Advanced Technologies for Security Applications

NATO Science for Peace and Security Series

This Series presents the results of scientific activities supported through the framework of the NATO Science for Peace and Security (SPS) Programme.

The NATO SPS Programme enhances security-related civil science and technology to address emerging security challenges and their impacts on international security. It connects scientists, experts and officials from Alliance and Partner nations to work together to address common challenges. The SPS Programme provides funding and expert advice for security-relevant activities in the form of Multi-Year Projects (MYP), Advanced Research Workshops (ARW), Advanced Training Courses (ATC), and Advanced Study Institutes (ASI). The NATO SPS Series collects the results of practical activities and meetings, including:

Multi-Year Projects (MYP): Grants to collaborate on multi-year R&D and capacity building projects that result in new civil science advancements with practical application in the security and defence fields.

Advanced Research Workshops: Advanced-level workshops that provide a platform for experts and scientists to share their experience and knowledge of security-related topics in order to promote follow-on activities like Multi-Year Projects.

Advanced Training Courses: Designed to enable specialists in NATO countries to share their security-related expertise in one of the SPS Key Priority areas. An ATC is not intended to be lecture-driven, but to be intensive and interactive in nature.

Advanced Study Institutes: High-level tutorial courses that communicate the latest developments in subjects relevant to NATO to an advanced-level audience.

The observations and recommendations made at the meetings, as well as the contents of the volumes in the Series reflect the views of participants and contributors only, and do not necessarily reflect NATO views or policy.

The series is published by IOS Press, Amsterdam, and Springer, Dordrecht, in partnership with the NATO SPS Programme.

Sub-Series

A.	Chemistry and Biology	Springer
B.	Physics and Biophysics	Springer
C.	Environmental Security	Springer
D.	Information and Communication Security	IOS Press
E.	Human and Societal Dynamics	IOS Press

- http://www.nato.int/science
- http://www.springer.com
- http://www.iospress.nl

Series B: Physics and Biophysics

Advanced Technologies for Security Applications

Proceedings of the NATO Science for Peace and Security 'Cluster Workshop on Advanced Technologies', 17–18 September 2019, Leuven, Belgium

edited by

Claudio Palestini

North Atlantic Treaty Organization

Springer

Published in Cooperation with NATO Emerging Security Challenges Division

Proceedings of the NATO Science for Peace and Security 'Cluster Workshop on Advanced Technologies',
17–18 September 2019,
Leuven, Belgium

ISBN 978-94-024-2023-4 (PB)
ISBN 978-94-024-2020-3 (HB)
ISBN 978-94-024-2021-0 (e-book)
https://doi.org/10.1007/978-94-024-2021-0

Published by Springer,
P.O. Box 17, 3300 AA Dordrecht, The Netherlands.

www.springer.com

Printed on acid-free paper

Contents

Chapter 1
Advanced Technologies at NATO: An Overview

Deniz Beten, Claudio Palestini, Ettore Marchesoni, and Marie-Anne Brouillon

Abstract Understanding and managing the political and practical implications of emerging and disruptive technologies is a priority for NATO and for the Science for Peace and Security (SPS) Programme. To take stock of the Programme's contribution to this field, NATO SPS organized the "SPS Cluster Workshop on Advanced Technologies" on 17 and 18 September 2019, under the auspices of the Katholieke Universiteit Leuven.

1.1 Science for Peace and Security: A History of Scientific Cooperation

Historically, technology was the spark that ignited NATO's interest and commitment to scientific advancement, a little over 60 years ago. Indeed, it was following the 1957 launch of Sputnik, the first artificial Earth satellite, that NATO realized the gap between Soviet and Allied technologies, and saw a need to act. This led to the creation of the NATO Science Committee in 1958, and to decades of NATO-supported practical cooperation in science and technology, through fellowships, research grants, knowledge sharing and capacity building, in the civilian and defence fields.

Science and technology in NATO is today a prominent effort brought forward by a number of programmes and stakeholders, which aims to sustain the technological advantage of the Alliance and its partners as described in the NATO Science & Technology Strategy.[1]

[1] NATO Science & Technology Strategy – Sustaining Technological Advantage. Approved by the North Atlantic Council on 27 July 2018

D. Beten · C. Palestini (✉) · E. Marchesoni · M.-A. Brouillon
NATO Emerging Security Challenges Division (ESCD), Brussels, Belgium
e-mail: palestini.claudio@hq.nato.int

C. Palestini (ed.), *Advanced Technologies for Security Applications*, NATO Science for Peace and Security Series B: Physics and Biophysics,
https://doi.org/10.1007/978-94-024-2021-0_1

Among all programmes, the Science for Peace and Security (SPS) Programme, established in 2006 as the merger of the Science Programme and the Committee on the Challenges of Modern Society (CCMS), is the living proof of NATO's longstanding commitment to science, which over the years brought about many outstanding results in a variety of fields. To mention a few: the project "SILK Afghanistan", which deployed a satellite-based network providing internet access to researchers in Afghanistan and other countries of the Southern Caucasus and Central Asia; the various projects developing real time explosive detection systems for public transportation; the multinational telemedicine system, enabling specialists to share real-time recommendations with first responders in emergency scenes or combat zones; or the establishment of a terahertz imaging laboratory in Algeria, contributing to the detection of explosive devices and other illicit substances – the first capability of this kind installed in North Africa.

Since its creation, SPS has demonstrated a flexible and versatile approach to practical scientific cooperation, and has proved instrumental to promoting knowledge-sharing, building capacity, and projecting stability outside NATO. The priorities addressed by the SPS Programme are aligned with NATO's Strategic Objectives, and aim to tackle emerging security challenges that require dynamic adaptation for the prevention and mitigation of risks. By addressing priorities such as advanced technologies, hybrid threats, and counter-terrorism, the Programme deals with new, contemporary challenges.

The SPS Programme has a unique way to tackle these challenges through partnerships, by engaging a network of scientists both from NATO and partner nations. As the threats it addresses are multidisciplinary in nature and often cross borders, international collaboration is essential to counter them. Bringing together researchers and institutions from Allied and partner nations, SPS helps finding solutions to challenges of mutual concern, at national and regional levels. In doing this, it plays a very important role in building capacity within and outside the Alliance, to deal with challenges affecting Allies and Partners alike.

In order to keep abreast of an ever-changing security landscape, it is in NATO's interest to understand the challenges and opportunities presented by disruptive and emerging technologies. The 1957 Sputnik launch may be considered the first of such challenges for the Alliance, and will not be the last. NATO's continuous adaptation and modernization have been essential to maintain its relevance through the years, which explains why innovation is such a critical element for its future. The future is also tied to advanced technologies, which will certainly have an impact on the security and defence landscape.

1.2 SPS Between Diplomacy and Technology

Science and technology can be a highly effective vehicle for international dialogue, due to their universality and dependence upon international networks. This is the *raison d'être* of the SPS Programme, which has been throughout its history a bridge between science, technology and public diplomacy.

In the early years, the Science Programme contributed to building a positive image of NATO as a security provider through engagement with Allied scientific and civilian communities based on the principles of solidarity. Over the years, and particularly at the end of the Cold War, this intra-alliance programme was transformed to reach out to and offer practical cooperation across NATO's partnership frameworks.

Today, the SPS Programme is as vital and necessary as ever before. NATO and Western societies needs to operate in times and environments that are Volatile, Uncertain, Complex and Ambiguous (VUCA), while innovation is following an unprecedented pace, with emerging technologies able to provoke disruption in the economy and societies in very short time and to radically change the security environment.

Additionally, we have experienced in the past few years a process of democratization of technology: cutting-edge technologies are becoming ubiquitous, affordable and easily accessible for everyone; their development has been more and more driven by commercial and civilian actors rather than the militaries, and as such emerging technologies are intrinsically more difficult to be framed, assessed and governed by institutions.

Today and in the future, emerging and disruptive technologies will not be developed in isolation, but they will be the result of collaborative works, merging ideas and experiences from different fields and contexts. That is why, in this environment, SPS is best placed to tackle these challenges and to bridge the gap between NATO nations and partner nations; academia, industry, end-users and governments; scientists, researchers and diplomats.

Since the SPS Programme's inception, a wide international network of scientist and experts from NATO member and partner countries has been established. More than 20 Nobel Laureates have been associated with the SPS Programme, a testament to the scientific excellence supported by SPS and its important role in innovation and advanced technologies.

1.3 SPS Cluster Workshop on Advanced Technologies

The topic of disruptive and emerging technologies has become more and more important for SPS. To take stock of the Programme's contribution to this field, NATO SPS organized the "SPS Cluster Workshop on Advanced Technologies" on 17 and 18 September 2019, under the auspices of the Katholieke Universiteit Leuven.

The event built upon the experiences from 26 SPS projects supported under the SPS key priority of "Security-related Advanced Technologies",[2] and promoted the exchange of knowledge and scientific know-how between NATO and partner nations. This was an opportunity for researchers to present their activities, to highlight future trends and to provide feedback on how SPS can contribute to scientific and technical development in this field. Additionally, the workshop was an opportunity to appreciate the high scientific value and educational success provided by the SPS Programme: in the context of these activities, prototypes were developed, patent applications prepared, and hundreds of scientific articles were published in high-ranked scientific journals and conference proceedings. Several young students and researchers (MSc and PhD students) were sponsored by SPS and played important roles in the projects.

Presentations and discussions revolved around four thematic areas, which are reflected in the structure of this volume: communication systems, innovative and advanced materials, sensors and detectors, and unmanned and autonomous systems.

In the field of communication systems, seven projects presented advanced technologies to:

- Detect intrusions in the electromagnetic spectrum by means of cooperative and intelligent detections systems;
- Predict and forecast cyber-attack trends through the analysis of big data and geopolitical strategic events;
- Deploy ad-hoc networks in crisis management and counter-terrorism operations;
- Establish a Quantum Key Distribution system over underwater fiber-optic communication networks;
- Develop post-quantum cryptographic algorithms, i.e. algorithms and protocols able to secure communications even in the case of attacks by a quantum computer.

Six SPS projects on innovative and advanced materials focused on the development of new materials (e.g. graphene, silicon carbide, ceramic, titanium, etc.) providing interesting properties for specific security applications. They also developed cutting-edge production and synthetization techniques, contributing to:

- Develop nuclear screening systems to detect illicit trafficking of nuclear materials;
- Develop radiation hardening techniques, i.e. making electronic components resistant to high energy radiation, especially for outer space vehicles and high-altitude flights;
- Develop novel armor concepts based on titanium, providing high strength, micro hardness, bending capabilities, etc.;
- Develop ceramic nanomaterials for infrared transparency.

[2]The list of SPS Key Priorities are reported in Annex II.

Eight SPS projects on sensors and detectors presented their achievements; they developed:

- Multi-static and multi-band radars for border security;
- Noise radar networks for the detection of low altitude Unmanned Aircraft Systems (UAS);
- Compact and portable LIDAR (Light Detection and Ranging) systems to be mounted on UAS for 3D imaging and mapping;
- Sensor fusion technologies to create common operational pictures and provide tactical information to security operators.

Finally, five activities on unmanned and autonomous systems were presented, covering the full domain spectrum (maritime and underwater, ground, air and space).

These projects aim to:

- Develop and deploy an underwater autonomous platform for situational awareness;
- Develop intelligent tyres to guarantee mobility in severe terrain environments;
- Deploy multiple cooperative UAS to monitor large areas;
- Deploy high altitude balloons in the stratosphere, providing high altitude radar coverage;
- Develop and demonstrate countering UAS capabilities, through the release of multiple cooperative drones attacking the enemy drone.

1.4 Outline

This book provides a summary of the projects presented. It is organized in four main Parts, divided according to the themes of the workshop. The full list of projects is reported in Annex I.

Note that the articles do not respect necessarily the opinion of NATO or the editors and are the full contribution of the authors.

Part I
Communication Systems

Chapter 2
Large Scale Collaborative Detection and Location of Threats in the Electromagnetic Space

Domenico Giustiniano, Vincent Lenders, and Sofie Pollin

Abstract In the twenty-first century, the security of the electromagnetic spectrum has tremendous strategic importance to society. In particular, the wireless infrastructure that carries vital services such as 5G cellular networks, communication to aircraft and Global Navigation Satellite System is especially critical. This rapid change is even more impressive considering that in the 1980s the only concern for spectrum management was mostly about radio/television broadcasting and military communications. The allocation of spectrum has become over the years more and more complex with different players and stakeholders that depend on largely of their correct operation. However, today, the cost of commodity radio technology prices is so low that access to it is no longer restricted to governments and network operators. It is now affordable to individuals, giving them the potential to become malicious intruders. More frequent and more sophisticated threats from such infiltrators could wreak havoc and are among the most serious challenges faced by society. Unauthorized transmissions could threaten the operation of networks used by air traffic control systems, police, security and emergency services in populated areas. The SOCRATES (Large Scale Collaborative Detection and Location of Threats in the Electromagnetic Space, Grant G5461) project started in June 2018 and aims to deliver a security system to protect our electromagnetic environment and the services and users that depend upon it. SOCRATES will provide an accurate, autonomous, fast and secure system based on a novel and disruptive IoT (Internet of Things) architecture. By detecting and locating unusual RF signal and source

D. Giustiniano (✉)
IMDEA Networks Institute, Madrid, Spain
e-mail: domenico.giustiniano@imdea.org

V. Lenders
Electrosense, Burgdorf, Switzerland
e-mail: vincent.lenders@electrosense.org

S. Pollin
KU Leuven, Leuven, Belgium
e-mail: sofie.pollin@esat.kuleuven.be

C. Palestini (ed.), *Advanced Technologies for Security Applications*, NATO Science for Peace and Security Series B: Physics and Biophysics,
https://doi.org/10.1007/978-94-024-2021-0_2

activity it will identify intruders in the electromagnetic space, before a threat can become serious, learning about its physical layer features and its geographic location. By providing the capability to detect, identify and locate potential threats to electromagnetic infrastructure security, SOCRATES represents an important step in ensuring society's readiness to respond effectively to them. SOCRATES will shield economic and social structures from those who would harm them. In this contribution we present a summary of published results achieved in the first year of the project, how funds from SOCRATES have foster the collaboration between NATO countries Spain and Belgium and partner country Switzerland, including the activities led by Electrosense as partner country.

2.1 Introduction

Protecting the wireless infrastructure and detecting malicious intruders in the spectrum are deemed essential for the society in the twenty-first century. In light of this problem, new technologies for detecting threats in the electromagnetic space must be devised to counteract possible catastrophic threats and take fast decisions. In the SOCRATES project we aim to deliver a security system to protect our electromagnetic environment and the services and users that depend upon it [1]. SOCRATES started in June 2018 and will be delivered in May 2021. Our system will blend new emerging and diverse disruptive technologies in order to provide accurate and fast detection, classification and location of intruders in the spectrum in the time, frequency and space domains. Using the innovative solutions developed in the project, stakeholders will be able to significantly reduce the danger of threats and have access to advanced methodologies for swiftly taking the correct actions. Our system builds on top of the emerging Electrosense (https://electrosense.org) initiative led by the partner country of the consortium. Electrosense is a non-profit association for distributed crowdsourcing, monitoring and storage of the wireless spectrum, which uses low-cost spectrum sensors, with its beta version already operative both for research and for providing a first set of applications to end users and stakeholders [4]. We will build on top of this network to address the most dangerous threats in the wireless electromagnetic space, with the objective of spotting and locate the spectrum saboteurs.

The paper is organized as follows. In Sect. 2.2 we present the overall goal of the project, followed by Sect. 2.3, that shows the results in this first year of the project, with particular emphasis on published material. Finally, in Sect. 2.4, we briefly present the deployment status of Electrosense through the funding received in SOCRATES, the key factor for collecting large scale data set as needed to execute the project.

2.2 Overall Goal

To counteract the threats in the electromagnetic space, there is the pressing need to design novel, flexible and autonomous methods to protect the wireless infrastructures from cyber-attackers and develop novel architectures and technologies to support cyber defence capabilities. In order to address this problem, the vision of SOCRATES is to create the foundations for an accurate, autonomous, fast and secure system that identifies intruders in the electromagnetic space, before the threat can become serious, learning about its physical layer features and its geographic location. SOCRATES is led by IMDEA Networks, Spain (NATO country), which works in this project with non-profit association Electrosense, Switzerland (partner country), and KU Leuven, Belgium.

In the project we will exploit novel techniques and new disruptive technologies emerging in diverse research sectors and their largely unexplored mutually interactions to find ground-breaking solutions to the grand challenges in security above described. We will start from the current Electrosense spectrum monitoring system, currently in its beta version, designed and deployed by project members for civilian applications, and investigate novel system concepts, methods, and algorithms to introduce the second generation of the Electrosense system that will target security applications. Electrosense uses off-the shelf software-defined low-cost spectrum sensors (but it can also operate with more expensive boards, such as Ettus), currently mainly located around Europe. It has a backend component to control the sensors, support low-latency stream processing and large-scale batch analyses at the same time. Spectrum data computation and storage in Electrosense is implemented using state-of-the-art big data lambda architecture [4].

The new system that will be developed in SOCRATES will process raw observables in the sensor, that is, in-phase and quadrature phase (I/Q) samples of spectrum received from the RF front-end, which we consider a key enabler for the security applications studied in the project. The availability of I/Q samples will unleash new opportunities to innovate on the methods to extract information in the data received by each spectrum sensor. In fact, while derived observables such as averaged power spectral density (PSD) measurement data suffices for coarse understanding of the usage of the spectrum, accessing and processing raw I/Q measurements is fundamental in security contexts.

The raw observables (I/Q data) can have a tremendous potential to provide greater performance in cooperatively detecting and locating RF intruders beyond state-of-the-art solutions. Yet, it first implies a deluge of data to be transmitted to the backend, which is not feasible to sustain for every users at their home with current Internet deployments. Therefore, it is not possible to collect continuously I/Q data. Second it requires more powerful machines to process data in the backend and novel algorithmic solutions that can scale well with the number of sensors. These algorithms must find the best solutions to process (at least partially) raw observables in the spectrum sensors while taking advantage of derived and lower-complexity observables such as PSD for a first level of detection of the intruder.

SOCRATES will dynamically select what observables to fetch and their structure in time, frequency and space dimensions. In turn, careful decision about what observables (raw, derived, type of derived, etc.) to fetch at any point in time from the spectrum sensors is needed.

Proper and flexible integration of new disruptive technologies will be used to investigate methods for geo-locating radio transmitters that can work with low-cost software-defined spectrum sensors distributed over the public Internet. In the envisioned system, we expect to concurrently localize and detect the anomaly, and we will look for novel approaches to perform these tasks jointly. At the end of the project, real experiments will showcase the system's ability to detect the waveforms and wireless technologies of adversaries who are misusing wireless resources. We will also demonstrate how the physical location of an intruder can be swiftly identified. Adopting an agile approach, SOCRATES will build, demonstrate and showcase early prototypes throughout the project.

2.3 Results in the First Year of the Project

In this section, we present the scientific contribution and other dissemination activities of the project. The three partners of the project are largely collaborating in the project, as the Electrosense infrastructure from the partner country Switzerland is used for experimentation. This is also demonstrated by several joint top publications among project partners, as well as two hackathons that have been organized in Spain and Germany as training activities. Members from all three organizations have participated in the hackathons (the results of these hackathons have resulted in papers that are currently under submission).

2.3.1 Scientific Contribution

In terms of scientific output, the project has resulted in the first year in one ACM/IEEE IPSN 2019, a top conference in networked embedded systems, one ACM Mobicom 2019, the premier conference in mobile computing, and one IEEE Dyspan 2019, a very well recognized international conference that mixes technology and policy issues, becoming the reference meeting point for sharing and exploring advanced spectrum technologies. Furthermore, one paper has been published in IEEE Transactions on Cognitive Communications and Networking (TCCN) (extension of the IEEE Dyspan 2019 paper).

In what follow, we describe in details the scientific output in the first year, mainly focusing of published material and open source code.

Fig. 2.1 Electrosense expansion board to operate from DC to 6 GHz

2.3.1.1 Sensor Architecture

In the first year of the project, we have designed an expansion dongle for Electrosense, to extend the sensing capabilities of the first generation dongle from almost DC up to 6 GHz, compared to 24 MHz to 1.7 GHz for the first generation dongles. The expansion board is illustrated in Fig. 2.1 and it is required for collecting spectrum data in the most significant frequency bands. The up and downconvertor board has three possible RF modes, operating at different bands:

- Up-conversion input: 0–30 MHz
- Direct input: 30 MHz–1.6 GHz
- Down-conversion input: 1.6–6 GHz

The output of the convertor board has a bandwidth of 20 MHz, even though the RTL-SDR device (used in the typical Electrosense spectrum sensor) only has 2.4 MHz bandwidth. Higher quality SDRs can be used to take advantage of the higher bandwidth. The convertor is controlled by an STM32 ARM based microcontroller, which exchanges data with the Raspberry PI through a USB port. The firmware running on the microcontroller exposes a text based interface. Both the convertor board and the firmware have been tested and calibrated. We have distributed the boards to project members and have started to distribute them to the Electrosense community for deployments. We have released the schematic (https://github.com/electrosense/hardware) as open source.

In addition, we have worked on a new version of the software code running in the software-defined spectrum sensor. This new version allows to fetch I/Q data (I/Q pipeline) on demand from the GUI interface at Electrosense. This is the first step for collecting I/Q data for detection and localization using the Electrosense system. The code has been also optimized for efficiently scanning the spectrum, reducing the time to scan the full spectrum by 25%. The code has been also publicly released as open source at https://github.com/electrosense/es-sensor. We have also introduced a *decoding pipeline* to allow to decode specific I/Q data directly in the spectrum sensor and provide this data directly to users in order to guarantee low latency, bypassing the backend processing. Currently, this decoded data can be requested by a user in near real time (see also work in Sect. 2.3.1.5). In the sensor architecture, we are now able to integrate different pipelines (PSD data, I/Q data, decoding pipeline), and that can be controlled seamlessly to deliver data requested by the backend and by the users.

2.3.1.2 Data Quality, Trustworthiness and Obfuscation Techniques

The work has focused on the investigation of techniques to transform spectrum data (PSD, I/Q) to a less complex space that preserves signal features, and send data in this domain to the backend. The motivation for this study is to limit the amount of network bandwidth needed to operate each spectrum sensor and perform computation extensively in this space. The majority of the work has been conducted to design a framework for analyzing the spectrum characteristics in large spatio-temporal scales. The framework operates in the SVD domains, where it extracts long-lived signal, short-lived signal, signal shape change, energy difference, etc. and it is characterized by high compression ratio, almost no information loss, and high run time efficiency [6]. However, the proposed framework still requires a high computational cost and it can work only on powerful embedded devices, we are also investigating different methodologies that can work on embedded boards.

2.3.1.3 Detection of Intruders

Related to intruder or anomaly detection, we have worked on a deep learning framework called SAIFE, an unsupervised wireless spectrum anomaly detection framework with interpretable features [5]. Demanding anomalous behaviour or intruders in wireless spectrum is a demanding task due to the sheer complexity of the electromagnetic spectrum use. Anomalies can take a wide range of forms from the presence of an unwanted signal in a licensed band, to the absence of an expected signal, which makes manual labelling of anomalies difficult and suboptimal. SAIFE is based on an adversarial autoencoder based anomaly detector using PSD data which achieves good anomaly detection and localisation (in the spectrum) in an unsupervised setting. It is able to learn interpretable features such as signal bandwidth, class and center frequency in a semi-supervised fashion. It has been tested on data from one of the distributed Electrosense sensors over a long term of 500 hours showing its anomaly detection capabilities.

2.3.1.4 Geo-Location of Intruders

We have worked on the problem of collaborative decoding to decode data from a transmitter of larger bandwidth than the one of the single receiver. In order to circumvent the hardware limitations of single receivers, we envision a scenario where non-coherent receivers sample the signal collaboratively to cover a larger bandwidth than the one of the single receiver and then, enable the signal reconstruction and decoding in the backend. In the work we have proposed and verified with real experiments a methodology to synchronize I/Q samples of multiple non-coherent spectrum sensors. In order to cover a large bandwidth, we suppose there exists only a very limited band in the spectrum where the signals received by multiple sensors overlap in frequency [3]. The work is an enabler for the investigation of localization techniques, as it allows to study to synchronize I/Q samples in a very challenging application (with very limited amount of information available).

2.3.1.5 System Solutions, Integration and Demonstrations

We have worked on a new architecture (Electrosense+ [2]) to give power to the users. We have found that this approach was necessary to provide incentives to users to install sensors, key for a crowdsourcing initiative. The idea has been to introduce the capability of decoding I/Q samples directly in the spectrum sensors, for broadcast and control data (and thus without privacy concerns), and provide this information to users in real time. The work has relied on open source implementations for the decoders, that has been optimized to work with low-cost embedded hardware. Large engineering effort has been done to integrate the proposed solution with the current Electrosense system in order to control the status of the sensor and to manage the signaling server needed by this new approach. In addition, we have integrated WebRTC as tool for direct streaming from the sensor. The system is already operative in test mode and it is currently able to decode FM and AM radio (with audio), Cell ID of LTE base stations, ADS-B, AIS and ACARS (with maps). It will be launched to the community in Fall 2019.

2.3.2 Other Dissemination Activities

Project members have also disseminated the results of the project in other activities. The NPD Dr. Giustiniano gave a keynote speech at EAI Crowncom 2018, the International Conference on Cognitive Radio Oriented Wireless Networks, in front of about 60 scientists. This has allowed to give high visibility to the research and the roadmap in Electrosense and SOCRATES. Dr. Giustiniano gave also an invited talk in April 2019, as part of the IEEE 5th World Forum on Internet of Things. Sofie Pollin gave a keynote speech during the workshop on IoT, Big Data & Smart Cities (ITCities'18, www.itcities.org) in June 2018 in Rabat, and explained how citizen science, and Artificial Intelligence all meet in Electrosense. Furthermore, following the agile approach in the project, we have shown early demonstrators of SOCRATES. In all events, the feedback has been very positive.

The press release of the start of the project has received significant attention from Spanish and international media. We highlight an interview in Spanish from the newspaper Vozpopuli and the international outreach in website such as phys.org, a very well know portal for news in science and technology.

2.4 Deployment

The approach of Electrosense is to use its own backend infrastructure as alternative to cloud services. Doing the processing in the cloud can be very expensive because of the egress costs and data management via public cloud. Furthermore, it forces to trust cloud providers and results in data loss/not more accessible after the end of the project, as leasing of space ends. As processing data for the applications envisioned in the process is a fundamental challenge for a large scale distributed system, Electrosense has upgraded its backend through the funding received in SOCRATES. Figure 2.2 shows a picture of the servers acquired by Electrosense within the project. Each server runs 2 × 12 Core Intel Xeon Silver 4116 2.1 GHz Processor.

In addition, SOCRATES requires sensors deployed in any location to collect data and train the models. For this reason, budget of the project has been used for making

Fig. 2.2 Electrosense servers

Electrosense sensors available to the community. In particular, part of these sensors operate up to 6 GHz (cf. Sect. 2.3.1.1). At the time of writing, we have started the distribution of sensors to interested applicants. Finally, project budget has been used in part to support the deployment of controlled testbeds for experimentation with higher-end software defined radios.

2.5 Conclusion

In this contribution, we have presented SOCRATES, Large Scale Collaborative Detection and Location of Threats in the Electromagnetic Space, with particular focus on the overall goal of the project and the results achieved in the first year of the project. The three partners of the project largely collaborate in the project, as the Electrosense infrastructure from the partner country Switzerland is used for experimentation. This is also demonstrated by several joint top publications among project partners. In the follow up of the project, we plan to massively deploy our sensors with expansion boards and send them to interested users in populated regions of the world. We also plan to release Electrosense+ to allow users to connect to any sensors and listen to a variety of events. This will provide incentives to users to deploy sensors on their own, further allowing to demonstrate the finding of the project in real conditions. With these new sensors, SOCRATES will be able to test in real life conditions new approaches for detecting and localizing intruders as investigated in the project.

References

1. (2019) SOCRATES project, Large Scale Collaborative Detection and Location of Threats in the Electromagnetic Space, Grant G5461. https://socrates.networks.imdea.org/
2. Calvo-Palomino R, Cordobés H, Engel M, Fuchs M, Jain P, Liechti M, Rajendran S, Schäfer M, den Bergh BV, Pollin S, Giustiniano D, Lenders V (2018) Electrosense+: empowering people to decode the radio spectrum, *arXiv*
3. Calvo-Palomino R, Cordobés H, Engel M, Fuchs M, Jain P, Liechti M, Rajendran S, Schäfer M, den Bergh BV, Pollin S, Giustiniano D, Lenders V (2020) Electrosense+: Crowdsourcing radio spectrum decoding using IoT receivers, 174, 107231, Computer Networks
4. Rajendran S, Calvo-Palomino R, Fuchs M, Van den Bergh B, Cordobes H, Giustiniano D, Pollin S, Lenders V (2018) Electrosense: open and big spectrum data. IEEE Commun Mag 56:210–217
5. Rajendran S, Meert W, Lenders V, Pollin S (2019) Unsupervised wireless spectrum anomaly detection with interpretable features. IEEE Trans Cogn Commun Netw 5(3):637–647. https://dblp.org/rec/bibtex/journals/tccn/RajendranMLP19
6. Zeng Y, Chandrasekaran V, Banerjee S, Giustiniano D (2019) A framework for analyzing spectrum characteristics in large spatio-temporal scales. In: Proceedings of the 25th annual international conference on mobile computing and networking, New York. ACM

Chapter 3
A Primer On Public Safety Communication in the Context of Terror Attacks: The NATO SPS "COUNTER-TERROR" Project

Muhammad Mahtab Alam, Yannick Le Moullec, Rizwan Ahmad, Maurizio Magarini, and Luca Reggiani

Abstract Terrorism is emerging as one of the most serious threats worldwide. Terrorist attacks are becoming more coordinated, sophisticated and hence more devastating. One of the important reasons for higher casualities is the "sluggish response time". In some of the recent attacks, police and law enforcement agencies even after many hours were unable to have some of the basic information such as how many people are inside the attacked premises, the number of seriously injured persons, the number of terrorists, what their location is and so on.

So, one of the most important questions is "How to provide the fundamental information to public safety agencies as quickly as possible to reduce the response time following terrorist attacks"? The NATO – Science for Peace (SPS) "COUNTER-TERROR" G5482 project (2019–2021) presented in this paper investigates and proposed innovative ideas and solutionsto address this question from an information and communication technology viewpoint, including the establishment of secure D2D communication networks to quickly gather information and the use of UAVs to discover and localize weak signals.

M. M. Alam (✉) · Y. LeMoullec
Thomas Johann Seebeck Department of Electronics, Tallinn University of Technology, Tallinn, Estonia
e-mail: muhammad.alam@ttu.ee; yannick.lemoullec@ttu.ee

R. Ahmad
School of Electrical Engineering and Computer Science, National University of Sciences and Technology (NUST), Islamabad, Pakistan
e-mail: rizwan.ahmad@seecs.edu.pk

M. Magarini · L. Reggiani
Dipartimento di Elettronica, Informazione e Bioingegneria, Politecnico di Milano, Milan, Italy
e-mail: maurizio.magarini@polimi.it; luca.reggiani@polimi.it

C. Palestini (ed.), *Advanced Technologies for Security Applications*, NATO Science for Peace and Security Series B: Physics and Biophysics,
https://doi.org/10.1007/978-94-024-2021-0_3

It is envisioned that "connecting" the on-scene available (OS-A) heterogeneous devices (through multi-hop device-to-device (D2D) communication) to the nearest mobile deployed command center in an efficient way cannot only rapidly disseminate fundamental information but also aid in significantly reducing the response times and, consequently, many lives and infrastructure.

The objectives of this project are to design and evaluate efficient heterogeneous resource management by adaptive power control, throughput enhancement and interference management for device-to-device (D2D) communication. The originality is to exploit machine learning (ML) techniques to improve the existing state of the art. Further, the use of unmanned aerial vehicle (UAV) for weak signal detection and devices accurate position evaluation is an important objective. The deployed setup with UAV assisted connectivity is one of the novel contributions of this project. In addition, context aware and reliable D2D multi-hop routing and network connections to ensure high end-to-end throughout and low end-to-end energy consumption and delay is another core objective.

In order to enable the emergence of future pervasive communication systems enabling life-critical, public safety and preparedness, D2D communication can be realized both in licensed (driven by cellular spectrum) and un-licensed spectrums (used by other wireless technologies e.g., WiFi Direct, long term evolution (LTE) Direct, Bluetooth). In the licensed spectrum, smart phones can connect through LTE-A and can create multi-hop communication by exploiting UAVs (i.e., relay nodes). Whereas, in the unlicensed spectrum, smartphones can connect through WiFi Direct, LTE Direct, or LTE-U. The fundamental aspect of this work is to connect the devices (i.e., establish and maintain reliable connections) in harsh environments and existing works have to be extended and tested so that they can deal with such conditions.

So far, there have been only a few research projects specific to the context of terrorist attacks.

To enable future pervasive public safety communication systems, this work will foster the vision to achieve beyond state-of-the-art ambitious, highly innovative and challenging research and development goals. For the given context, in particular, connecting the OS-A devices in an efficient way to optimize their heterogeneous resources and improve reliability is key factor to the innovation of the proposed system.

By implementing the proposed D2D-based solution, the average response time shall be reduced by an average of 8 to 10 times. The direct positive consequences are that more lives will be saved, the number and severity of injuries will be reduced, and damages to infrastructure will be decreased.

Keywords Device-to-device communication · On-scene available (OS-A) devices · Beamforming · Localization · Machine learning · Software defined networking

3.1 Introduction

The increasing number of terrorist attacks and fatalities have raised serious concerns for the entire world. One of the fundamental issues in most of the terrorist attacks is the *slow response time* as it is observed that, even after many hours, police and other law enforcement agencies are often unable to take actions against terrorists due to the lack of basic information such as the number of terrorists and that of people trapped inside the attacked zone, their location and identity, etc.

This NATO-SPS project shall provide innovative answers and solutions to the problem "how to reduce the response time in the context of terrorist attacks?"

As of today, from an information and communication technology (ICT) viewpoint, classical public safety networks (PSNs) do not provide the appropriate infrastructure to deal with such attacks. This is because such networks are not designed (i) for the specific context of terrorist attacks, and (ii) to exploit the devices (in particular user equipment) and networks, which are active, and already inside the terror zone (e.g., shopping mall, theatre etc.). These requirements are different from classical PSNs since a) in typical natural disasters, the emergency rescue teams can gain access to the disaster site during or after the emergency incident with only slight (if not without) direct life threat and can start the operations immediately, whereas in the case of terrorist attacks, rescue and law enforcement teams remain unable to immediately step-in because of the unclear information and the situational facts (i.e., number of terrorists, their positions, number and type of weapons used, etc.), and b) in terrorist attacks, the devices held by the people are critical (for example, their personal ID number and name can be traced through the device ID and this can further help to provide the exact location of the persons). In addition, other important information about the devices such as battery levels, devices radio-link conditions, signal strengths and connectivity patterns, as well as mobility profiles can be traced autonomously. In this work, we are developing methods and techniques to exploit smartphones and/or on-scene available (OS-A) devices, which have enabled direct communication feature in the affected zone, to get the fundamental and critical information to reduce the response time.

To achieve this, research and development in Pervasive Public Safety Communication (PPSC) is carried out to enable connectivity-aware, autonomous, cooperative, reliable, and real-time communication with constrained resources.

The objectives of this project are to design and evaluate efficient heterogeneous resource management by adaptive power control, throughput enhancement, and interference management for D2D communication. The originality is to exploit machine learning (ML) techniques to improve the existing state of the art. Further, by exploiting unmanned aerial vehicle (UAV) for weak signal detection, devices' accurate position evaluation is an important objective. The deployed setup with UAV assisted connectivity is one of the novel contributions of this project. In addition, the context aware and reliable D2D multi-hop routing and network connections to ensure high end-to-end throughout and low end-to-end energy consumption and delay is another core objective. In order to enable the emergence of future pervasive communication systems enabling life-critical, public safety and preparedness, D2D communication can be realized both in licensed (driven by cellular spectrum) and un-licensed spectra (used by other wireless technologies e.g., WiFi Direct (WiFi-D), long term evolution (LTE) Direct, Bluetooth). In the licensed spectrum, smartphones can connect through LTE-A and can create multi-hop communication by exploiting UAVs (i.e., relay nodes), whereas, in the unlicensed spectrum, smartphones can connect through WiFi-D or LTE Direct (LTE-D) and LTE-U. The fundamental aspect of this work is to connect the devices (i.e., establish and maintain reliable connections) in harsh environments and existing works must be extended and tested so that they can deal with such conditions.

In this paper, we highlight the achieved results for (i) D2D communication based on LTE-D using ML techniques, (ii) innovative techniques for localization and beamforming for the given context, and (iii) new software defined raadio (SDN) based energy efficient routing techniques.

The rest of the paper is organized as follows. First, we present LTE-D based on uncentralized D2D communication architecture and machine learning based D2D communication. Then we present the localization and beamforming results using UAVs and, finally, relaible networking techniques. Conclusions are drawn at the end.

3.2 Reliable and Resource-Efficient D2D Communication Using Machine Learning Techniques

D2D devices generate interferences while reusing the spectrum both in in-band and out-band modes. Typically, a base station (BS) allocates the resources to cellular users and D2D users dedicatedly. The interference can be controlled if dedicated resource allocation and proper power allocation are applied at the D2D devices while communicating at short range (overlay mode). However, in out-band mode and underlay modes, due to absence of base-station control, interference management is an open research and development challenge. Our objective is to create multi-hop D2D communication, and through relaying, establish a link to the deployed external command center. Consequently, controlling the resources (i.e., maximize throughput, increase energy efficiency of the OS-A devices) as well as minimizing the interference is paramount.

The resource scheduler needs to perform a task for the power allocation at each time step. The amount of power to be allocated is essential for the interference level and the quality-of-service (QoS) of the network. In this regard, several works exist in the literature; these include random power allocation [1], which seems not an appropriate solution given the dynamic environment in our case.

Efficient resource and power allocation assignment helps to maximize the resources. The existing literature [1–3] lacks in exploiting online mechanisms to learn the resource allocation adaptively, which would improve the overall network performance and provide efficient management of the resources. Recent advances in online learning algorithms, e.g., reinforcement learning (RL) algorithms and its variants i.e., SARSA (lambda), exp3 and multi-armed bandit solver, create an enormous scope for applying them to a number of applications. In this project, we apply online learning algorithms for optimizing resource utilization among devices in D2D communication.

At first, we describe some classical approaches for resource optimization in D2D communication followed by existing RL-based resource allocation algorithms and their limitations, followed by our preliminary results.

In [1], an efficient resource allocation technique is proposed for multiple D2D pairs which maximizes the system throughput. The proposed solution is interesting; however, it does not consider interference metrics in the system model for D2D pairs. Feng et al. [2] explore a resource allocation technique by considering the QoS of cellular users (CU) and D2D pairs simultaneously to improve the system performance. Further, in [3], a centralized heuristic is proposed where the resources of cellular users and D2D pairs are synchronized considering the interference link gain from D2D transmitter to the BS. Zhao et al. [4] propose a resource allocation method for the D2D pairs to enhance the system sum-rate. In [5], Min et al. propose an interference restricted region where multiple cellular users and a single pair D2D exist. In [6], the authors propose the improved target rate of cellular users for maximizing the system throughput. In [7], the authors propose a centralized resource allocation method for the PSN where transmitting devices close to each other based on communication range are assigned orthogonal transmission opportunities. A common limitation of these works is that they are fully centralized i.e., the base-station requires full knowledge of the link state information that creates redundant information over the network and consequently causes network congestion as well as rapid depletion of battery-limited devices (in disaster situations, this is one of the main problem in centralized D2D communication).

In addition to above-mentioned works, An et al. [8] propose a heuristic resource allocation technique. They implement an adaptive interference restricted region for multiple D2D pairs that share the resources. However, their proposed method is not adaptive for the power allocation to the users. In [9], the authors propose an adaptive resource allocation method based on two phase optimization. They apply Lagrangian dual decomposition which is computationally complex. Wang et al. [10] propose a joint scheduling strategy for the resource allocation where the average D2D throughput is improved. Here, only one D2D pair reuse the channel assigned to the cellular users. The co-tier interference is not considered in this work. In [11], Graziosi et al. propose a distributed resource allocation method where minimum rates of cellular users and D2D pairs are maintained. However, this approach shows low spectral efficiency.

Motivation for using and improving ML – Recent studies show significant gains in performance and improved resource utilizations and, for this reason, our novel approach will build-upon these preliminary studies. There are only a few works based on ML for resource allocation in D2D communications [12–13]. These works focus on online learning techniques for optimal power allocation to the resource blocks (RB), which increases the system throughput. However, these preliminary studies are limited (i) on the modeling of reinforcement learning and lack in exploitation, particularly to maximize the observation space to explore the full potential of the overall system, (ii) the reward function is composed of only signal to interference noise ratio (SINR) threshold values, which does not reflect the impact on the channel gains and link quality measures between users, and (iii) these works do not consider the task scheduling, energy efficiency, and the fairness issues for resource allocation.

Thus, we investigate an online task scheduling strategy for adaptive resource allocation using reinforcement learning methods such as SARSA (lambda) and multi-armed bandit solver method by considering improved state space and a reward function. Our proposed resource allocation methods helps to provide mitigated interference level, increased D2D throughput, and consequently improved system throughput, energy efficiency, and fairness.

Preliminary Simulation Results In our on-going research (where the base-station is partially available), preliminary results show that cooperative reinforcement learning using SARSA (lambda) method can yield better D2D throughput in comparison to random and distributed approaches while operated in overlay mode. Figure 3.1 shows the average D2D throughput, average cellular user throughput and the average system throughput over the number of D2D users. We can observe that D2D throughput increases with the increase of D2D users, but on the other hand,

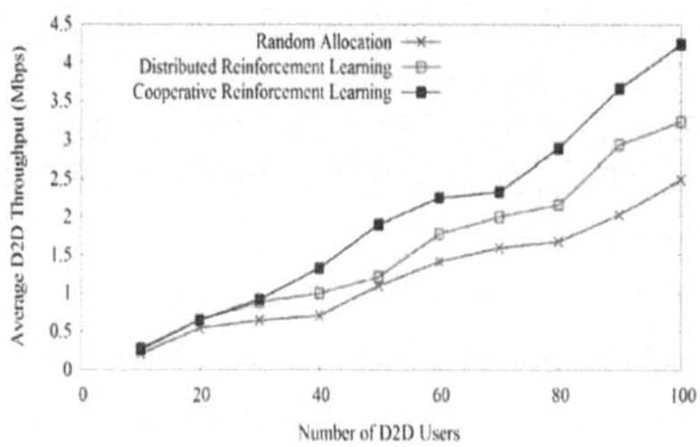

(a) Average D2D user throughput over number of D2D users

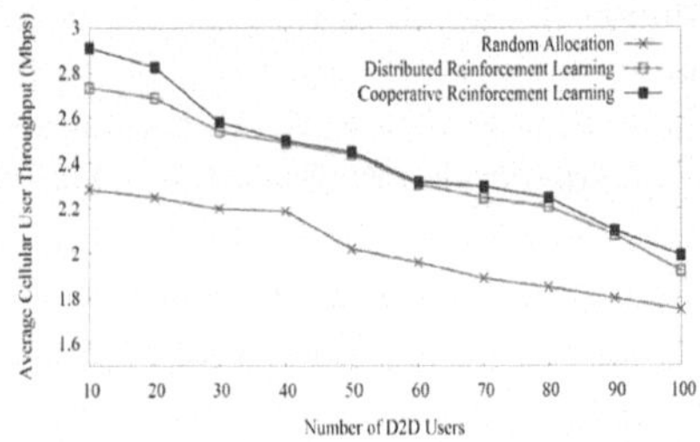

(b) Average cellular user throughput over number of D2D users

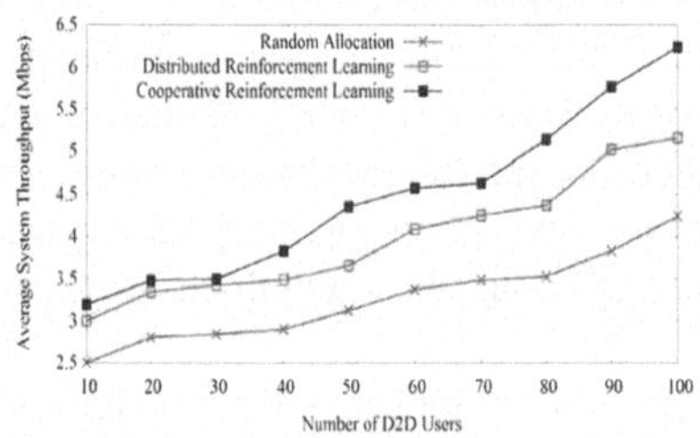

(c) Average system throughput over number of D2D users

Fig. 3.1 Analysis of average D2D (**a**), Cellular User (**b**), and System (**c**) throughputs over number of D2D users

cellular user throughput decreases. To balance the trade-off, system throughput i.e., the summation of D2D and the cellular user throughput is analysed. All methods show these same trends over the number of D2D users. From this experiment, we can also investigate the issue about the appropriate number of D2D users, which provides the better trade-off between D2D and cellular user throughput in a single cell scenario. Here, we can observe that a moderate number of D2D users, for example 50, provide a suitable amount of D2D and cellular user throughputs. The proposed cooperative reinforcement learning outperforms the other methods regarding D2D throughput, cellular user throughput and overall system throughput for every number of D2D users.

In this NATO-SPS project, we are building upon these preliminary results and extend them for non-assisted underlay mode, as well as for the out-band modes. For these modes, heterogeneous resource optimizations are critical which will be achieved by adaptive power control, throughput enhancement and interference management for D2D communication using novel cooperative reinforcement learning approach.

3.3 UAV Assisted Relay for Reliable Communication

In the 4G context (LTE, LTE-Advanced), there has been significant attention towards the use of UAVs as flying base stations to provide better cellular coverage and boost the capacity at the mobile user stations on the ground [13–17], using either low (LAP) or high altitude aerial platforms (HAP). Small UAVs such as quadcopters are easy to deploy and can fly at low altitude, providing better services to the users on the ground. One of the main advantages of having BS onboard a UAV is that of dynamically changing the coverage area according to traffic intensity. This is cost effective and it is also highly efficient in situations where a large mass of users confined in a small area demand high speed data as in stadiums, public rallies, festivals, concerts or areas interested by terrorist attacks, as considered in this project. In this context, some of the project partners already have a 2-year experience in High Altitude Platforms and UAV-to-ground communications [18, 19].

Aerial wireless sensor networks (WSNs) have also been successfully employed in many non-critical application domains ranging from agriculture and habitat and smart buildings monitoring. Often, they employ multi-hop communication and routing protocols over lossy channels with varying levels of reliability. However, a recent survey on reliability in WSNs indicates that by carefully selecting link metrics and power consumption levels, WSN-based systems can often provide a high level of reliability [20]. Some of the project partners in this NATO-SPS project have proposed a novel technique in [21] which employs heartbeats, algorithms and redundant gateways to improve failure detection of primary gateways in aerial WSNs and providing a fast recovery mechanism using a secondary gateway to continue transmitting sensor data to the cloud with no sensor data loss. Performance measurements using gateway prototypes built using low-cost commercial off-the-shelf (COTS) hardware demonstrate the feasibility of the approach. The redundancy on the sensor node level is provided by redundant sensor nodes, which in turn are capable of connecting to multiple gateways. For ZigBee we have addressed the problem of the network coordinator being a single point of failure by using the method described in [22].

Even if the use of UAVs in military communications and applications has a long history (see, for example, [23, 24] for studies on UAV relay communications in the military field), more recently, in the civilian context the use of access point (AP)/UAV for enhancing communications by means of relaying techniques has acquired a great interest and several works are present in the literature, especially for allowing out-of-range communications. In addition, LTE, from Release 10 (LTE-Advanced), has included capabilities to enable fast deployment of UAVs to restore or extend coverage of the network. In this project, UAVs are intended to work primarily as relays between terminals inside a definite, emergency area and an ECC. In general, while the scenario of a single UAV relay has been considered in several studies in the literature, the scenario with multiple UAVs needs still research efforts in several directions, and also in the parts concerning the cooperative resource management and routing, previously discussed. One of the classical issues of relaying, particularly important for the UAV case, is the computation of the optimal position of the relays for maximizing the network performance (e.g. minimizing the outage probability of the devices in the covered area) and this has been studied in [25–27]. On the other hand, in the literature specifically devoted to UAVs, this project initially benefit from several contributions [28–30]. In [28], it is developed an algorithm in which the heading angle of UAV relays is controlled for optimizing the performance of the ground-based terminals to the network base station relay links. In [29], a particular path planning of UAVs, also able to provide collision avoidance, is investigated and discussed, with the aim of achieving a continuous long-range communication relaying and, in [30], it is proposed another, optimal UAV deployment algorithm, based on the received signal strength and distance between nodes and UAV.

Among the studies in the context of emergency scenarios, in [31] it is presented a study on the use of UAVs for public safety communications during natural disasters, in case of damages on the network communication infrastructure and, in [32], a drone-assisted multi-hop D2D communication system is analysed in order to extend the network coverage in regions interested by disasters.

With an UAV acting as a relay, a twofold advantage could be provided by adopting an antenna array on the UAV, possibly with a beamforming (BF) mechanism: (i) an additional antenna array gain for enhancing weak signals coming from the emergency area (e.g. from indoor or shadowed environments), (ii) an additional information source regarding the angle of arrival of the communication link, useful for estimating the positions of the transmitting devices inside the emergency area. BF consists of a set of techniques for electronic steering of antenna beams, exploiting linear signal combinations at the input (transmitting) or output (receiving) of the antenna elements [33, 34]. In the project, one of the technical challenges is the integration of an antenna array in the UAV, because of its large size.

On the other hand, the use of relaying UAVs could be integrated also in a positioning system for enhancing the localization of emitting devices inside the emergency area. In all the wireless systems, including the current and future cellular mobile systems (4G–5G), the localization of fixed or moving devices is obtained by exchanging radio signals among reference base stations placed in known positions (called beacons or anchors) and the targets in unknown positions. In general, the target coordinates are computed by using a preliminary estimation of some physical parameters that can be derived from the received signals: (i) Time of arrival (TOA) and time difference of arrival (TDOA), (ii) Angle of arrival (AOA) or direction of arrival (DOA), (iii) Received signal strength (RSSI). RSSI measures are inexpensive and simple while the high-resolution TOA-based measures can require relevant resources in terms of sampling rate and computational complexity. Angles of arrival can be estimated by using antenna arrays, which search the angle with the best RSSI and this approach is the basis of the simulations and experiments in this project, together with the analysis of the same RSSI measures. A detailed review of the localization algorithms can be found in [35–43].

In the context of localization with UAVs, in [42] a TDOA cooperative approach, among a set of UAVs, is used for localizing a signal source with an explicit optimization based on Fisher Information Matrix. Also [43] presents a method based on TDOA for estimating the positions of a set of radars with a cooperative fleet of UAVs. Finally, [44] proposes a method for localization of slow-moving radio frequency sources with unknown transmitted strength of signal using an UAV also in absence of GPS signal or when the GPS signal is poor.

3.4 Network Connectivity and Routing

In emergency situations, the traditional telecommunication infrastructure is typically not functional due to: (1) deliberate target by the terrorists, (2) indirectly by some explosion and (3) overuse [45–47]. Sometimes the security/ rescue forces make it non-functional on purpose to isolate the terrorists. Therefore, a temporary communication network to ensure smooth and reliable communication between rescue operators and the victims is required.

Historically, at the network level this was realized by Mobile Ad hoc Networks (MANETs) and WSNs. Emergency MANETs (eMANETs) are deployed in emergency cases with the goal to provide communication for emergency workers with intelligent devices such as smartphones and Personal Digital Assistants (PDAs) [46]. Recently, an integration of UAVs in MANETs for routing to increase the coverage area is proposed [48]. The authors in [49] provided a comprehensive survey of multi-hop routing protocols for different classes of MANETs and integration of networking technologies for disaster response scenarios. Similarly, [50] highlighted the merits and demerits of MANETs and Delay Tolerant Networks

(DTNs) and presented the case of integrated MANET and DTN for improving the performance in dynamic environments. This survey also highlights the lack of realistic simulations models for the disaster environment. Kawamoto et al. presents the case of hybrid MANET and DTN implementation [51].

A single technology is not able to provide a complete solution. Therefore, a fusion of MANET based technologies such as MANETs, Vehicular Ad hoc Networks (VANETs), Flying Ad Hoc Networks (FANETs), WSNs and DTNs is a suitable choice. Authors in [52] reiterate the fact that most of the research work is simulation based. However, they have tried to present a detail of the real experimental work in this domain and concluded that it is feasible provided the interoperability issues are resolved. In [53], a survey of routing algorithms and mobility models proposed for MANETs, DTNs, VANETs, WSNs for communication under disaster scenarios is presented. It also highlighted the challenges, gaps between applications, protocols evaluations and mobility models. In another effort, dynamic routing in FANETs is discussed in detail for the case of self-organizing wireless networks in [54].

The success of any networking technology is also dependent on the underlying access technology. These days, user devices are equipped with multiple access and network technologies which allow these to communicate on multiple interfaces. This makes it more feasible to enable and facilitate D2D communication. Figure 3.2 shows a proposed framework for multi-hop D2D communication [49]. As seen below, both MANET and DTN are available choices for networking technologies. In addition, a gateway function is proposed which is responsible for filtering of redundant messages. An amalgam of access technologies is also available.

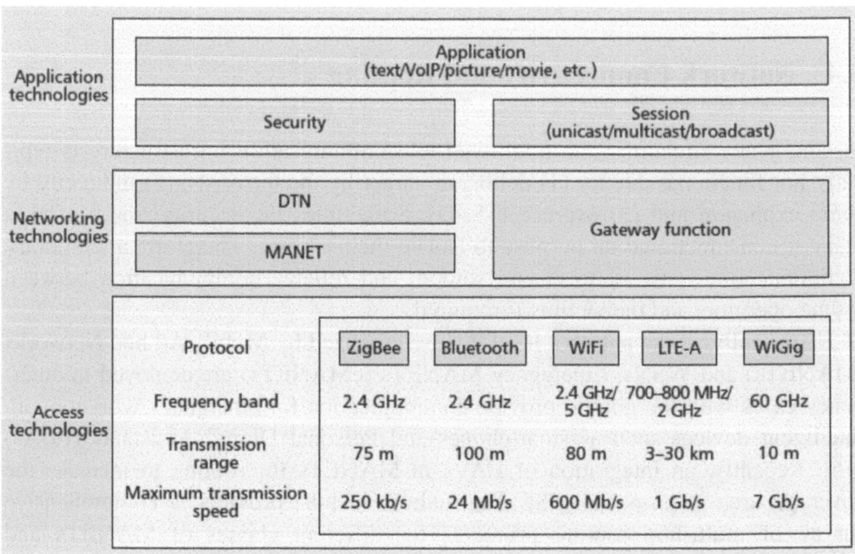

Fig. 3.2 Important factors for multi-hop D2D communications [45]

In [51], the authors propose a hybrid MANET and DTN approach with MANETs for static and dense scenario and DTNs for saving resources. These two paradigms work in a completely different way and a hybrid approach takes advantage of both. Integration of MANETs and DTNs is also discussed in [51, 52]. The performance of routing protocols based on this integration from the perspective of simulations and experimental testbed are shown.

MANETs based solutions are good but they suffer from configuration problems for the users. Authors in [55] highlighted the real-life deployment issues of existing multi-hop ad hoc networks in emergency scenarios. They targeted requirements for dedicated hardware, IP address allocation, network configuration and user's inability (due to not being familiar or in panic) to change their devices in to ad hoc mode and configure ad hoc networks. They also proposed a fast and simple internet connectivity solution based on Wireless multi-hop communication abstraction (WMCA) and the practical tree-based disaster recovery network (TDRAN) scheme which utilizes the software-based access node (SAN).

The emergency route selection scheme for D2D cellular communications during an urban terrorist attack is presented in [56]. It dynamically selects multi-hop routes for D2D communications in spectrum co-existence with completely congested conventional cellular network (CCN). In that work, different routing algorithms, namely, shortest path-routing (SPR), interference-aware-routing (IAR), and broadcast-routing (BR) are investigated. The comparison results show that there is a trade-off between different algorithms, for example, for a small D2D communication distances, both SPR and BR achieve slightly higher packet reception ratio (PRR) than IAR, whereas as the distance increases the impact of interference increases and therefore IAR scheme performs much better. Another interference-aware routing scheme is proposed in [53]. The scheme minimizes the hop-count in wireless D2D networks which can decrease not only the delay for D2D connections, but also reduces the power consumption. The proposed approach jointly takes the geometric information, interference constraint, and D2D rate requirements into account, and yields low computational complexity. Several routing algorithms are cited in [57] with emphasis on interference aware routing protocol; for example, shortest path, or shortest hop-counts and farthest neighbour routing approaches are discussed. In these approaches, the algorithm reaches the destination as fast as possible from the current node position with a condition that it achieves the data rate requirements of the D2D communications while maintaining the required QoS [57]. In [58] an energy-efficient heuristic multi-hop D2D routing algorithm was proposed which uses both channel reusing and power allocation policies to maximize the energy-efficiency. The performance analysis shows that the proposed algorithm outperforms Dijkstra, Closest to Destination (CD) routing and Ad-hoc on-demand distance vector (AODV). A multi-hop D2D communications scenario is presented in [59] and results show that the average energy-efficiency and spectral efficiency are enhanced when the number of hops used increases. A device-aware routing and scheduling algorithm (DARS) that takes into account device capabilities including computing power, energy limitations, and incentives is proposed [60]. The protocol is tested on a multi-hop D2D testbed using Android-based smartphones and tablets by exploiting Wi-Fi Direct and legacy Wi-Fi connections. The results show that DARS significantly improves throughput as compared to state-of-the-art routing protocols.

A centralized flow-based routing for D2D users in the dead spot is proposed in [61]. It takes into consideration the limited energy resources and computational complexity of relay nodes and tries to use their resources efficiently. The motivation comes from the fact that the distributed routing protocols waste a lot of resources in broadcast/multicast of control messages. It improves the utilization of resources by using an SDN solution to cater for specific service needs of the flow. Park [62] proposed a fast and energy efficient D2D multi-hop routing framework. It advocates the use of central presence server for exchange of management packets and also admits its limitation as a single point of failure. The proposed D2D routing method operates based on the geographic locations of nodes. The data packets are delivered through D2D mode. Additionally, a reliable data transfer algorithm is also proposed for scenarios when no geographic information is available. The proposed method results in reduced end to end delay compared to maximum progress method (MaxP), cost over progress method (CoP), and congestion-aware forwarder selection method (CAFS).

In [63] a Load Balancing Based Selective Ad hoc On Demand Multipath Distance Vector (LBSAOMDV) scheme for disaster recovery is discussed. It is an enhancement of the Ad hoc On Demand Multipath Distance Vector (AOMDV) protocol, which reduces the control traffic by decreasing the number of nodes while maintaining the QoS. Further, it intelligently selects the nodes which receives the route requests (RREQ) and optimizes the bandwidth utilization in comparison to AOMDV. Recently, the concept of multi-hop smart phone networks based on WiFi-Direct has been proposed in [64]. An energy efficient cluster-based routing protocol, called Quasi Group Routing Protocol (QGRP) is developed to address the energy issue which is critical due to high energy costs of the smart phones. In addition, a virtual hierarchical distributed cluster algorithm for smart phone networks is introduced. The simulations demonstrate that QGRP can save significant amounts of energy compared to the cases without QGRP. A framework for publishing content for D2D multi-hop routing is proposed in [65]. The framework is validated through android implementation.

D2D communication over multi-hop networks is very rare (especially for multi-standard), it requires multiple dimensions of complexity and cross layer features which are yet to be explored. The desired features the D2D multi-hop routing are high end to end throughput, low end to end delay and energy consumption. This coupled with technologies like machine learning and SDNs is deemed to fetch the desired results. As opposed to disaster management scenarios, in a terrorist attack DTN concept would be a strong candidate due to its store-and-forward capabilities in volatile connectivity conditions and would be addressed.

3.5 Conclusion

This paper has introduced the motivation, concept, challenges, and innovative solution ideas explored in the NATO – SPS "COUNTER-TERROR" G5482 project. At the time of writing, the project is well on track; ongoing and future work are related to e.g. joint performance evaluation of energy-spectral efficiency, end-to-end delay, impact of power allocation on D2D discovery processes, weak signal detection, user equipment localization, routing framework with clusters, energy efficiency of clustering vs. non-clustering, as well as integration and prototyping.

References

1. Zhao J, Chai KK, Chen Y, Schormans J, Alonso-Zarate J (2015) Joint mode selection and resource allocation for machine-type d2d links. Trans Emerg Telecommun Technol
2. Kai Y, Zhu H (2015) Resource allocation for multiple-pair d2d communications in cellular networks. In: Communications (ICC), 2015 IEEE International Conference on IEEE, 2015, pp 2955–2960
3. Feng D, Lu L, Yuan-Wu Y, Li GY, Feng G, Li S (2013) Device-todevice communications underlaying cellular networks. IEEE Trans Commun 61(8):3541–3551
4. Zhao J, Chai KK, Chen Y, Schormans J, Alonso-Zarate J (2015) Joint mode selection and resource allocation for machine-type d2d links. Trans Emerg Telecommun Technol
5. Min H, Lee J, Park S, Hong D (2011) Capacity enhancement using an interference limited area for device-to-device uplink underlaying cellular networks. IEEE Trans Wirel Commun 10(12):3995–4000
6. Yu G, Xu L, Feng D, Yin R, Li GY, Jiang Y (2014) Joint mode selection and resource allocation for device-to-device communications. IEEE Trans Commun 62(11):3814–3824
7. Zhao Y, Pelletier B, Pani D (2014) Resource allocation schemes for d2d communications. In: Systems, Applications and Technology Conference (LISAT), 2014 IEEE Long Island. IEEE, 2014, pp 1–5
8. An R, Sun J, Zhao S, Shao S (2012) Resource allocation scheme for device-to-device communication underlying lte downlink network. In: Wireless Communications & Signal Processing (WCSP), 2012 International Conference on. IEEE, 2012, pp 1–5
9. Esmat HH, Elmesalawy MM, Ibrahim II (2016) Adaptive resource sharing algorithm for device-to-device communications underlaying cellular networks. IEEE Commun Lett 20(3):530–533
10. Wang F, Song L, Han Z, Zhao Q, Wang X (2013) Joint scheduling and resource allocation for device-to-device underlay communication. In: Wireless Communications and Networking Conference (WCNC), 2013 IEEE. IEEE, 2013, pp 134–139
11. Luo Y, Shi Z, Zhou X, Liu Q, Yi Q (2014) Dynamic resource allocations based on q-learning for d2d communication in cellular networks. In: Wavelet Active Media Technology and Information Processing (ICCWAMTIP), 2014 11th International Computer Conference on IEEE, 2014, pp 385–388
12. Nie S, Fan Z, Zhao M, Gu X, Zhang L (2016) Q-learning based power control algorithm for d2d communication. In: Personal, Indoor, and Mobile Radio Communications (PIMRC), 2016 IEEE 27th Annual International Symposium on IEEE, 2016, pp 1–6
13. Mozaffari M, Saad W, Bennis M, Debbah M (2016) Efficient deployment of multiple unmanned aerial vehicles for optimal wireless coverage. preprint arXiv:1606.01962

14. Mozaffari M, Saad W, Bennis M, Debbah M (2016) Unmanned aerial vehicle with underlaid device-to-device communications: performance and tradeoffs. IEEE Trans Wirel Commun 15(6)
15. Kong M, Yorkinov O, Tran THV, Shimamoto S (2011) Elevation angle-based diversity access employing high altitude platform station and unmanned aerial vehicle (UAV) for urban area communications
16. Li Y, Cai L (2017) UAV-assisted dynamic coverage in a heterogeneous cellular system. IEEE Netw
17. Cileo D, Sharma N, Magarini M (2017) Coverage, capacity and interference analysis for an aerial base station in different environments. In: Proceedings of International Symposium on Wireless Communication Systems (ISWCS 2017), August 2017, Bologna, Italy
18. Sudheesh PG, Magarini M, Muthuchidambaranathan P (2016) Achieving maximum system capacity in multiple-high altitude platforms through interference alignment. In: Proceedings of International Conference on Industrial and Information Systems (ICIIS 2016), December 2016
19. Sudheesh PG, Sharma N, Magarini M, Muthuchidambaranathan P (2017) Interference alignment in multiple-high altitude platforms based communication with a generalized long distance line-of-sight channel model. In: Proceedings of ICCMIT, April 2017
20. Mahmood MA, Seah WKG, Welch I (2015) Reliability in wireless sensor networks: a survey and challenges ahead. Comput Netw 79:166–187., ISSN 1389-1286
21. Scazzoli D, Mola A, Silverajan B, Magarini M, Verticale G (2017) A redundant gateway prototype for wireless avionic sensor networks. In: Proceedings of IEEE International Symposium on Personal, Indoor and Mobile Radio Communications (PIMRC 2017, Special Session on Mission Critical Communciations: Advancements, Challenges, and Opportunities), October 2017, Montreal, QC, Canada
22. Scazzoli D, Kumar A, Sharma N, Magarini M, Verticale G (May 2017) Fault recovery in time-synchronized mission critical ZigBee-based wireless sensor networks. Int J Wireless Inf Networks:1–10
23. Pinkney FJ, Hampel D, Di Pierro S (1996) Unmanned Aerial Vehicle (UAV) communications relay. Proc Milcom 96:47–51
24. Orfanus D, de Freitas EP, Eliassen F (2016) Self-organization as a supporting paradigm for military UAV relay networks. IEEE Commun Letters
25. Sadek AK, Han Z, Liu KR (2010) Distributed relay-assignment protocols for coverage expansion in cooperative wireless networks. IEEE Trans Mobile Comput 9(4):505–515
26. Feng H, Cimini LJ (2012) On optimum relay deployment in a multihop linear network with cooperation. In: IEEE Military Communications conference
27. Bao VNQ, Thanh TT, Nguyen TD, Vu TD (Jun. 2013) Spectrum sharingbased multi-hop decode-and-forward relay networks under interference constraints: performance analysis and relay position optimization. J Commun Netw 15(3):266–275
28. Zhan PC, Yu K, Swindlehurst AL (2011) Wireless relay communications with unmanned aerial vehicles: performance and optimization. IEEE Trans Aerosp Electron Syst
29. Cetin O, Zagli I (2012) Continuous airborne communication relay approach using unmanned aerial vehicles. J Intell Robot Syst
30. Hanif U, McClean S, Nixon P, Parrt G, Luo C (2017) An optimal UAV deployment algorithm for bridging communication
31. Merwaday A, Guvenc I (2015) UAV assisted heterogeneous networks for public safety communications. In: IEEE Wireless Communication and Networking Conference. Workshops (WCNCW), March 2015
32. Li X, Guo D, Yin H, Wei G (2015) Drone-assisted public safety wireless broadband network. WCNC
33. Godara LC (1997) Applications of antenna arrays to mobile communications, part I: performance improvement, feasibility, and system considerations. Proc IEEE 85(7)
34. Godara LC (1997) Application of antenna arrays to mobile communications, part II: beamforming and direction-of-arrival considerations. Proc IEEE 85(8)

35. Patwari N, Ash J, Kyperountas S, Hero A, Moses R, Correal N (2005) Locating the nodes: cooperative localization in wireless sensor networks. Signal Process Mag IEEE 22(4)
36. Mao G, Fidan B, Anderson BD (2007) Wireless sensor network localization techniques. Comput Netw 51(10)
37. Wang J, Ghosh R, Das S (2010) A survey on sensor localization. J Control Theory Appl 8(1)
38. Guvenc I, Chong C-C (2009) A survey on TOA based wireless localization and NLoS mitigation techniques. IEEE Commun Surv Tutorials 11(3):107–124
39. Macagnano D, Destino G, Abreu G (2012) A comprehensive tutorial on localization: algorithms and performance analysis tools. Int J Wireless Inf Networks 19(4)
40. Gezici S (2008) A survey on wireless position estimation. Wirel Pers Commun 44(3):263–282
41. Reggiani L, Fiorina J, Gezici S, Morosi S, Najar M (2013) Radio context awareness and applications. J Sens
42. Purvis KB, Astrom KJ, Khammash M (2008) Estimation and optimal configurations for localization using cooperative UAVs. IEEE Trans Control Syst Technol 16(5)
43. Mallick P, Routray A, Mohapatra J, Jana K (2014) Minimum variance estimators for enemy radar localization by unmanned aerial vehicles. In: IEEE International Conference on Electronics Computing and Communication Technologies (IEEE CONECCT), 2014
44. Dehghan SMM, Moradi H (2016) SLAM-inspired simultaneous localization of UAV and RF sources with unknown transmitted power. Trans Inst Meas Control 38
45. Legendre F, Theus H, Felix S, Bernhard P (2011) 30 years of wireless ad hoc networking research: what about humanitarian and disaster relief solutions? What are we still missing? In: Proceedings of the 1st ACM international conference on wireless technologies for humanitarian relief
46. Panaousis EA, Ramrekha TA, Millar GP, Politis C (2010) Adaptive and secure routing protocol for emergency mobile ad hoc networks. Int J Wireless Mob Netw 2(2)
47. Nishiyama H, Ngo T, Oiyama S, Kato N (2015) Relay by smart device: innovative communications for efficient information sharing among vehicles and pedestrians. IEEE Veh Technol Mag 10(4):54–62
48. Reina DG, Askalani M, Toral SL, Barrero F, Asimakopoulou E, Bessis N (2015) A survey on multihop ad hoc networks for disaster response scenarios. Int J Distrib Sens Netw 2015, Article ID 647037, 16 pages
49. Nishiyama H, Ito M, Kato N (2014) Relay-by-smartphone: realizing multihop device-to-device communications. IEEE Commun Mag 52(4):56–65
50. del Pilar Salamanca M, Camargo J (2016) A survey on IEEE 802.11-based MANETs and DTNs for survivor communication in disaster scenarios. In: 2016 IEEE Global Humanitarian Technology Conference (GHTC), Seattle, WA, pp 197–204
51. Kawamoto Y et al (2013) Toward terminal-to-terminal communication networks: a hybrid MANET and DTN approach. (Invited Paper). In: IEEE 18th International Workshop on Computer Aided Modeling and Design of Communication Links and Networks, Berlin, Germany, September 2013, pp 228–232
52. Reina DG et al (2014) A survey on ad hoc networks for disaster scenarios. In: International Conference on Intelligent Networking and Collaborative Systems, Salerno, pp 433–438
53. Rosas E, Hidalgo N, Gil-Costa V, Bonacic C, Marin M, Senger H, Arantes L, Marcondes C, Marin O (2016) Survey on simulation for mobile ad-hoc communication for disaster scenarios. J Comput Sci Technol 31(2):326–349
54. Rosati S, Kruželecki K, Heitz G, Floreano D, Rimoldi B (2016) Dynamic routing for flying ad hoc networks. IEEE Trans Veh Technol 65(3):1690–1700
55. Minh QT, Shibata Y, Borcea C, Yamada S (2016) On-site configuration of disaster recovery access networks made easy. Ad Hoc Netw 40:46–60
56. Hu Y, Guo W, Wang S (2014) "Emergency route selection for D2D cellular communications during an urban terrorist attack. In: IEEE International Conference on Communications (ICC)
57. Ren P, Du Q, Sun L (2013) Interference-aware routing for hopcount minimization in wireless d2d networks. In: IEEE/CIC International conference on communications in China – workshops (CIC/ICCC)

58. Babun L (2015) Extended coverage for public safety and critical communications using multi-hop and D2D communications. M.Sc dissertation, Florida International University
59. Jingyi Z, Xi L, Quansheng X (2015) Multi-hop routing for energy-efficiency enhancement in relay-assisted device-to-device communication. J China Univ Posts Telecommun 22(2):1–51
60. Ali K, Nguyen HX, Shah P, Vien QT, Ever E (2017) D2D multi-hop relaying services towards disaster communication system. In: 24th International Conference on Telecommunications (ICT), Limassol, pp 1–5
61. Xing Y, Seferoglu H (2017) Device-aware routing and scheduling in multi-hop device-to-device networks. 2017 Information Theory and Applications Workshop (ITA), San Diego, CA, USA, pp 1–7
62. Tanha M, Sajjadi D, Tong F, Pan J (2016) Disaster management and response for modern cellular networks using flow-based multi-hop device-to-device communications. In Proceedings of the IEEE Vehicular Technology Conference, VTC-Fall'16, pp 1–7
63. Park J (2016) Fast and energy efficient multihop D2D routing scheme. Int J Distrib Sens Netw 12(5)
64. Tata C, Kadoch M (2014) Multipath routing algorithm for deviceto-devicecommunicationsforpublic safety over lte heterogeneous networks. In: 1st International conference on information and communication technologies for disaster management (ICTDM).
65. El Alami M, Benamar N, Younis M, Shahin AA (2017) A framework for hotspot support using Wi-Fi direct based device-to-device links. In: Proceedings of the. 13th International Wireless Communications and Mobile Computing Conference (IWCMC), Valencia, pp 552–557

Chapter 4
Flash Crowd Management via Virtualized Network Resources (FALCON)

Liljana Gavrilovska, Alberto Leon-Garcia, Valentin Rakovic, Daniel Denkovski, Simona Marinova, Vladimir Atanasovski, Thomas Lin, and Hadi Bannazadeh

Abstract The FALCON project, presented in this paper, focuses on dynamic on-demand virtual resource allocation for wireless network environments, leveraging a highly flexible and adaptable wireless network to provide reliable and efficient communications in flash crowd scenarios and emergency situations.

4.1 Introduction

Network virtualization has been identified as one of the potential technologies that can meet the underlying 5G technical requirements in radio access and core networks, fostering efficient network and resource sharing and real-time and flexible scheduling. By exploiting solutions of Software Defined Everything (SDx) and virtualization, *network virtualization* can increase the wireless system scalability and flexibility and can overcome several challenges in the current legacy wireless systems, such as interference and high power consumption [16]. Network virtualization enables networks to become programmable, adaptable, centrally managed and cost effective. It is identified as one of the most impactful technologies for future networks, especially addressing emergency and flash crowd scenarios.

Flash crowd scenarios are characterized by a significant upsurge of the traffic demand and/or user connectivity due to either high concentration of users per unit area (e.g. sport events, concerts etc.) or high volume of users connection demands

L. Gavrilovska (✉) · V. Rakovic · D. Denkovski · V. Atanasovski
Faculty of Electrical Engineering and Information Technologies, Ss. Cyril and Methodius University in Skopje, Skopje, N. Macedonia
e-mail: ljana@feit.ukim.edu.mk

S. Marinova · A. Leon-Garcia · T. Lin · H. Bannazadeh
Department of Electrical and Computer Engineering, University of Toronto, Toronto, ON, Canada

C. Palestini (ed.), *Advanced Technologies for Security Applications*, NATO Science for Peace and Security Series B: Physics and Biophysics,
https://doi.org/10.1007/978-94-024-2021-0_4

as a result of emergency scenarios arising out of natural (e.g. earthquakes, floods) or man-made (e.g. terrorist attacks, industrial accidents, transportation failures) disasters.

The latest research developments in the area [10] focus on introducing elastic virtualization and processing that adapt to the fluctuations in capacity demand and analyze the radio remote head density. The work in [8] introduces the notion of efficient on-demand resource provisioning, and proposes innovative techniques where clustering, consolidating, and cooperation of virtual baseband processing can improve the overall system performance. The authors in [2, 9, 14, 15] focus on improving the joint virtual computing and resource allocation in *network virtualization* by introducing advanced cloud-based solutions and optimization algorithms. However, most of the current works are strictly theoretical and none of them provide functional solutions for emergency and flash crowd situations in general.

This paper presents the **FALCON project approach** [13] for building an agile, flexible and scalable network architecture that is capable of addressing the demands in flash crowd scenarios, with emphasis on emergency situations. The FALCON architecture leverages existing technologies and exploits the virtualization of wireless network resources. The project also focuses on developing and designing a novel self-autonomous network entity called Virtual Resource Manager (VRM), which can orchestrate the operation of various underlying wireless networks by optimally mapping cloud hardware resources, such as CPU and memory, based on the existing user requirements and demands. The role of the VRM becomes especially important in scenarios with uneven spatio-temporal traffic distribution and user connectivity demands.

The paper is organized as follows. Section 4.2 discusses the FALCON's system requirements and compares the FALCON system to legacy solutions. Section 4.3 gives an overview of the FALCON architecture and deployment. Section 4.4 presents set of specific prove of concept results. Section 4.5 concludes the paper.

4.2 FALCON Requirements and Technology Comparison

Conventional network systems are not capable, or only partially capable, of meeting the emergency scenario system requirements. These requirements clearly call for new networking solutions, such as SDN/NFV that leverage highly flexible, efficient, and robust network deployments. The FALCON architecture builds upon these solutions and provides a novel network infrastructure also capable of addressing the flash crowd scenario requirements.

In emergency and flash crowd scenarios, the first responder teams are the focal end users. They pose strict service requirements with respect to network availability, scalability, reliability and latency. Several distinct system requirements are specific to flash crowd scenarios and emergency related scenarios, such as *Agile network service provisioning, Disaster resilience, Network access priority, Mission-critical*

Network technology	Agile service provisioning	Disaster resilience	Network access priority	Mission-critical service support	Backhaul scalability and robustness	Cybersecurity
Conventional networks	Hours	Very limited capabilities for static and mobile deployments	Partial. Requires significant network configuration	Fragmented. Services are technology/ system dependent	Relying only on node redundancy	System dependent. Limited possibilities for extensions
SDN/NFV	Enabled by software implementation on NFs	On-the-fly and efficient static and mobile deployment	Different levels of priority to different user groups	Can support all services within the same system	Can facilitate dynamic reconfiguration of the backhaul	Highly flexible and extendable for new security mechanisms
FALCON	Seconds	Supported	Ongoing work	Yes	Supported	Ongoing work

Fig. 4.1 Technology comparison

service support, Backhaul robustness and scalability, Cybersecurity. Figure 4.1 compares the capabilities of conventional network systems and network systems based on SDN/NFV versus the FALCON design. FALCON strives to address the given system requirements by introducing a novel network infrastructure based on state of the art concepts, utilizing network virtualization.

4.3 FALCON Approach

The FALCON project focuses on dynamic on-demand virtual resource allocation for wireless network environments, leveraging a highly flexible and adaptable wireless network [11]. The architecture is able to provide reliable and efficient communications in flash crowd scenarios and emergency situations by taking into consideration the required QoS profile. In particular, FALCON focuses on developing and designing a novel software-defined and virtualized cloud-based system that provides efficient and flexible adaptation of the underlying communication infrastructure.

4.3.1 FALCON Architecture

The FALCON system aims to support the operation of various underlying wireless technologies and the dynamic and optimal allocation of cloud resources (CPU and RAM), to augment the operation of available wireless network resources according to the wireless environment and its behavior in case of the traffic demands. Figure 4.2, depicts the generic FALCON architecture. It consists of *FALCON RAN*, i.e. the virtualized C-RAN (comprised of the RRHs and VBUs) and the *FALCON CORE*, represented by the Cloud Core datacenter. Moreover, FALCON leverages

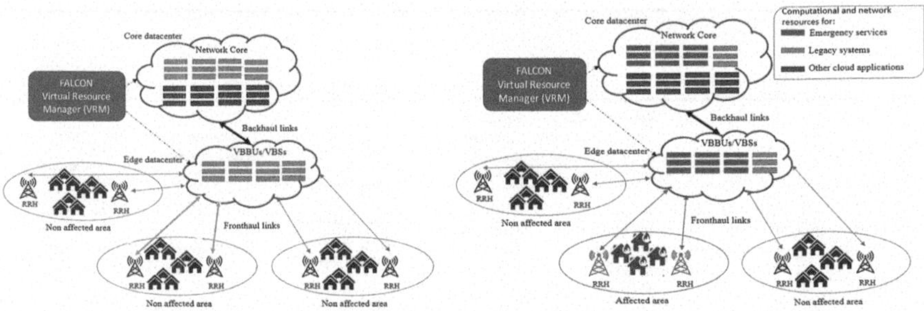

(a) (b)

Fig. 4.2 FALCON generic system architecture. (**a**) Disaster free scenario. (**b**) Disaster scenario

the aspects of Multi-Access Edge Computing (MEC) [5], by placing the latency sensitive services at the network edge. FALCON's two-tier solution mitigates the practical deployment and complexity issues related to an single-tier end-to-end solution by organizing the system capabilities and functionalities into two physically and logically complementary entities, i.e. the FALCON RAN and the FALCON CORE.

Figure 4.2, depicts a generic example of an emergency scenario, where the FALCON system dynamically adapts to the underlying conditions. Specifically, the available computational resources (e.g. CPU, RAM, etc.) are dynamically reallocated in order to foster the operation of the FALCON system in the affected area. In conventional systems, this would have not been possible, because they rely on static configuration and deployment. Moreover, the system uses the available resources in the most efficient manner in order to foster the required service in the affected and in the non-affected areas. This decision making is vital for optimal performance of the system.

The *FALCON RAN* has two fundamental components, Remote Radio Heads (RRHs) and Virtual BaseBand Units (VBBUs)/Virtual Base Stations (VBSs). The RRHs contain the RF hardware of the system and are located at remote sites. They transmit and receive wireless signals, but they do not perform any baseband signal processing. By utilizing Software Defined Radios (SDRs) as the RRHs, the FALCON RAN decouples the RAN and underlying signal processing from the physical RF hardware. Hence FALCON fosters the possibility to efficiently virtualize and deploy different types of wireless technologies, such as, GSM, LTE, LTE-M, SigFox, etc [6, 12]. This decoupling enables dynamic and flexible implementation of different RANs without any hardware extensions. The RAN-specific baseband processing is performed in a softwarized virtualized environment of VBBUs that are housed separately in centralized processing pools located within edge datacenters, and are connected to the RRHs via high speed/low latency fronthaul links. Furthermore, the decoupling of the functionalities of the RRHs and the VBBUs/VBSs enables remote control and configuration as well as the independent evolution of each component. It must be noted that the optimal performance of the RAN heavily depends on the fronthaul design.

The *FALCON CORE* exploits the benefits of NFV and SDN to facilitate the core network capabilities of the cellular systems. It offers fully flexible and programmable network, which is able to respond to the surges in traffic demands. FALCON CORE can customize the placement and configuration based on the service required by the focal users in the emergency scenarios. For instance, delay sensitive services (i.e., mission critical delay sensitive signalling such as push-to-talk signalling and video) will require values less than 60 ms [4], which can be achieved by placing the FALCON CORE in proximity to the RAN. However, if the requirements are not very stringent, a distributed cloud deployment (i.e., FALCON RAN in one datacenter, FALCON CORE in another), which further improves the performance and scalability of the platform, is desired. Furthermore, the communication between the components in the architecture (i.e., RRHs to VBBUs, and VBBUs to FALCON CORE) is leveraged by SDN equipment. Hence, FALCON can adapt to changing underlying scenario requirements and conditions using dynamic QoS enforcement.

The communication between the components in the architecture (i.e., RRHs to VBBUs, and VBBUs to FALCON CORE) leverages SDN. Hence, FALCON can adapt to changing underlying scenario requirements and conditions using dynamic QoS enforcement. Moreover, the architecture supports a distributed cloud deployment (i.e., FALCON RAN in one datacenter, FALCON CORE in another), and thus improves the performance, scalability and agility of the C-RAN platform by bringing the centralized processing pools closer to the wireless nodes [3].

4.3.2 FALCON Network Virtualization Approach

FALCON stems from two fundamental virtualization techniques, applied for the RAN and CORE virtualization. Specifically, FALCON utilizes the *container-based virtualization* for the RAN and the *hypervisor-based virtualization* for the CORE.

The hypervisor-based framework runs complete isolated guest operating systems (OS) as virtual machines (VMs) which results in lengthy deployments and slow booting up. Also, hypervisor-based virtualization provides scaling by modifying the VM's computational resources, CPU and memory. However, the necessity to suspend the VM during vertical rescaling results in a delay (service unavailability) in order of seconds up to several minutes.

Container-based virtualization provides on-the-fly reallocation of computational resources of a single container without suspending the operation of the container. This provides significantly faster scaling of computational resources, in order of milliseconds. As shown in [5] container-based resource scaling of a virtualized LTE system results in delays less than 20 ms using a moderate computing hardware. Due to the dynamic nature of C-RAN operations and strict low-latency and high reliability requirements, especially in flash crowd scenarios, containerization is a more suitable solution to RAN virtualization, whereas the CORE leverages both VM and container virtualization technologies. In FALCON, containers are operated as isolated instances within the OS. They do not deploy a full OS as they share the host OS kernel, which makes them significantly lighter and faster to boot up. Using orchestration engines for container management can further enhance the system by providing: the option to specify the amount of computational resources, monitoring, automated resource scaling, and custom-made schedulers to achieve efficient resource allocation. FALCON uses its own developed orchestration engine, in order to minimize the workload from operating such softwares. This makes FALCON's RAN virtualization more agile and resource efficient.

4.4 Experimental Results

This section presents experimental results that demonstrate the operational proof of concept of the FALCON system and its advantages. The results focus on several system aspects such as *Service Deployment Time, Reconfiguration*, and *Programmable Network Management*. These system features are especially important in emergency scenarios where prolonged system outages and/or high service blocking rate can have detrimental consequences.

The experimental FALCON platform, comprises three core logical entities, Remote Radio Heads (RRHs) and Multi-access Edge Computing (MEC) segment, and Core segment implemented on commercial-grade hardware components. The RRHs are implemented on a Software Defined Radio (SDR) platform. SDRs foster the implementation of specific hardware capabilities using only software tools. The platform, uses the Universal Software Radio Peripheral (USRP) X310 devices, developed by National Instruments. The RAN-specific baseband processing is performed in a softwarized virtualized environment of BBUs. The virtualization of the BBU's is provided on the MEC segment of the platform that incorporates a container-based virtualization and utilizes the Docker framework. It runs on DELL T640 servers. FALCON uses a commercial LTE BBU software, Amarisoft that implements a commercial-grade full stack LTE Rel. 14. The Core segment is hosted on the multi-region and multi-tenant SAVI testbed [7]. SAVI is fully SDN-enabled, featuring advanced network management, which is indispensable for FALCON in order to enable various QoS requirements and backhaul robustness. The emergency scenario of interest focuses on a homogeneous wireless system setup, executed over the FALCON system with full stack LTE capabilities. Table 4.1 presents the experiment parameters of interest.

Table 4.1 Experiment parameters

Experiment parameters	Parameter value
Channel bandwidth [MHz]	5, 10, 20
No. of eNB	1
Antenna mode	SISO, 2 × 2-MIMO
LTE band number	26
Modulation	64 QAM
UE model	Samsung Galaxy S9

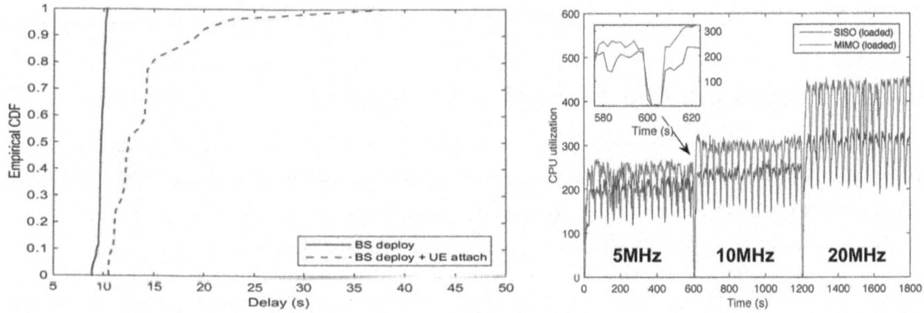

(a) (b)

Fig. 4.3 Experimental results. (**a**) Empirical CDF of LTE deployment time. (**b**) LTE BS reconfiguration during flash crowd

4.4.1 Service Deployment Time and Reconfiguration

Figure 4.3a presents the empirical cumulative distribution function of the FALCON system deployment time, including the UE attachment time. The results show that the FALCON system is deployed in a period of approximately 10 s for 100% of the cases and it takes additional 10 s for the UE to find the newly deployed base station and attach (90% of the cases). This is significantly less than the 5GPP recommendation that the average creation time must not be higher than 90 min [1]. Thus, the results in Fig. 4.3 prove that the FALCON system implementation is capable to address the 5GPP performance requirements and respond to the dynamism in the emergency situations and minimize the network outage time.

Figure 4.3b depicts the time series of the CPU utilization and the transitions between LTE bandwidths of 5, 10 and 20 MHz, for the SISO and the MIMO case, in case of flash crowd occurrences. Specifically, the scenario focuses on the case when the traffic demand is significantly increased (flash crowd occurring on 600 and 1200 s) such that the active LTE configuration is not able to support the aggregate traffic demand. Thus, the system needs to reconfigure using a higher channel bandwidth. In conventional systems, this will require prolonged periods of time, in the order of minutes, for the system to reconfigure. In the case of the FALCON system, the reconfiguration delay is in order of few seconds. Moreover, conventional systems require manual intervention for the reconfiguration process, while the FALCON system is envisioned to use the VRM module that tracks the underlying traffic conditions and automatically adapts. The presented results prove the flexibility and the swiftness of the FALCON system and its ability to adapt to the underlying communication demands, which is especially important in emergency situations.

4.4.2 Programmable Network Management

For emergency scenarios it is important to enable coexistence of multiple mission-critical services, which can be achieved by proactively changing the QoS parameters in the network. Figure 4.4 shows a guaranteed bit rate (GBR) service that provides information exchange between the first responders and affected users. Another service (e.g., video stream) appears around the 10th and the 20th second. To successfully admit the new users, FALCON reduces the throughput of the active GBR service. The result shows that in emergency scenarios, the FALCON system will able to adapt to the users' requests, and support the requirements for mission-critical services.

The presented results show that the developed FALCON system is capable of agile and dynamic reconfiguration that leverages reliable communication in variety of scenarios such as emergency and flash crowd situations.

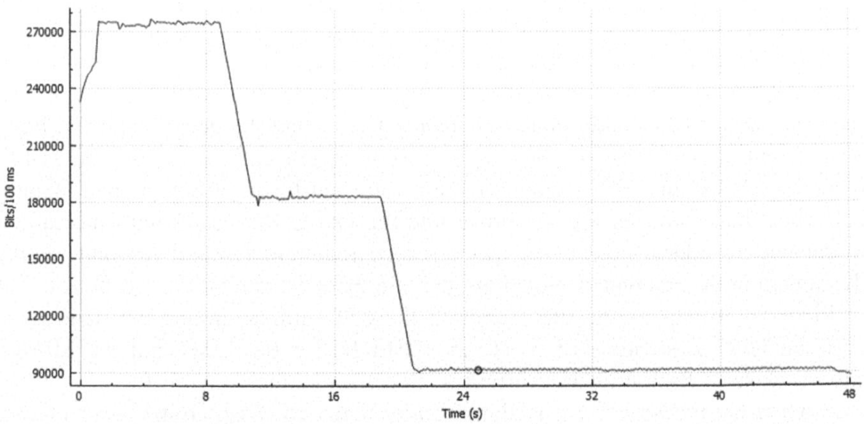

Fig. 4.4 Dynamic QoS enforcement for FALCON

4.5 Conclusion

This paper presented the FALCON project that aims to establish a novel networking and computing infrastructure that will be resilient to flash crowds and common attacks (e.g. terrorist attacks, emergency situations etc.) by virtualizing the network resources and adaptively dedicating them to the entities of interest in a secure and real-time manner. The developed FALCON system is capable of on-the-fly adaptations of the network resources in order to foster the best communication performances to all affected users. The FALCON project strongly focuses on increased reliability and robustest to efficiently support efficient communication in flash crowd scenarios, thus alleviating the loss of communication or coordination among the users.

The FALCON project is still ongoing and continues to build upon the developed system. Future work will focus on introducing an AI-based virtual resource orchestrator that will be able to efficiently map the cloud computing resources onto the physical network resources and predict possible flash crowd occurrences. Moreover, the future work will also test the developed FALCON system in different real-world scenarios focusing on aspects such as scalability, network robustness, fronthaul design, etc.

References

1. 5GPPP (2019) Creating a Smart Ubiquitous Network for the Future Interne, https://5g-ppp.eu/wp-content/uploads/2014/02/Advanced-5G-Network-Infrastructure-PPP-in-H2020_Final_November-2013.pdf
2. D'Oro S, Marotta MA, Both CB, DaSilva L, Palazzo S (2019) Power-efficient resource allocation in C-RANs with SINR constraints and deadlines. IEEE Trans Veh Technol 68:6099–6113
3. ETSI (2018) MEC in 5G networks. https://www.etsi.org/images/files/ETSIWhitePapers/etsi_wp28_mec_in_5G_FINAL.pdf
4. For 4G LTE Q (2019) QoS for 4G LTE. https://5g-tools.com/qos-for-4g-lte/. Online; accessed 16 Aug 2019
5. Gavrilovska L, Rakovic V, Denkovski D (2018) Aspects of resource scaling in 5G-MEC: technologies and opportunities. In: 2018 IEEE Globecom workshops (GC Wkshps), pp 1–6
6. Gavrilovska L, Rakovic V, Ichkov A (2018) Virtualization approach for machine-type communications in multi-RAT environment. Wirel Pers Commun 100:67–79
7. Lin T, Park B, Bannazadeh H, Leon-Garcia A (2018) Deploying a multi-tier heterogeneous cloud: experiences and lessons from the SAVI testbed. In: NOMS 2018–2018 IEEE/IFIP network operations and management symposium, pp 1–7
8. Liu Q, Han T, Ansari N (2019) Energy-efficient on-demand resource provisioning in cloud radio access networks. IEEE Trans Green Commun Netw, 3(4): 1142–1151, Dec 2019, https://doi.org/10.1109/TGCN.2019.2926287.
9. Luong P, Gagnon F, Despins C, Tran L (2018) Joint virtual computing and radio resource allocation in limited fronthaul green C-RANs. IEEE Trans Wirel Commun 17:2602–2617
10. Pompili D, Hajisami A, Tran TX (2016) Elastic resource utilization framework for high capacity and energy efficiency in cloud RAN. IEEE Commun Mag 54:26–32

11. Rakovic V, Denkovski D, Gavrilovska L (2018) Practical implementation of Cloud-RAN: FALCON's approach. In: European conference on networks and communications
12. Rakovic V, Ichkov A, Marinova S, Todorovski D, Atanasovski V, Gavrilovska L (2017) Dynamic virtual resource allocation in virtualized multi-RAT cellular networks. Wirel Pers Commun 97:1677–1692
13. SpS F (2017) FALCON Project. http://falcon.feit.ukim.edu.mk
14. Tran TX, Le DV, Yue G, Pompili D (2018) Cooperative hierarchical caching and request scheduling in a cloud radio access network. IEEE Trans Mobile Comput 17:2729–2743
15. Vu TT, Ngo DT, Dao MN, Durrani S, Nguyen DHN, Middleton RH (2018) Energy efficiency maximization for downlink cloud radio access networks with data sharing and data compression. IEEE Trans Wirel Commun 17:4955–4970
16. Wang K, Yang K, Magurawalage CS (2018) Joint energy minimization and resource allocation in C-RAN with mobile cloud. IEEE Trans Cloud Comput 6:760–770

Chapter 5
ThreatPredict: From Global Social and Technical Big Data to Cyber Threat Forecast

Jérôme François, Frederic Beck, Ghita Mezzour, Kathleen M. Carley, Abdelkader Lahmadi, Mounir Ghogho, Abdellah Houmz, Hicham Hammouchi, Mehdi Zakroum, Narjisse Nejjari, and Othmane Cherqi

Abstract Predicting the next threats that may occurs in the Internet is a multifaceted problem as the predictions must be enough precise and given as most as possible in advance to be exploited efficiently, for example to setup defensive measures. The ThreatPredict project aims at building predictive models by integrating exogenous sources of data using machine learning algorithms. This paper reports the most notable results using technical data from security sensors or contextual information about darkweb cyber-criminal markets and data breaches.

Keywords Threat · Prediction · Darknet · Darkweb · Data breaches

5.1 Introduction

Predicting future cyber-threats with reasonable lead-time and accuracy can give security practitioners sufficient time to prepare for upcoming major attacks. For example, these practitioners can increase network provisioning to deal with an

J. François (✉) · F. Beck · A. Lahmadi
Inria – LORIA – University of Lorraine, Nancy, France
e-mail: jerome.francois@inria.fr; frederic.beck@inria.fr; abdelkader.lahmadi@loria.fr

G. Mezzour · M. Ghogho · A. Houmz · H. Hammouchi · M. Zakroum · N. Nejjari · O. Cherqi
International University of Rabat, Rabat, Morocco
e-mail: ghita.mezzour@uir.ac.ma; mounir.ghogho@uir.ac.ma; abdellah.houmz@uir.ac.ma; hicham.hammouchi@uir.ac.ma; mehdi.zakroum@uir.ac.ma; narjisse.nejjari@uir.ac.ma; othmane.cherqi@uir.ac.ma

K. M. Carley
Carnegie Mellon University, Pittsburgh, PA, USA
e-mail: kathleen.carley@cs.cmu.edu

C. Palestini (ed.), *Advanced Technologies for Security Applications*, NATO Science for Peace and Security Series B: Physics and Biophysics,
https://doi.org/10.1007/978-94-024-2021-0_5

upcoming major Denial-of-Service (DoS) attack or purchase security insurance. These preventive measures can stop future attacks or at least reduce their impact.

However, most prior work focused on detecting attacks instead of predicting them. A notable exception [12] predicts future cyber rates by taking into account statistical temporal properties of cyber attacks. Unfortunately, it only achieves reasonable accuracy for a limited prediction time horizon (a few hours). Liu et al. [8] forecasts security incidents on organizations by using security features about the organization such as network mismanagement symptoms. Unfortunately, the authors were unable to predict geo-politically motivated attacks.

A promising direction to improve these prediction models is to take into account exogenous global social and technical data such as Twitter sentiments and Darkweb trades and discussions. Social and traditional media data has been successfully used to predict civil unrest [10], disease outbreaks [7], and the stock market [1]. Exogenous technical events such as the discovery of new software vulnerabilities are also known to affect cyber attack rates.

ThreatPredict aims so at developing machine-learning models that predict future cyber threat rates up to a few days in advance. Our models will take into account past cyber attack temporal evolution as well as global social, geo-political, economical, and technical events that are known to affect cyber attack rates.

To reach our ultimate goal of predicting cyber threats days in advance, three steps must be achieved:

- Data collection: the goal is to collect all relevant data and to preprocess them to clean and index them for further correlation;
- Data analysis: based on collected data, the first analysis consists into refine them in order to have a more fine-grained view about activities we can observe. For example, a security sensor can emit a large number of alerts even if they all concern a single attack. This is usually the case for scanning or DDoS attacks that are launched from multiple hosts.
- Threat prediction: once all data is indexed and refined, we can pursue the analysis by mining historical data to predict and characterize the next threats or attacks to occur will be.

This paper focuses on describing relevant exogeneous data sources for cyber-security in Sect. 5.2 but also on an initial sets of data-analysis results. First, darknet data will be investigated to show how such raw attack traffic data can be better integrated into machine learning algorithms to support predicting scanning activities of cyber-criminals (Sect. 5.3). We will also leverage contextual information such as twitter data to increase our predictive model for attack rates. Finally, Sects. 5.4 and 5.5 focus on contextual data, darkweb market trends and data breaches respectively.

5.2 Overview

As briefly mentioned in Introduction, we rely on various sources of data to infer the future threats of attacks. Figure 5.1 summarizes them. The initial set of data are collected by security sensors in our own infrastructure. They provide very fine-grained information about attacker activities and with a large history over multiple years. We can differentiate the darknet and the honeypots. Honeypots are artificial systems with vulnerabilities the attackers can interact with them. They allow to observe attacker behaviors on different services such as SSH, telnet, web or even industrial systems. However, this paper mostly focuses on darknet data analysis. A darknet or telescope is using a whole subnetwork, which is announced over Internet such that packets sent to the IP addresses are properly routed over. However, in a darknet, this subnetwork does not host any services or devices and so no legitimate traffic is supposed to reach it. The entity hosting the darknet is then silently collecting all incoming packets, i.e. without replying to any of them. Such an infrastructure is mainly characterized by the size of the subnetwork defined by the prefix length. In our case, we host a darknet of a /20 network meaning that we monitor more than 4000 adresses.

Darknet data consists so in network traffic flow, which are monitored in real-time. It is completed by other security technical data that are freely accessible over Internet as highlighted on top of Fig. 5.1. They are used for two purposes. First, data collected by the darknet are raw data. So, all traffic is abnormal but classifying and grouping observed attacks is not trivial. We rely on external information such as blacklists or attack and threats reports. The latter contains information about hosts (IP addresses) which have been involved in attacks. These IP addresses are matched against our darknet data and serve as a refinement of data but also to validate our classification algorithms. Indeed, darknet data can be only partially enriched or annotated through public information. We thus develop methods to automatically perform this task (mainly to group traffic flows by attack instances

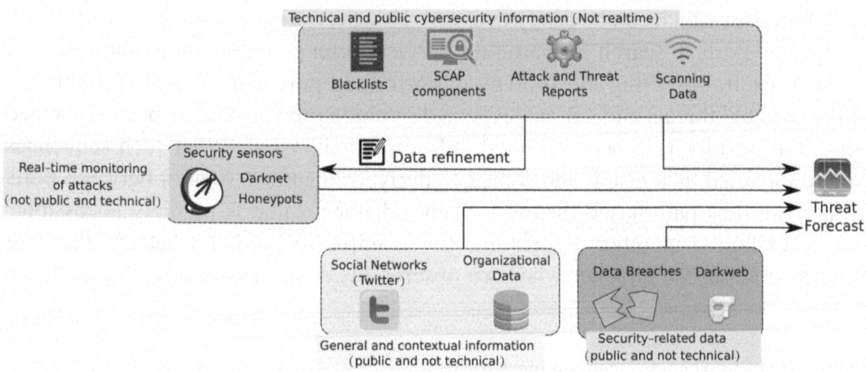

Fig. 5.1 ThreatPredict overview

and label them with the attack type). Other technical and public data refer to SCAP (Security Content Automation Protocol) components such as Common Vulnerabilities and Exposures (CVE) or Common Attack Pattern Enumeration and Classification (CAPEC) provided by MITRE and to scanning data like Rapid7 sonar project[1] or Censys [3]. Scanning data allows to evaluate the exposure of certain devices or services and indirectly their risk of being attacked if there exist past or new threats against them, which can be discovered in SCAP components.

All sources previously described can be qualified as technical. However, the added value of ThreatPredict is to leverage non technical data. Firstly, evidences about past attacks can be found in PRC[2] (Privacy Rights Clearinghouse) data breaches providing evidences of security issues. They can be used to assess the risk of exposure of a given organization to a data breach. this risk can be measured by the configuration of the network, which can be used to forecast future attacks. Finally, to build these predictive models contextual information is considered. For example, analysis of sentiment or trendy topics over Twitter are good indicators to identify potential victims of attacks, e.g. major events usually attracts attackers. Also, Darkweb tradings and discussion could reveal important indicators on hackers interests and future attacks. Moreover, Organizational data such as economic value would give insights about an organization and could be a valuable predictor in our predictive models.

5.3 Darknet Monitoring and Prediction

5.3.1 Predicting TCP Scans

When performing an attack, the first stage usually consists in identifying the potential targets. Discovering accessible machines and services often relies on IP sweeping and/or scanning TCP and UDP ports. Naive approach testing all ports numbers and all IP addresses of a targeted subnetwork is time-consuming and has a large footprint which can be easily detected. However, smarter attackers would look for specific ports to search for particular services with potential vulnerabilities.

As a result, observing the strategy of performed port scans revealed the nature of the targeted threats and can be leveraged to predict the next sequence of scanned ports. Full results have been reported in [4] and demonstrated that such sequences can be modeled as a graph and mined to derive semantic distances between ports using a shortest path-based distance. Such a distances that is publicly accessible[3] can be exploited in other algorithms, for example to predict what are the next targeted ports by an attacker when we observe his or her first scans. We analyzed

[1] https://www.rapid7.com/research/project-sonar/
[2] https://www.privacyrights.org/
[3] http://port2dist.lhs.inria.fr/

around 815 million of darknet packets collected during 9 months to infer the values of distances. Based on them, K-nearest neighbors (with K = 3) has been applied to the MAWI Labs dataset [9] to deduce future port scans. We consider the period from 2 to 9 September 2015 (except the 5th and 7th of September because of dataset unavailability) with a total of 590,173,645 IP packets with a mean of 98,362,274 packets a day. Our technique has been successfully applied since 99.98% (True Positive Rate) of scans have been predicted and would have been ideally proactively blocked while the false positive is limited 0.0015%. The interested reader is encouraged to read [4] for all methodological details and results.

5.3.2 Prediction Attack Rates

In this section we study different aspects of the traffic captured by the /20 network telescope including the prediction of the traffic rate.

As a first step, we performed an exploratory data analysis of 3 years of traffic. We found that 90% of probing activities are targeting only 550 ports of the port's space. The latter includes remote access services which are the most sought by network probers, followed by database management systems, web services and miscellaneous services as well.

Network probers tend to scan only a limited number of ports. Thus, extracting the network probers exploration patterns reveals their interest in terms of services and also the sets of related ports. We model these behaviors using transition graphs decorated with probabilities of switching from a port to another. Each graph represents the scanning activities of one prober, where a node represents a port targeted by the prober, and the edges represent the probability of transition between ports. The results of the analysis showed strong relations between ports of similar services, as depicted in Fig. 5.2, such as the ports used for remote access including the port 23 (TelNet), 22(SSH) and their alternatives 2323 and 2222, and the same is found for database systems, web services etc.

Besides the descriptive analysis, observing a significant disparity between the predicted probing rate and the actual value may help detecting an imminent threat. Furthermore, exploiting the similarities found between ports can improve the predictions. Thus, we extracted time series of traffic targeting each port in the darknet, then we assessed the capacity of Non-stationary Autoregressive (AR) and Vector Autoregressive (VAR) models in predicting port probing rates. We applied the two models on the extracted time series of probing rates for each port. The result in Fig. 5.3 represents the R^2 score of the models, applied on different ports' time series with different time resolutions. We conclude that VAR models generally give better results than AR models, as they use data of related ports to predict rates on the target port. We found that the non-stationary AR model produces satisfying results for services exhibiting low short term probing rates variability such as the ports 23, 2323 and 445, with an R^2 of more than 0.90 for time resolutions ranging

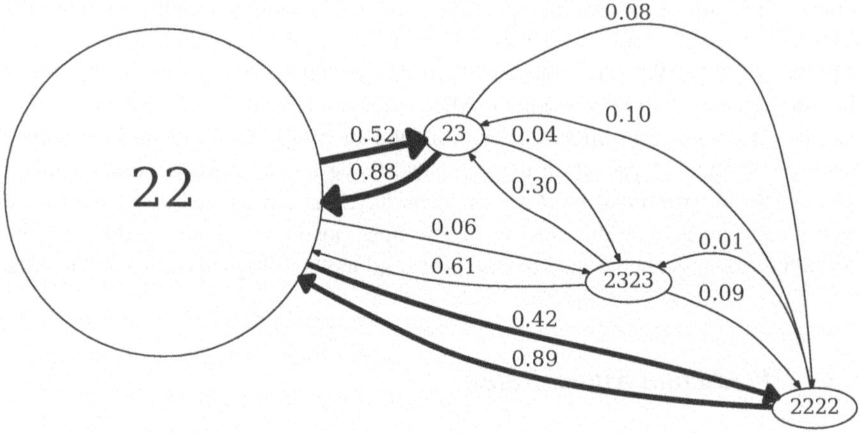

Fig. 5.2 Targeted probing pattern (remote access services)

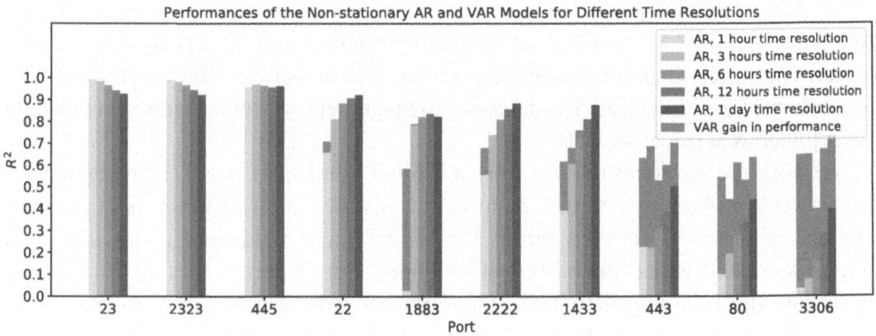

Fig. 5.3 Comparison of performances of the non-stationary AR and VAR models

from one hour to one day. Where non-stationary VAR model consistently produces better results for services exhibiting high probing rate variability. More details could be found in the paper [11].

5.3.3 Probing Rate Severity Prediction

Since we analyzed the probing activities and performed probing rates prediction hours in advance, we explore in this section the aspect of predicting probing rate severity by leveraging twitter sentiments as exogenous data [6]. We investigate the relationships between probing rates and Twitter sentiments towards France through a statistical analysis. It revealed a significant correlation between the darknet attack rates and negative sentiments which further motivated the use of these sentiments as additional features for prediction. We cluster the probing rates into four homogeneous levels of severity using spectral clustering, and we cast the prediction as a classification problem. Then, we predict future rate levels using several classifiers. Incorporating negative sentiments to our models allows to predict the next day severity with 81% accuracy using AdaBoost classifier. It is equivalent to a 5% improvement compared to only using past probing rates for prediction. This demonstrates the importance of incorporating social signals in models when predicting future probing rates.

5.4 Darkweb Analysis

Another research aspect to consider is Darkweb since it contains important amounts of information on hackers tradings and discussions. The Darkweb is a subset of the Deep Web, which cannot be accessed using standard web browsers. Notably, tt is used by cyber-criminals to offer all kinds of illegal services and goods.

In this section, we analyze the illegal trade in the darkweb's marketplaces to provide a cyber threat intelligence. The analysed services include the most common and severe malwares, zero-day threats, exploits, DDOS services etc. Full details are available in [2].

We focused on analyzing four of the most popular and well known marketplaces: Agora, Silkroad 2.0, Alphabay and Nucleus. We retrieved the advertisements that have been published and extracted all information about the offers. We were able to collect a total of 972.655 items for our analysis, starting from 7 July 2013 until 20 December 2015.

The main information we collected about the products are: Vendor's alias, Item's Category, Item's description, Price of the article, Seller's location, Destinations vendor ships to, Rating given by consumers to vendors based on a stars rating scale.

Our descriptive analysis highlights the large extent of this phenomenon with markets worth up to millions of USD each year. It is a scale of organized crime, well anchored in the field and whose demand and supply is booming. Also we were able to highlight the existence of well organized cells that monopolize the market. Indeed, 98% of the market is controlled by only 20% of sellers. In addition to this, it is a market that is based mainly on the reputation of sellers and the feedback of customers. Users are generally satisfied by the outcome of their transactions.

Being proactive is the best and most effective defense. By gathering threat intelligence from a largely anonymous and underground network, we could obtain targeted, relevant dark web intelligence, and facilitate the understanding of threats in order to prevent or mitigate the damages. For example, getting alerts if a breached data is found on the Dark Web, allows to take active measures to reduce the legal liability.

5.5 Data Breaches Analysis

Cyber attacks are diverse and having a bigger picture on cyber threats requires analyzing different sources of data. We explore in this section another perspective of attacks that is not purely technical but allows to give context about attacks and threats. Indeed, data breaches represent a permanent threat to all types of organizations since they disclose personal and confidential information to the public. In accordance with the new regulations such the EU General Data Protection Regulation (GDPR), and with the objective of predicting these breaches in advance to prevent the damage, we conducted an exploratory analysis to study the frequency, trends, and the causes of these leaks. In this work [5], we considered a dataset of data breaches mainly encountered in the USA, consisting of about 9,000 breaches involving organizations in different sectors, collected in the period between 2005 and 2019. These breaches led to the loss of 11.5 Billion individual records, which have a significant financial loss and technical impact. There exist several methods that involves a breach, from insiders (INSD) and unintended disclosure (DISC) to hacking activities (HACK) as illustrated in Fig. 5.4. Hacking is the most prevalent method with 29% and also the most devastating in terms of breached records and financial impact. For example, the Yahoo breach as the biggest reported breach was

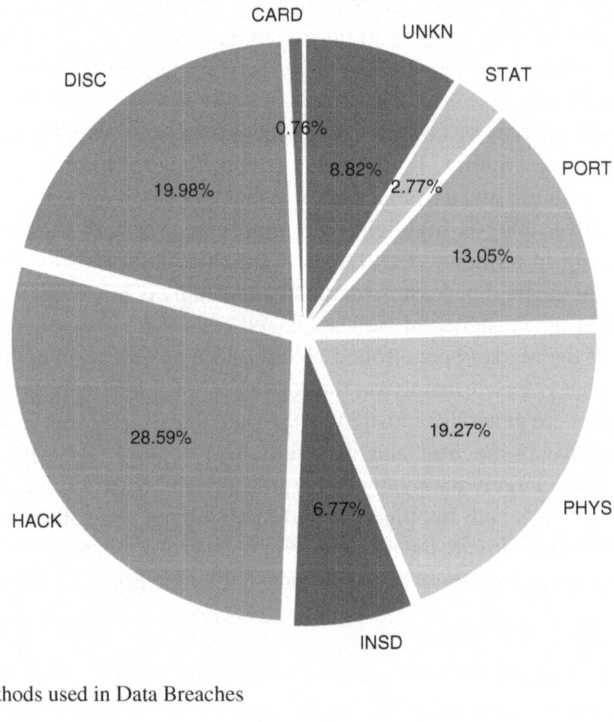

Fig. 5.4 Methods used in Data Breaches

estimated to have a financial loss of **$444B**. In addition, we found that the most targeted sectors are medical institutions and online services and platforms, which could be explained by the amount of sensitive and personal data they hold on their users which makes them the top target to the attackers. From 2014, we observed that the frequency of attacks has increased significantly in medical and online sectors with an interarrival time (time between two consecutive attacks) of 4 and 7.5 days respectively.

As breaches due to hacking are the most interesting to research on, we are currently working on predicting these breaches by analyzing an organization's network and looking at several aspects such as DNS misconfigurations, IP blacklists, and expired certificates in order to assess the risk of exposure to breach attacks.

5.6 Conclusion

In this paper, we introduced the ThreatPredict project that pursues the goal to provide an accurate forecasting of internet threats day in advance. It resides in co-jointly analyzing various sources of security and contextual data such as darknet, honeypots, darkweb, data breaches... Although the sources of data can should be constantly updated to capture the multiple sides of the security threats, we showed in this paper the veracity of some of them, in particular darknet, Twitter, darkweb and data breaches. Although the analysis is mostly limited to a single or two sources of data, the next stage of the project will be focused on improving the joint analysis of multiple sources of data.

Acknowledgments This work has been partially supported by the NATO Science for Peace and Security Programme under grant G5319 Threat Predict: From Global Social and Technical Big Data to Cyber Threat Forecast and by the project SecureIoT, funded from the European Union's Horizon 2020 research and innovation programme under grant agreement no. 779899.

It is also supported by the High Security Lab (https://lhs.loria.fr) providing the French darknet data and we gratefully thank its technical director, Frederic Beck from Inria Nancy Grand Est. We also thank the NICT providing the Japan darknet data (https://www.nicter.jp).

We acknowledge the support from the National Center of Scientific and Technical Research (CNRST), Rabat, for the grant of an excellence scholarship.

References

1. Bollen J, Mao H, Zeng X (2011) Twitter mood predicts the stock market. J Comput Sci 2:1–8
2. Cherqi O, Mezzour G, Ghogho M, El Koutbi M (2018) Analysis of hacking related trade in the darkweb. In: 2018 IEEE international conference on intelligence and security informatics (ISI), pp 79–84
3. Durumeric Z, Adrian D, Mirian A, Bailey M, Halderman JA (2015) A search engine backed by internet-wide scanning. In: Proceedings of the 22Nd ACM SIGSAC conference on computer and communications security

4. Evrard L, François J, Colin J-N (2019) Attacker behavior-based metric for security monitoring applied to darknet analysis. In: Proceedings of IFIP/IEEE international symposium on integrated network management, Washington, DC
5. Hammouchi H, Cherqi O, Mezzour G, Ghogho M, El Koutbi M (2019) Digging deeper into data breaches: an exploratory data analysis of hacking breaches over time. Proc Comput Sci 151:1004–1009
6. Hammouchi H, Mezzour G, Ghogho M, El Koutbi M (2019) Predicting probing rate severity by leveraging twitter sentiments. In: 2019 15th international wireless communications & mobile computing conference (IWCMC), pp 883–888
7. Lazer D, Kennedy R, King G, Vespignani A (2014) The parable of Google flu: traps in big data analysis. Science 343:1203–1205
8. Liu Y, Sarabi A, Zhang J, Naghizadeh P, Karir M, Bailey M, Liu M (2015) Cloudy with a chance of breach: forecasting cyber security incidents. In: 24th USENIX security symposium (USENIX security 15), Washington, DC
9. Johan M, Romain F, Kensuke F (2014) A taxonomy of anomalies in backbone network traffic. In Proceedings of 5th International Workshop on TRaffic Analysis and Characterization (TRAC), pp. 30–36
10. Ramakrishnan N, Butler P, Muthiah S, Self N, Khandpur R, Saraf P, Wang W, Cadena J, Vullikanti A, Korkmaz G, Kuhlman C, Marathe A, Zhao L, Hua T, Chen F, Lu CT, Huang B, Srinivasan A, Trinh K, Getoor L, Katz G, Doyle A, Ackermann C, Zavorin I, Ford J, Summers K, Fayed Y, Arredondo J, Gupta D, Mares D (2014) 'Beating the news' with EMBERS: forecasting civil unrest using open source indicators. In: Proceedings of the 20th ACM SIGKDD international conference on knowledge discovery and data mining
11. Zakroum M, Houmz A, Ghogho M, Mezzour G, Lahmadi A, François J, El Koutbi M (2018) Exploratory data analysis of a network telescope traffic and prediction of port probing rates. In: 2018 IEEE international conference on intelligence and security informatics (ISI). IEEE, pp 175–180
12. Zhan Z, Xu M, Xu S (2015) Predicting cyber attack rates with extreme values. IEEE Trans Inf Forensics Secur 10:1666–1677

Chapter 6
Analysis, Design and Implementation of an End-to-End QKD Link

Marina Mondin, F. Daneshgaran, F. Di Stasio, S. Arnon, J. Kupferman, M. Genovese, I. Degiovanni, F. Piacentini, P. Traina, A. Meda, M. Gramegna, I. Bari, O. Khan, and M. Khan

Abstract This manuscript discusses the most relevant aspects of the practical implementation of a long-range Quantum Key Distribution (QKD) link with trusted nodes, achieving the highest possible secret key rate generation within the security and system level constraints. To this purpose, we report on recent pilot studies for the measurements of detection efficiency and source photon statistics for validating the calibration facilities (i) at telecom wavelength for realistic quantum backbone implementation through standard telecommunications grade optical fiber, and (ii) for the telecom and VIS-NIR regime. In addition, since there are circumstances when a fiber optical link may not be available, we will also discuss the characterization of a Free Space Optics (FSO) QKD link. Finally, the manuscript also discusses the problem of information reconciliation in Continuous Variable QKD (CV-QKD) scenarios.

Keywords Quantum key distribution · Detection efficiency · Free space optics · Information reconciliation

---

M. Mondin (✉) · F. Daneshgaran · F. Di Stasio
California State University Los Angeles, Los Angeles, CA, USA
e-mail: marina.mondin@calstatela.edu

S. Arnon · J. Kupferman
Ben Gurion University of the Negev, Beer-Sheva, Israel

M. Genovese · I. Degiovanni · F. Piacentini · P. Traina · A. Meda · M. Gramegna
INRIM, Torino, Italy

I. Bari · O. Khan · M. Khan
NUCES, Peshawar, Pakistan

C. Palestini (ed.), *Advanced Technologies for Security Applications*, NATO Science for Peace and Security Series B: Physics and Biophysics,
https://doi.org/10.1007/978-94-024-2021-0_6

6.1 Calibration Facilities for Realistic Quantum Backbone Implementation: Detection Efficiency and Source Photon Statistics

The development of fiber based point-to-point quantum networks for absolute secure quantum communication requires the exploitation of the most advanced available quantum technology on the market and of the use of innovative diagnostic tools to prevent possible quantum hacking attacks.

A possible solution for the development of a Quantum Key Distribution (QKD) [1, 2] network is the installation of QKD devices on existing optical fiber links connecting metropolitan areas and the secure communication, both classical and quantum, between the most important cities.

The point–to-point connection can be then ensured by the realization of QKD trusted nodes. Nevertheless, metropolitan fiber links are subject to very high losses. It is of utmost importance to exploit state-of-the-art measurements techniques that metrology for quantum technology [3, 4] is developing in order to pave the way to implementation of secure practical QKD systems and networks.

In this section, we report on pilot studies for the measurements of detection efficiency and source photon statistics for validating the calibration facilities at telecom wavelength, together with an analogue comparison characterisation in the VIS-NIR regime, and the realisation of pilot measurement comparisons to validate the techniques developed.

As a route to the validation of the measurement facilities, four European NMIs (INRIM, PTB, NPL and CMI) started two pilot studies on two key measurands in the 1550 nm region correlated with fibre-based QKD systems, i.e. the detection efficiency of single-photon detectors, and the Glauber second-order auto-correlation function of a pseudo single-photon source. Measurement protocols and procedures were developed on purpose.

Firstly, the pilot study towards a comparison on the measurements of detection efficiency of SPADs in the 1550 nm region, exploited a free-running InGaAs/InP SPAD-based detector. Single-photon avalanche diodes (SPAD) based on InGaAs/InP semiconductor materials are the most exploited detectors in many quantum technologies [5, 6]. The successful development of such new technologies and products requires the solution to a number of metrological challenges; for this reason a metrological characterization in terms of detection efficiency, dead time, after pulsing and dark counts of single photon detectors is mandatory. A pilot study to compare different detection efficiency measurement strategies at the wavelength of 1550 nm was performed by four European National Metrology Institutes: CMI, INRIM, NPL and PTB. The device under test was a commercial free-running fibre-coupled InGaAs/InP single-photon detector. The setup and the reference standard used as well as a detailed estimation of the measurement uncertainty of the detection efficiency was compared. The DUT was a fiber pigtailed free running SPAD (Id Quantique ID220), with nominal detection efficiency of 10% and dead time D of 10 μs. All the participants provided quantum efficiency measurement with the detector illuminated by a pulsed laser source, a commercial short-pulse laser source (ID Quantique, id300), which is based on a Distributed-feedback laser diode at 1550 nm. The measurement was carried out with the common repetition rate of 110 KHz. The exact wavelength of the source is measured with an optical spectrum analyzer (Anritsu MS974 OA). The measurement principle used by all participating laboratories for determining the detection efficiency of the InGaAs/InP SPAD detector was based on the substitution method. In a general scheme, adopted by all the participants, the output of the laser was sent to a device that provide a variable calibrated attenuation to attenuate light at single photon level. The detection efficiency was estimated by comparing the optical power measured by the DUT with the incident mean optical power per laser pulse measured with an analogue calibrated detector. Data are still under analysis but it seems that an excellent agreement (within the uncertainty) is obtained.

Secondly, the pilot study towards a comparison on the measurements of the Glauber second order autocorrelation function in the telecom range achieved a good agreement within the uncertainty. The source used for this test was a CW heralded single-photon source emitting real single photons at 1550 nm. The Pilot study was carried on jointly in the INRIM labs.

Single-photon sources represent one of the fundamental tools for quantum information, quantum metrology and related quantum technologies. Because of this, the need for their characterization and standardization is becoming of the utmost relevance. Here, we show a procedure providing a quantitative estimate of the multi-photon component of a CW single-photon source, showing the results achieved in a pilot study [7] for the measurement of the second-order autocorrelation function $g^{(2)}(0)$ of a low-noise heralded single-photon source (HSPS) prototype operating at telecom wavelength , i.e. 1550 nm.

In our setup (Fig. 6.1), our source hosts a CW laser (532 nm) pumping a 10 mm $\times$ 1 mm $\times$ 10 mm periodically-poled lithium niobate (PPLN) crystal to produce non-degenerate SPDC. Our signal and idler photons, respectively with wavelengths 1550 nm and 810 nm, are filtered and coupled to single-mode fibers (SMF). The detection of an idler photon by the SPCM-AQR silicon single-photon avalanche detector heralds the arrival of a 1550 nm signal photon, addressed to a 20 m long single-mode optical fiber connected to an electro-optical shutter (OS) operated by a fast pulse generator controlled by a field programmable gate array (FPGA). For each heralding signal, the FPGA operates the pulse generator in order to open our HSPS output channel, i.e. OS channel A, for a time window of 7 ns in correspondence of the passage of a 1550 nm photon.

For this joint measurement, the HSPS output is addressed to a 50:50 fiber beam splitter (FBS) whose outputs are sent to two Hanbury Brown & Twiss (HBT) interferometers, one belonging to INRiM and the other to the guest NMI (PTB or NPL), allowing simultaneous data collection between INRiM and the guest NMI to avoid mismatch due to some drift in the HSPS output over time. Every HBT is composed of two infrared InGaAs-InP SPADs, be they free-running or gated (when gated, the SPADs are triggered by the same FPGA signal opening the OS), whose outputs are sent to time-tagging coincidence electronics. Actually, the parameter

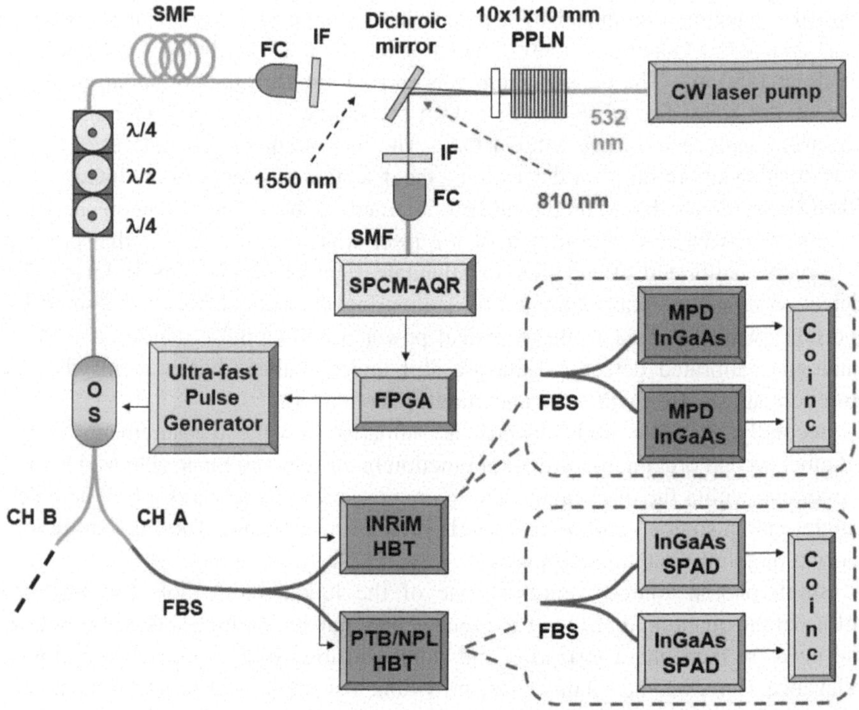

Fig. 6.1 Experimental setup for the joint estimation of the multi-photon component in the emission of a low-noise CW single-photon source in the infrared (1550 nm)

Table 6.1 Comparison of the three measurement campaigns

Measurement session	α (INRiM)	α (NPL)	α (PTB)
INRiM – PTB	0.016 ± 0.006	–	0.04 ± 0.05
INRiM – NPL	0.013 ± 0.008	0.02 ± 0.02	–

we evaluate to characterize the emission of our source is $\alpha = \frac{P_{12}^{(ph;ph)}}{P_1^{(ph)} \cdot P_2^{(ph)}} \cong g^{(2)}(0)$, where $P_{12}^{(ph;ph)}$ and $P_i^{(ph)}$ $(i = 1, 2)$ are, respectively, the probability of a coincidence count between the two HBT SPADs and the photon count probability for each of the HBT SPADs (dark counts subtracted). After a careful analysis of the setup parameters for a proper evaluation of the uncertainties associated to our measurements, we obtained the results below.

As visible in the Table 6.1 above, the results of the whole measurement campaign (involving different measurement setups and data collection methodologies) are all in agreement within the experimental uncertainties reported (coverage factor: k = 1).

This pilot study, involving INRiM (also responsible for the source realization), NPL and PTB, represents a step forward towards a robust procedure for the characterization of this kind of single-photon sources.

Concerning with the pilot study [8, 9] towards a comparison on the measurements of the Glauber second order autocorrelation function in the VIS-NIR regime, the source used for this test was a pulsed single photon source based on nitrogen-vacancy in diamond emitting single photons.

In order to establish standard techniques in single-photon metrology, device-independent and reproducible methods for source characterization are needed. Measurement of the $g^{(2)}(0)$ parameter is of utmost importance in characterizing and understanding single-photon sources emission.

In this paragraph we report on the pilot studies, performed by INRIM, NPL and PTB, on the complete characterization of a test source based on a single colour centre in diamond excited in pulsed regime and emitting radiation in the visible range. More details can be found in [7].

This comparison was hosted at INRIM from October 16 to October 29 2017 and was composed of two joint measurements of $g^{(2)}(0)$, on the same emitter: one performed by INRIM and PTB and the other one by INRIM and NPL.

The experimental setup [9] is composed of a laser-scanning confocal microscope whose signal is split by a 50:50 beam-splitter and connected to two measurement devices, i.e. two single-photon sensitive Hanbury Brown & Twiss (HBT) interferometers. The excitation light, produced by a pulsed laser (48 ps FWHM, 560 pJ per pulse) emitting at 532 nm with a repetition rate $R = 2.5$ MHz was focused by a 100X oil-immersion objective on the nano-diamond (ND) sample hosting an SPS based on a single Nitrogen Vacancy (NV) center of negative charge, with emission in a broad spectral band starting approximately at 630 nm and ending at 750 nm. The optical filters used were a notch filter at 532 nm and two long-pass filters (FEL600 and FEL650). The photoluminescence signal (PL), thus occurring in a 650–750 nm spectral range, was collected by a multimode fiber and split by a 50:50 beam-splitter (BS). As stated above, each end of the BS was connected to a separate HBT setup used for the joint measurement. In particular each faciliy was composed of:

- INRiM: a fused 50:50 fibre beam-splitter connected to two Excelitas SPCM-AQR-14-FC single-photon avalanche detectors (SPADs). Single and coincidence counts were sampled via ID Quantique *ID800* time-to-digital converter (60 ps time resolution).
- NPL: a fused 50:50 fibre beam-splitter connected to two Perkin-Elmer SPCM-AQR-14-FC single-photon avalanche detectors (SPADs). Coincidence counts were sampled via PicoQuant *HydraHarp 400* multichannel picosecond event timer (1 ps time resolution).
- PTB: a fused 50:50 fibre beam-splitter connected to two Excelitas SPCM-AQR-14-FC single-photon avalanche detectors (SPADs). Single and coincidence counts were sampled via PicoQuant *HydraHarp 300* multichannel picosecond event timer (4 ps time resolution).

Table 6.2 Comparison of results obtained by the three partners (INRIM, PTB and NPL)

	INRIM	PTB
$g^{(2)}(0)$	0.079 ± 0.009	0.076 ± 0.007
	INRIM	**NPL**
$g^{(2)}(0)$	0.065 ± 0.005	0.068 ± 0.005

The validity of the technique is demonstrated by compatibility of the results obtained by the three partners (see Table 6.2), demonstrating a system-independent (and unaffected by the non-ideality of the apparatus), estimate of $g^{(2)}(0)$, emission lifetime ($\tau = 15.34 \pm 0.08$) ns and their uncertainty.

This study will greatly benefit the single-photon metrology community, as well as rapidly-growing quantum-technology-related industries. The main results of this study was the development of a standardized measurement technique as well as an uncertainty estimation procedure.

6.2 Hacking and Security in Free-Space QKD System

This section reviews the challenge that arise in a realistic implementation of a free space quantum key distribution (QKD) system. QKD exploits quantum mechanics to achieve secure communication, by enabling two parties to produce and to share secret key to encrypt and decrypt messages. Practical QKD systems have been constructed deployed all over the world. However, while in theory QKD systems are secure there are still opportunities for attacks on the system, since actual devices have non ideal physical characteristic hat can be exploited. We focus here on one attack that exploits a detector non ideal physical characteristic, called backflash or Breakdown flash hacking. Many QKD systems use very sensitive detector, which detection is typically done using single photon avalanche photodiodes (SPADs). In this type of system Alice sends a photon to Bob and he use his SPAD detector to detect the photon. However, due to unwanted phenomena or side effects, sometime the detector also emit a secondary photon which the eavesdropper (Eve) could detect and achieve information from it. This secondary emission is called backflash, and its physical characteristics depend on the structure of the detector. This is not a new phenomenon and has already been identified in the past for silicon and InGaAs/InP detectors [10, 11]. A single photon avalanche photodiodes (SPADs) detector is operated in a Geiger mode. In Geiger mode, the photo diode detector is reverse-biased above the breakdown voltage. As a result, signal photoelectrons can create an avalanche in high probability. In addition in low probability around 0.4% backflash photon is created [12]. In conventional SPAD the nominal detection efficiency is about 10%, as a results the backflash could contain at least 0.04 photons emerging from the devices. This may result in a considerable amount of information leakage that has to be considered in practical QKD implementations generate a self-sustaining discharge.

The Backflash spectral distribution, timing and information leakage for the different types of detectors has been measured and characterized in various detectors. The backflash was examined for InGasAs/InP SPADs at telecom wavelength, and for visible silicon SPADs in. Their setup used an attenuated laser sending photonsss at 15550 nm. The back reflected light was analyzed and for the prototype detector this was nearly 10%, for the commercial one it was around 6%. Information leakage was affected by detector voltage bias (since higher voltage bias means more carriers in the avalanche. The spectral distribution between 1000 and 1600 *nm* of Backflash for InGaAs SPADs is examined in [12]. In [13] the authors tested commercial silicon SPADs, and demonstrated that an eavesdropper can distinguish detector clicked. Base on this work it seems that backflash can reveal information about the optical electrical system that the photon propagates One solution to the reduce information leakage by backflash is to insert proper characterized passive optical devices into the system, in particular spectral filters (isolators, spectral filters, circulators), and use short gates and small avalanches. The best solution would be use of superconducting nanowire QKD detector, which would probably have less backflash, but this is not commercially in usc at this point.

6.2.1 Backflash Principle

Backflash is a results of secondary photons that are emitted during the avalanche process, due to recombination of electrons and holes in the APD junction [14]. The backflash emission timing and spectrum depend on the detector, so Eve can gain information about the detector and the optical path the photon pass and use it to exploit other weaknesses. For example in a BB84 protocol Eve could know which detector the photon came from [13] because some photons from the vertical polarization detector will be emitted back through the vertical polarizer, and will themselves be vertically polarized.

In Fig. 6.2 we can see the typical scenario under consideration: Alice sends a signal to Bob, and Bob's detector emits backflash through free space, which can then be utilized by Eve. For now, we ignore other optical elements of the system. To study the backflash, we suggest the following a mathematical model describe in the next lines. The probability for Eve to receive a photon due to backflash as a function of pointing direction error angle α and the distance separation between the Bob and Eve and is

$$P_E(\alpha) = \eta_q \frac{K_1 L(\alpha)}{Z^2}$$

where K_1 is a constant, η_q is a quantum efficiency and $L(\alpha)$ is the pointing loss factor (both defined below). The constant K_1 is

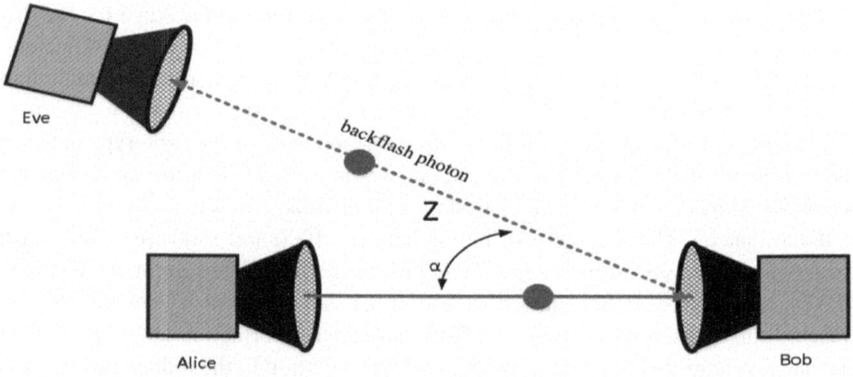

Fig. 6.2 General scheme for use of backflash in eavesdropping

$$K_1 = N_B\, \eta_B\, \eta_E \left(\frac{\lambda}{4\pi}\right)^2 L_A G_B G_E$$

L_A represents the atmospheric attenuation, η_E, η_B and G_B, G_E are the optical efficiencies and the telescope gain of the Bob and Eve, respectively. The backflash wavelength is λ, N_B is the probability of backflash due to absorption of photon, and $L(\alpha)$ is the pointing loss factor. The gain of Eve's telescope is $G_E = \left(\frac{\pi D_E}{\lambda}\right)^2$, where D_E represents the aperture diameter for Eve's telescope. The gain of Bob's telescope is given by

$$G_B \approx \phi^{-2} = K_2 \left(\frac{D}{2f}\right)^{-2}$$

D is the detector diameter and f is the equivalent focal length of the optical system and K_2 is a constant. The pointing loss factor is given by

$$L(\alpha) = \exp\left\{-G_B\, \alpha^2\right\}$$

We conclude that the closer Eve is, the higher probability to gain a photon. The other factor controlling Eve's probability is pointing loss multiple by gain. This is symmetric around zero, and the probability for Eve to gain a photon increases strongly as this goes to zero. The larger $G_B\alpha^2$, the larger the pointing loss (equation above).

It is easy to see that in order to minimize the possibility for an eavesdropper to gain information by means of backflash, Eve should be substantially kept as far as possible, and at as wide an angle as possible away from the line of sight between Alice and Bob.

6.3 Information Reconciliation Considerations

Continuous Variable (CV) Quantum Key Distribution (QKD) can be a viable alternative to its Discrete Variable (DV) counterpart, since CV-QKD does away completely with the requirement of operating with single or at least very low mean photon count per pulse. In CV-QKD Alice and Bob (and possibly Eve) share a set of correlated Gaussian samples. The trick is then to convert the samples into binary bits and use error correction techniques to ensure Alice and Bob's copies of the corresponding label sequence match.

At very low SNR, very little information is carried by the magnitude of the Gaussian samples is extremely low, while the sign contains almost all the information. Quantization of magnitudes of the Gaussian samples at Bob in reverse reconciliation (RR) can then be used to provide side information to Alice, which may be used in an unequal error protection scheme [15].

6.4 Conclusions

In this paper, we have discussed relevant aspects of the practical implementation of a long-range Quantum Key Distribution (QKD) link with trusted nodes. In particular, we have initially discussed on recent pilot studies for the measurements of detection efficiency and source photon statistics for validating the calibration facilities (i) at telecom wavelength for realistic quantum backbone implementation through standard telecommunications grade optical fiber, and (ii) for the VIS-NIR regime. Second, we discussed the problems that arise in a realistic implementation of a free space QKD system. Lastly, we briefly commented on information reconciliation problems in CV-QKD systems.

Acknowledgements This research was supported by NATO under the SPS program, project "Analysis, design and implementation of an end-to-end 400 km QKD link". This work received funds also from the projects EMPIR 14IND05 MIQC2, EMPIR 17FUN06 SIQUST, EMPIR 17FUN01 BECOME (the EMPIR initiative is co-funded by the European Union's Horizon 2020 research and innovation program and the EMPIR Participating States).

References

1. Bennett C, Brassard G (1984) Quantum cryptography: public key distribution and coin tossing. In: Proceedings IEEE International Conference on Computers, Systems and Signal Processing (Bangalore, India), pp 175–179.
2. Gisin N, Ribordy G, Tittel W, Zbinden H (2002) Quantum cryptography. Rev Mod Phys 74:145–195
3. https://www.euramet.org/european-metrology-networks/quantum-technologies/

4. Gramegna M et al (2018) European coordinated metrological effort for quantum cryptography. In: Proceedings of the SPIE 10674, Quantum Technologies 2018, 106741K (21 May 2018). https://doi.org/10.1117/12.2307841
5. Hadfield RH (2009) Nat Photonics 3:696–705
6. Stucki D et al (2001) J Mod Opt 48(13):1967–1981
7. Rebufello E et al (2019) Towards a standard procedure for the measurement of the multi-photon component in a CW telecom heralded single-photon source. Metrologia 56:025004
8. Reichert S (2019) Common ground. Nat Phys 15:110
9. Moreva E et al (2019) Feasibility study towards comparison of the $g^{(2)}(0)$ measurement in the visible range. Metrologia 56:015016
10. Spinelli A, Lacaita AL (1997) Physics and numerical simulation of single photon avalanche diodes. IEEE Trans Electron Dev
11. Meda A et al (2017) Quantifying backflash radiation to prevent zero-error attacks in quantum key distribution. Light Sci Appl Nat Group 6:e16261
12. Shi Y et al (2017) Breakdown flash at telecom wavelengths in InGaAs avalanche photodiodes. Opt Express 25(24):30388–30394
13. Pinheiro PVP et al (2018) Eavesdropping and countermeasures for backflash side channel in quantum cryptography. arXiv preprint arXiv:1804.10317
14. Lacaita A et al (1993) Photon-assisted avalanche spreading in reach-through photodiodes. Appl Phys Lett 62(6):606–608
15. Daneshgaran F, Mondin M, Olia K (2017) Permutation modulation for quantization and information reconciliation in CV-QKD systems. In: Proceedings of the SPIE, vol. 10409, id. 104090J 10 pp

Chapter 7
Secure Communication in the Quantum Era: (Group) Key Establishment

Christian Colombo, María Isabel González Vasco, Rainer Steinwandt, and Pavol Zajac

Abstract This paper gives a brief overview of NATO SPS project G5448. This project, which involves partners from four countries, focuses on the task of establishing a cryptographic key among a group of users over a (classic) public communication network. It is assumed that the adversary may eventually obtain access to quantum computers, i. e., standard cryptographic hardness assumptions like factoring integers being asymptotically hard may be invalidated. Next to laying out the basic project structure, we describe some of the work that has already been done since the beginning of the project, including in particular some of the work done by students.

Keywords Quantum computing · Cryptography · Runtime verification · Group key establishment

7.1 Introduction

At this point in time, quantum computing is widely considered more than a mere distant possibility. Substantial resources areds currently invested into bringing this technology to the level where it can be commercially used. Unsurprisingly,

C. Colombo
University of Malta, Msida, Malta
e-mail: christian.colombo@um.edu.mt

M. I. González Vasco
Universidad Rey Juan Carlos, Móstoles, Spain
e-mail: mariaisabel.vasco@urjc.es

R. Steinwandt (✉)
Florida Atlantic University, Boca Raton, FL, USA
e-mail: rsteinwa@fau.edu

P. Zajac
Slovak University of Technology in Bratislava, Bratislava, Slovakia
e-mail: pavol.zajac@stuba.sk

C. Palestini (ed.), *Advanced Technologies for Security Applications*, NATO Science for Peace and Security Series B: Physics and Biophysics,
https://doi.org/10.1007/978-94-024-2021-0_7

the implications of large-scale quantum computers on the security of today's communication networks receive a lot of attention at this point: From Shor's seminal work [20], we know that a large-scale quantum computer has the ability to invalidate assumptions that underlie widely deployed cryptographic solutions and to render these protocols insecure. Several popular cryptographic primitives for digital signatures and key establishment can no longer be considered secure in such a quantum era. With a large-scale quantum computer, secret keys in these protocols can be easily recovered from public data alone.

A particularly prominent effort to address this uncomfortable situation is NIST's ongoing effort to standardize some fundamental post-quantum cryptographic schemes. The process is remarkably complex in that there are no obvious drop-in replacements for currently deployed schemes. Even in the current (Round 2) stage of this effort, there is still quite some movement in the details of the proposed schemes. Our NATO SPS project G5448 already helped to establish a "PQC Wiki," which gives easy access to the current status of the considered candidates [16].

Overarching goal of this NATO SPS project is to provide a robust solution for (group) key establishment in the quantum era. In other words, we tackle the task that two or more parties want to establish a high-entropy secret over a public, insecure, communication network, and the adversary must be expected to invoke a quantum computer to undermine the security of the protocol. The project's scope includes the design of an authenticated group key establishment protocol and its theoretical analysis, but extends beyond this: to improve resistance against implementation-level attacks, we want to leverage runtime verification. In this way, we can add protections at execution time of a protocol and go beyond what is covered in a standard cryptographic protocol analysis.

7.2 Project Structure and Activities

To provide all necessary expertise, four different partners are involved in the research.

7.2.1 Project Partners

7.2.1.1 Florida Atlantic University (US)

Here, Rainer Steinwandt's team offers the needed expertise in using quantum computers for cryptanalysis. A member of the this team (Edoardo Persichetti) is a co-author of multiple active proposals in NIST's current post-quantum standardization effort, so that it can be ensured that state-of-the-art cryptographic tools are used.

7.2.1.2 Slovak University of Technology in Bratislava (SK)

Otokar Grošek serves as NPD for this project, and his team brings many years of experience in the secure implementation of cryptographic schemes to the project. The Slovak partner can in particular help with the detection (and prevention) of side-channel attacks, e.g., based on timing behavior or power consumption of an implementation.

7.2.1.3 University of Malta (MT)

Christian Colombo and his team bring extensive expertise in runtime verification to the project. Once the cryptographic protocol details have been determined, the Malta partner will take the lead in specifying and deriving a software implementation that offers robust security guarantees against implementation-level attacks at runtime.

7.2.1.4 Universidad Rey Juan Carlos (ES)

Under the leadership of María Isabel González Vasco, this project partner takes the lead in the protocol design phase. Owing to a strong track record in group key establishment, it is ensured that state-of-the-art security models and solutions are leveraged. The Spanish partner is also communicating with an end user, helping to ensure that practical needs are met.

7.2.2 *Initial Work*

In the initial project phase, we arranged for several in-person meetings among all project teams and of subsets of the project teams – the implementation specialists from Malta and Slovakia met, and the partners from Spain and the US met to work on protocol design and the security model. Both on the theoretical/conceptual side (with the partners in Spain and in the US taking the lead) and on the implementation side (with the partners in Malta and Slovakia taking the lead), the project has been active.

7.2.2.1 Theoretical/Conceptual Thrust

Much effort went into laying foundations – clarifying the security model and understanding classical and quantum resources for attacking relevant cryptographic primitives. A presentation on a Grover-based key search for AES at *Quantum Resource Estimation 2019* gives an example of our work in this line [13]. For the cryptographic tools to be used in our protocol solution, our focus is on code-

and lattice-based approaches, and this is reflected in the papers published in the project so far [14]. While a signature-based authentication would conceptually be the simplest approach to ensure authentication in a group-key establishment protocol, especially in a post-quantum scenario, such an approach comes with some (performance) challenges. Consequently, we looked in the initial project phase also at a different authentication mechanism, using passwords. It is our aim to use, whenever possible, non-expensive cryptographic primitives (e.g., message authentication codes and key encapsulation mechanisms), focusing on those whose resilience to quantum-attacks is better understood.

7.2.2.2 Implementation Thrust

From the more practical side, an initial design of a runtime verification setup for security protocol implementation has been drawn up and experiments are underway to validate the concept. A position paper is being drafted in this direction with the aim of disseminating the idea and getting feedback. We also involved students in our research, which has the beneficial side effect of increasing awareness of the NATO SPS Programme. We start the discussion of our practical efforts by highlighting some results of relevant student projects.

Post-quantum Digital Signatures on Android Phones

The transfer to post-quantum cryptography is also an engineering challenge that requires to adapt the used cryptographic infrastructure to new parameters and constraints enforced by post-quantum algorithms. The MS thesis [22] adapts currently available APIs and libraries to implement an application that can be used to sign (PDF) documents on Android devices with the post-quantum algorithm SPHINCS [4].

The main challenge was the key management, as the standard Android APIs only accommodate classical algorithms such as RSA. We have tested an experimental application on a Samsung Galaxy S10+ phone, and compared the signature and verification times with standard RSA-3076 (with equivalent pre-quantum security level as SPHINCS-256, which we used). While verification times are similar (in milliseconds), the SPHINCS-256 signature algorithm is relatively slow (0.5 s, compared to 5 ms of an RSA signature). However, from the user perspective, 0.5 s is still below noticeable delay. The main problem is SPHINCS-256's signature size of 41 kB (compared to 0.4 kB of RSA). This is partially solved in some proposed post-quantum candidates in NIST's standardization effort, but unfortunately most of the candidates still lack implementations in Android libraries.

Authenticating Ephemeral Post-quantum Public Keys with a Merkle Tree

Client-server communication is commonly initiated using public-key cryptography, and *forward secrecy* is a popular security requirement: old sessions are not to be compromised when long term secrets are revealed. This can be accomplished by using ephemeral (single-use) key pairs. Ephemeral keys are also useful to prevent some types of statistical attacks on post-quantum cryptosystems such as [7].

When client-server communication is initiated, typically at least server-side public keys need to be authenticated to prevent man-in-the-middle attacks. Unlike standard RSA/DSA/ECDSA signature schemes, post-quantum schemes have often large public keys and/or large signatures. Adapting specific post-quantum signatures can thus significantly slow down the protocol execution and consume resources on constrained devices.

In the MS thesis [15] a candidate for a quantum-safe key authentication proposed by one of the authors [23] was implemented and tested. In this system, a set of ephemeral key pairs[1] is generated in advance in a secure device on the server. A hash tree is constructed (see Fig. 7.1), and the root of the hash tree serves as a (short) public key for verifying authenticity of each key in the tree structure. A key is signed by providing a valid signature path from the leaf (containing the

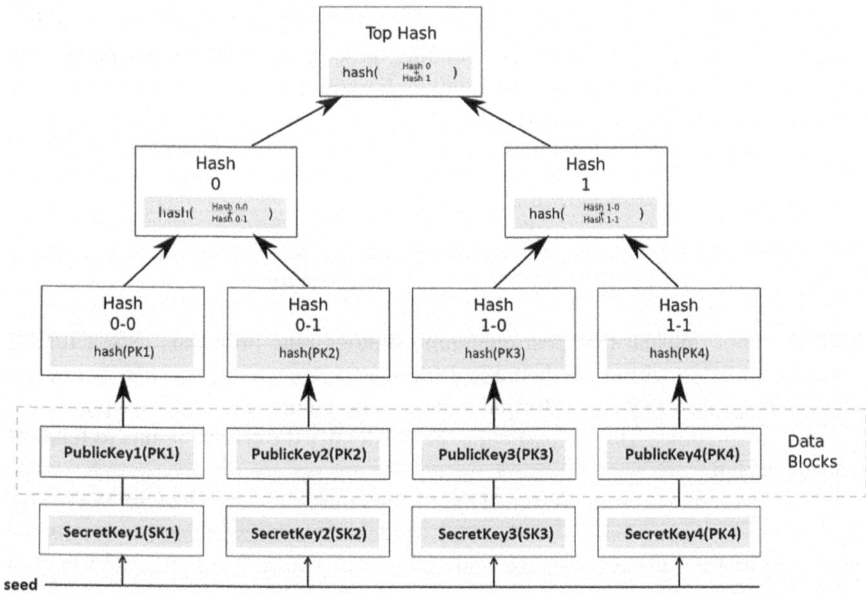

Fig. 7.1 Tree authenticated public keys

[1]For the tests, BIKE [1] was used.

authenticated ephemeral key) to the root of the tree. Experiments show that both signing and verification of ephemeral keys is fast (microseconds per keypair), and the signature size can be relatively short (up to 1 kB, depending on the number of pre-authenticated keys). In our stress test, a standard notebook was able to handle a data structure with 32 million key pairs. This suffices for the communication needs of a standard intra-net application. However, a more complex solution is required to cover the needs of large internet sites.

Systems Based on QC-MDPC and QC-LDPC Codes

Low and medium density parity check (LDPC/MDPC) codes are codes based on sparse parity-check matrices that allow to repair a large number of transmission errors. Structured quasi-cyclic (QC) versions of these codes have a very compact representation that makes them attractive for use in various post-quantum cryptosystems [2]. Classical decoding for LDPC and MDPC codes is based on bit-flipping. The main disadvantage for cryptographic use is its probabilistic nature, as the algorithm does not have an assured success. There are various attacks on QC-MDPC and QC-LDPC code-based systems that exploit the correlation between decoding failure rate and a structure of the secret key [6–8].

The MS thesis [12] successfully tested all proposed attacks on QC-MDPC code based systems in practice. Furthermore, unlike some theoretical predictions, these attacks were confirmed to work against QC-LDPC systems as well. On the other hand, the MS thesis [3] focused on defending these types of cryptosystems. It was experimentally verified that the Miladinovic–Fossorier decoding algorithm can be parametrized in such a way that the decoding failure rate is not correlated with the secret parameters. These algorithms and selected others were added to our BitPunch library [11].

Deeper Understanding of Post-quantum Tools and Resources

While it is not unusual that post-quantum cryptographic proposals lack a formal security evaluation within the theoretical framework known as *provable security*, it is indeed helpful to understand formally the reasons behind successful cryptanalysis of such constructions. This, in particular, helps identify the critical points to fend off in a new design.

In [10] we report on the Walnut Digital Signature Scheme (*WalnutDSA*). This paper is the first contribution of our Phd. student Jose I. Escribano; it details the original proposal and describes the main attacks that have been presented against this construction. Furthermore we discuss several modifications of the proposal, currently under debate, towards a secure implementation of this signature scheme.

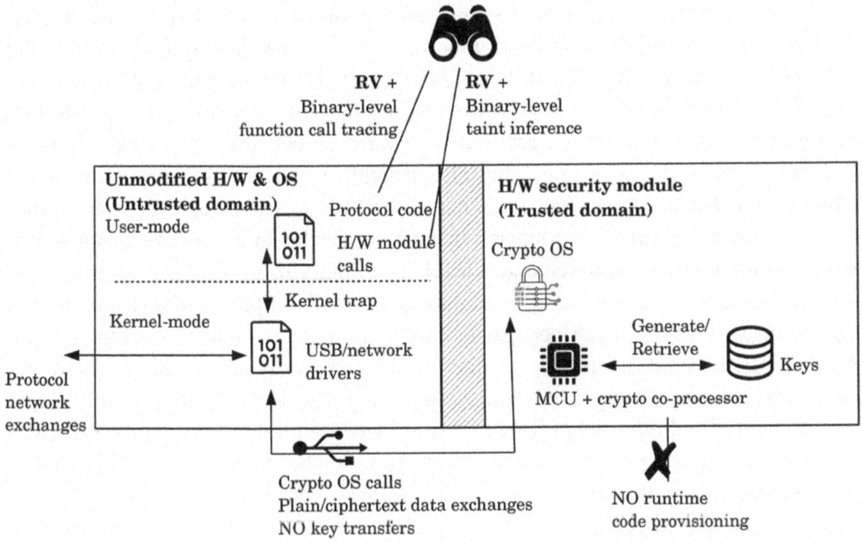

Fig. 7.2 RV-centric comprehensive security for cryptographic protocol implementations (USB stick example)

Runtime Verification (RV)

Figure 7.2 shows a proposed RV-centric solution of protocol implementation security. This setup requires no special hardware or OS modifications, mitigates threats related to hardware issues, including side channel attacks on ciphers, while keeping runtime overheads to a minimum. The primary components of this design are two RV monitors executing within the untrusted domain and a hardware security module (HSM) providing the trusted domain of the trusted execution environment (TEE). The chosen example HSM is a USB stick, comprising a micro-controller (MCU), a crypto co-processor providing h/w cipher acceleration and true random number generation (TRNG), as well as flash memory to store long term keys. In this manner, the cryptographic primitive and key management code are kept out of reach of malware that can potentially infect the OS and applications inside the untrusted domain. The co-processor in turn can be chosen to be one that has undergone extensive side-channel security analysis, thus mitigating the remaining low-level hardware-related threats (e.g., [5]). The *Crypto OS* is executed by the MCU, exposing communication and access control interfaces to be utilized for HSM session negotiation by the protocol executing inside the untrusted domain, after which a cryptographic service interface becomes available (e.g., PKCS#11). In a typical TEE fashion, cryptographic keys never leave the HSM. The proposed setup forgoes dealing with the verification of runtime provisioned code since the cryptographic services offered by the HSM are expected to remain fixed for long periods.

The RV monitors verify correct implementation of protocol steps and inspect all interactions with the hardware module, both of which happen through the network and external bus OS drivers respectively. Verifying protocol correctness leverages the high-level flavors of RV, checking that the network exchanges follow the protocol-defined sequence and that the correct decisions are taken following protocol verification steps (e.g., digital certificate verifications). Inspecting interactions with the HSM, on the other hand, requires a low-level approach similar to the Frama-C/SecureFlow plug-in. In both cases the monitors are proposed to operate at the binary (compiled code) level. The binary level provides opportunities to secure third-party protocol implementations, as well as optimized instrumentation applied directly at the machine instructions level. Overall, binary instrumentation is a widely-adopted technique in the domain of software security, including the availability of widely used frameworks (e.g., Frida [17]) that simplify tool development. The higher-level RV monitor is tasked with monitoring protocol steps and as such instrumentation based on library function hooking suffices. This kind of instrumentation is possible to deploy with minimal overheads.

In contrast the lower-level RV monitor has to rely on monitoring information flows, specifically, untrusted flows [18]. The main limitation is presented by impractical overheads [9]. Our proposed solution concerns inferring, rather than tracking, taint [19] and which takes a black-box approach to taint flow tracking, trading off between accuracy and efficiency. This method only tracks data flows at sources/sinks and then applies approximate matching in order to decide whether tainted data has propagated all the way in-between. With slowdowns averaging only $0.035\times$ for fully-fledged web applications, this approach seems promising. In fact we propose that this approach requires the same library function hooking type of instrumentation as with the higher-level RV monitor. Crypto OS calls may be considered both taint sources and sinks. In the case of data flowing into Crypto OS call arguments originating from suspicious sites, e.g., network input, interprocess communication or dynamically generated code, the Crypto OS calls present the sinks. All these scenarios are candidates of malicious interactions with the HSM. In the reverse direction, whenever data flows resulting from Crypto OS call execution and that end up at the same previously suspicious sites, the calls present the tainted sources while the suspicious sites present the sinks. In this case these are scenarios of malicious interactions targeting leaks of cryptographic keys/secrets, timing information or outright plaintext data leaks. Whichever the direction of the tainted flows, the same approximate matching operators can be applied between the arguments/return values of the sources/sinks.

A nonce-based remote-code attestation, e.g., [21], can optionally close the loop of trust, executed by the Crypto OS, with the HSM performing the tasks intended to be executed by a trusted platform module in such protocols.

7.3 Outlook and Opportunities for Cross-Project Collaboration

The decision on a target protocol is expected to be made as planned. Then the transition from theoretical protocol analysis to a software implementation and implementation-specific security aspects is to be tackled next, and we expect the project to proceed as planned.

Some of the research in this project may well be of interest when combining two-party solutions for *quantum key distribution* to larger networks: While the focus in our project is on classical communication networks, it is common to use a two-party solution as a primitive, and one could potentially explore to what extent non-classical (quantum) solutions could be used here as well, possibly leading to a hybrid protocol for group key establishment. Moreover, it is common to use a classical authenticated channel in quantum key distribution, and when transitioning to group communication, the solutions considered here might be of interest to establish such authenticated channels.

Acknowledgments This work has been supported through NATO SPS Project G5448.

References

1. Aragon N, Barreto P, Bettaieb S, Bidoux L, Blazy O, Deneuville J-C, Gaborit P, Gueron S, Guneysu T, Melchor CA et al (2017) BIKE: Bit Flipping Key Encapsulation. Submission to the NIST post quantum standardization process
2. Baldi M (2014) QC-LDPC code-based cryptography. Springer Science & Business, Cham
3. Baraniak T (2019) Decoding of QC-MDPC codes. Master's thesis, Slovak University of Technology in Bratislava
4. Bernstein DJ, Hopwood D, Hülsing A, Lange T, Niederhagen R, Papachristodoulou L, Schneider M, Schwabe P, Wilcox-O'Hearn Z (2015) SPHINCS: practical stateless hash-based signatures. In: Annual international conference on the theory and applications of cryptographic techniques, pp 368–397
5. Bollo M, Carelli A, Di Carlo S, Prinetto P (2017) Side-channel analysis of SEcube™ platform. In: 2017 IEEE east-west design & test symposium (EWDTS), pp 1–5
6. Eaton E, Lequesne M, Parent A, Sendrier N (2018) QC-MDPC: a timing attack and a CCA2 KEM. In: International conference on post-quantum cryptography, pp 47–76
7. Fabšič T, Hromada V, Stankovski P, Zajac P, Guo Q, Johansson T (2017) A reaction attack on the QC-LDPC McEliece cryptosystem. In: International workshop on post-quantum cryptography, pp 51–68
8. Guo Q, Johansson T, Stankovski P (2016) A key recovery attack on MDPC with CCA security using decoding errors. In: International conference on the theory and application of cryptology and information security, pp 789–815
9. Jee K, Portokalidis G, Kemerlis VP, Ghosh S, August DI, Keromytis AD (2012) A general approach for efficiently accelerating software-based dynamic data flow tracking on commodity hardware. In: NDSS
10. Pablos JIE, María Isabel González Vasco MEM, del Pozo ALP (2019) The cracking of WalnutDSA: a survey. Symmetry 11:9

11. Klein M (2016) Side channels in SW implementation of the McEliece PKC. Infocommun J 8:10–16
12. Kováč J (2019) Reaction attack on the QC-MDPC McEliece cryptosystem. Master's thesis, Slovak University of Technology in Bratislava
13. Langenberg B, Pham H, Steinwandt R (2019) Reducing the cost of implementing AES as a quantum circuit. Presentation at Quantum Resource Estimation QRE 2019, slides available at https://www.quantumresource.org/pdfs/pham.pdf
14. Marak P (2019) Secure communication in the quantum era. http://re-search.info/node/27. Web site for NATO SPS Project G5448. Section on *publications and project dissemination*
15. Novotný M (2019) Implementation of the experimental post-quantum protocol. Master's thesis, Slovak University of Technology in Bratislava
16. Persichetti E, Bai S, Karabina K, Ngo T, Steinwandt R, Catalano Gonzaga di Cirella M (2019) PQC WIKI. A platform for NIST post-quantum cryptography standardization. https://pqc-wiki.fau.edu
17. Ravnås OAV (2019) FЯIDA. Dynamic instrumentation toolkit for developers, reverse-engineers, and security researchers. https://frida.re/
18. Schwartz EJ, Avgerinos T, Brumley D (2010) All you ever wanted to know about dynamic taint analysis and forward symbolic execution (but might have been afraid to ask). In: 2010 IEEE symposium on security and privacy (SP), pp 317–331
19. Sekar R (2009) An efficient black-box technique for defeating web application attacks. In: NDSS
20. Shor P (1997) Polynomial-time algorithms for prime factorization and discrete logarithms on a quantum computer. SIAM J Comput 26(5):1484–1509
21. Stumpf F, Tafreschi O, Röder P, Eckert C et al (2006) A robust integrity reporting protocol for remote attestation. In: Second workshop on advances in trusted computing (WATC'06 Fall), pp 25–36
22. Pernický Ľ (2019) Post-quantum cryptography on Android. Master's thesis, Slovak University of Technology in Bratislava
23. Zajac P (2019) Tree authenticated ephemeral keys, Cryptology ePrint Archive, Report 2019/921. https://eprint.iacr.org/2019/921

Part II
Advanced Materials

Chapter 8
Engineering Silicon Carbide for Enhanced Borders and Ports Security

Ivana Capan, Tomislav Brodar, Zoran Ereš, Robert Bernat, Željko Pastuović, Adam Sarbutt, José Coutinho, Vitor Torres, Vladimir Radulović, Luka Snoj, Klemen Ambrožič, Takeshi Ohshima, Yuichi Yamazaki, and Takahiro Makino

Abstract Developing new state-of-the-art, low-cost and radiation hard detectors is an extremely difficult challenge which can be tackled only by a multisciplinary group of scientists and engineers from various fields having access to different infrastructure.

In our project, Engineering silicon carbide for enhanced borders and ports security (-SiCure) funded by the NATO SPS programme [project number G5215] five partners from Australia (ANSTO), Croatia (RBI), Japan (QST), Portugal (UA) and Slovenia (JSI) have created a team whose main goal is to develop a SiC detector of special nuclear materials.

Keywords Radiation detector · Silicon carbide · Neutron · Nuclear material · Schottky barrier diodes

I. Capan (✉) · T. Brodar · Z. Ereš · R. Bernat
Rudjer Bošković Institute, Zagreb, Croatia
e-mail: capan@irb.hr

Ž. Pastuović · A. Sarbutt
Australian Nuclear Science and Technology Organisation, Lucas Heights, NSW, Australia

J. Coutinho · V. Torres
Department of Physics and I3N, University of Aveiro, Aveiro, Portugal

V. Radulović · L. Snoj · K. Ambrožič
Jožef Stefan Institute, Ljubljana, Slovenia

T. Ohshima · Y. Yamazaki · T. Makino
National Institutes for Quantum and Radiological Science and Technology, Takasaki, Gunma, Japan

C. Palestini (ed.), *Advanced Technologies for Security Applications*, NATO Science for Peace and Security Series B: Physics and Biophysics,
https://doi.org/10.1007/978-94-024-2021-0_8

8.1 Introduction

Increasingly complex risks like geopolitical instability or decentralized terrorism threats, have led to the urge for deploying nuclear screening systems for detection of illicit trafficking of nuclear materials, and from that, to a growing interest in the field of research and development of new radiation detection technologies suitable for homeland security applications.

Recent progress in the manufacturing of high-quality bulk and epitaxial silicon carbide (SiC) and processing technologies for fabrication of SiC-based electronic devices, could enable unprecedented detection of neutron and alpha-particle sources. Unlike existing and commonly used neutron gas-based detectors, SiC-based devices have the potential to be simultaneously portable, operable at room temperature and radiation hard.

Our main objective is to combine theoretical, experimental and applied research towards the development of radiation-hard SiC detectors of special nuclear materials (SNM), and therefore to enhance border and port security barriers.

8.2 Experimental and Theoretical Methods

Two main components of any detector system are (1) the detector of particle events and (2) the electronic system for processing and recording of events.

Neutron detectors were made from n-type silicon carbide Schottky barrier diodes (SBDs), which were fabricated onto nitrogen-doped epitaxial grown 4H-SiC single crystal layers with different thicknesses (from 25 up to 170 μm) [1]. The lateral dimensions of the diodes were 1 mm by 1 mm or 2 mm by 2 mm. The Schottky barriers were formed by evaporation of nickel through a metal mask with patterned square apertures, while Ohmic contacts were formed on the backside of the SiC substrate by nickel sintering at 950 °C in Ar atmosphere [2]. Figure 8.1 shows typical SiC SBD used in the neutron detector. The quality of the SBDs was assessed by current-voltage (I-V) and capacitance-voltage (C-V) measurements.

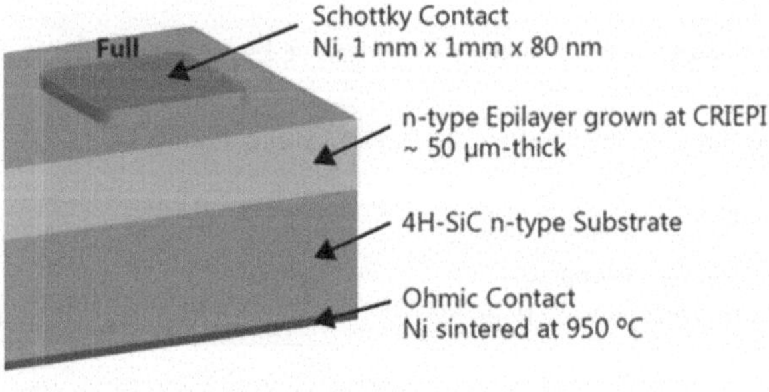

Fig. 8.1 Schematic drawing of the n-type 4H-SiC Schottky barrier diode

The detectors were equipped with converting layers consisting of ^{10}B or ^{6}LiF enriched powders, for more efficient thermal neutron detection.

Defects in as-grown and in irradiated diodes were investigated by highly sensitive space-charge methods, namely DLTS and high-resolution Laplace-DLTS [3] to determine their concentrations, activation energies for carrier emission and respective capture cross sections. For studying the capture kinetics, capacitance transients were measured with different pulse widths, while keeping the other parameters constant [4].

Defect modelling was carried out by first-principles using density-functional theory methods. Core and valence states were described by pseudo-potentials and a plane wave basis set, respectively. The many-body electronic potential was evaluated using the hybrid density functional of Heyd-Scuseria-Ernzerhof [5]. To large extent, this approach mitigates the well-known underestimated gap syndrome affecting conventional density-functional calculations. Defects, vacancy-related and interstitial-related defects were inserted in 576-atom hexagonal supercells. The outcome from the calculations were the electrical levels of defects and respective capture barriers. These quantities can be directly compared to the DLTS observations. In fact, by combining theory with experiments, we were able to come up with a detailed model of the defect with strongest detrimental impact to the performance of the detectors, namely the carbon vacancy [4].

Figure 8.2 displays a schematic diagram of the electronic data acquisition system that was assembled for the tests. The electronic system for particle event processing and recording consisted of a preamplifier followed by shaping amplifier and a multichannel analyser (MCA), operated by a laptop computer. In order to avoid noise from the mains power, the system was powered by a standalone battery voltage source. The latter also provided power to a separate high voltage module (not shown in Fig. 8.2), which was used to apply a reverse bias to the SBDs.

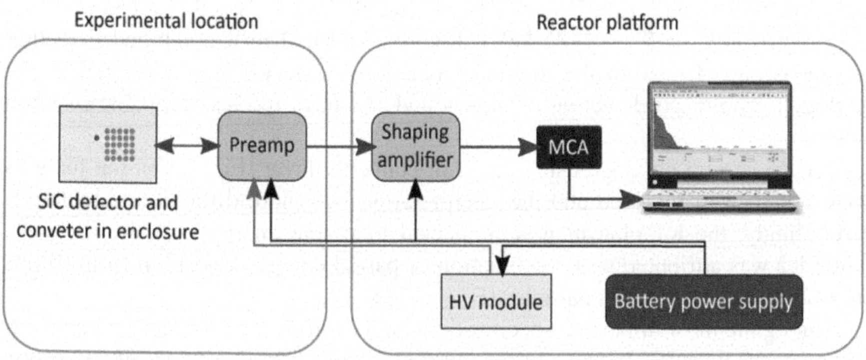

Fig. 8.2 Schematic diagram of the particle event acquisition system

8.3 Detector Prototype Testing

Neutron irradiations were performed in the Dry Chamber of the JSI TRIGA reactor. The Dry Chamber is a large irradiation room in the concrete body of the reactor, connected to the reactor core by means of a graphite thermalizing column [6]. It has been used for radiation hardness testing of detectors, electronic components and systems [7, 8]. In order to determine the neutron flux, activation dosimetry measurements were carried out using foils of certified reference material Al-0.1wt.%Au, obtained from the IRMM (now Joint Research Center – JRC, Geel, Belgium). These were irradiated to measure the $^{197}Au(n,\gamma)^{198}Au$ reaction rate, the cross-section for this nuclear reaction being a standard. The neutron flux is linearly proportional to the reactor power, which was varied between 10 and 250 kW (maximum steady state power). At full power the total neutron flux was 1.6 \times 10^7 $n \cdot cm^{-2} \cdot s^{-1}$. Further details may be found in Ref. [9].

Figure 8.3 shows the measured spectra of 4H-SiC detectors placed in the Dry Chamber of the reactor at different power levels (along with respective flux values). Spectra for detectors with either a ^{10}B or ^{6}LiF converter layers are reported. In all the recorded spectra a significant number of counts at higher energy channels was observed. Distinctive structures were also obtained, depending on the converter layer employed.

The spectra of Fig. 8.3 display a distinctive structures, which can be assigned to the recording of secondary charged particles originating from $^{10}B(n, \alpha)$ and $^{6}Li(n, t)$ reactions. The main features (regions) in the spectra are denoted as "R0-4" and "R0-2" in Fig. 8.3a, b, respectively. The sharp peak R0 that is present in both spectra at low energy channels is attributed to electronic noise. Inspection of the kinetic energy of the reaction products against the spectra allowed a tentative assignment of the spectral features.

For the ^{10}B-covered detectors (Fig. 8.3a), R1 was connected to hits bt ^{7}Li (E = 1013 keV) and $^{7}Li^{*}$ (E = 840 keV) ions, R2 was assigned to alpha particles (E = 1776 keV or E = 1472 keV), R3 was assigned to a combined detection of alphas and $^{7}Li_{*}$ from the dominant reaction, while R4 was attributed to the analogous combined detection of alphas and ^{7}Li from the less probable reaction branch.

For the spectra obtained using the ^{6}LiF converter layer (Fig. 8.3b), the features were not so well resolved and the interpretation was more difficult and tentative. Accordingly, the R1 plateau was connected to partial energy deposition events, while R2 was attributed to a combination of partial energy deposition from tritons (E = 2731 keV) and alpha particles (E = 2052 keV).

Among the most important specifications of a neutron detector are its sensitivity and response linearity. In order to minimize electronic noise, in a *real-life* detection system the signal recorded should be derived from the total counts above a certain channel number (energy threshold). Radulović and co-workers [9] set this threshold at around 600 keV, which is definitely above the R0 peak of Fig. 8.3a, b, and obtained total detected count rates as a function of reactor power and corresponding

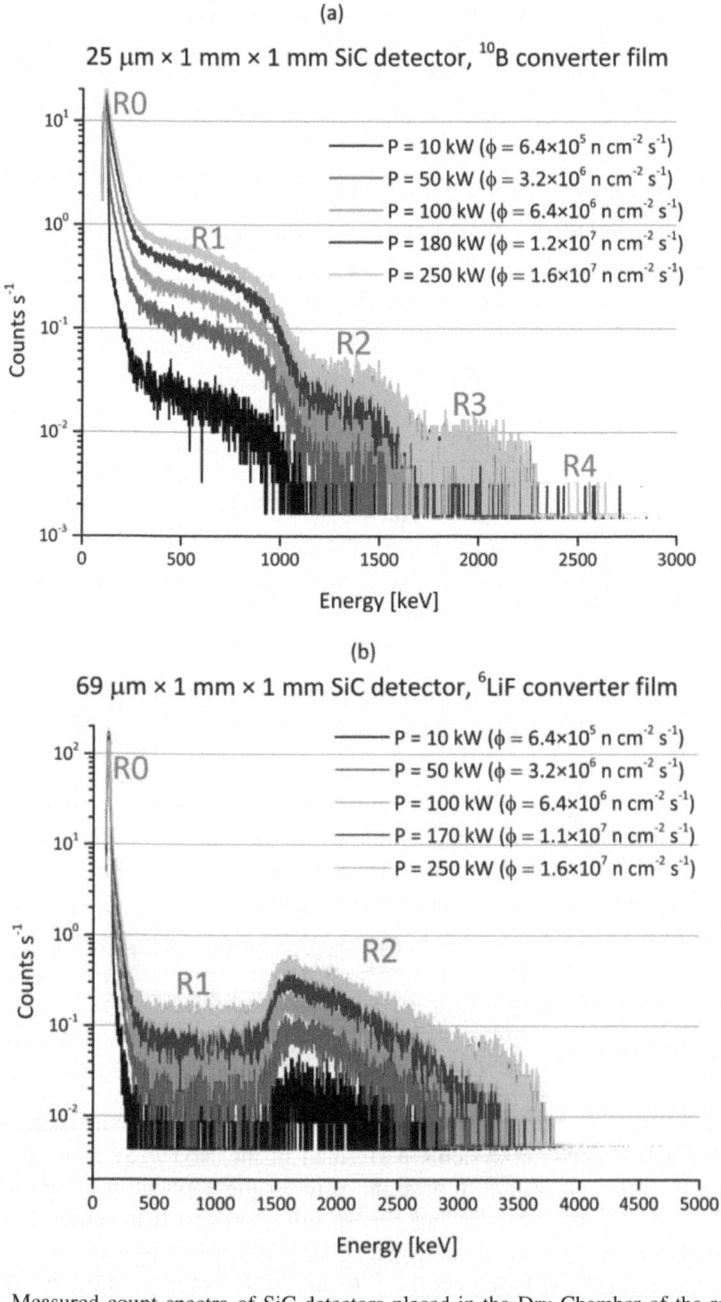

Fig. 8.3 Measured count spectra of SiC detectors placed in the Dry Chamber of the reactor at different reactor power/neutron flux. (**a**) Spectra of a detector with a ^{10}B converter layer and (**b**) Spectra of a detector with a ^{6}LiF converter layer

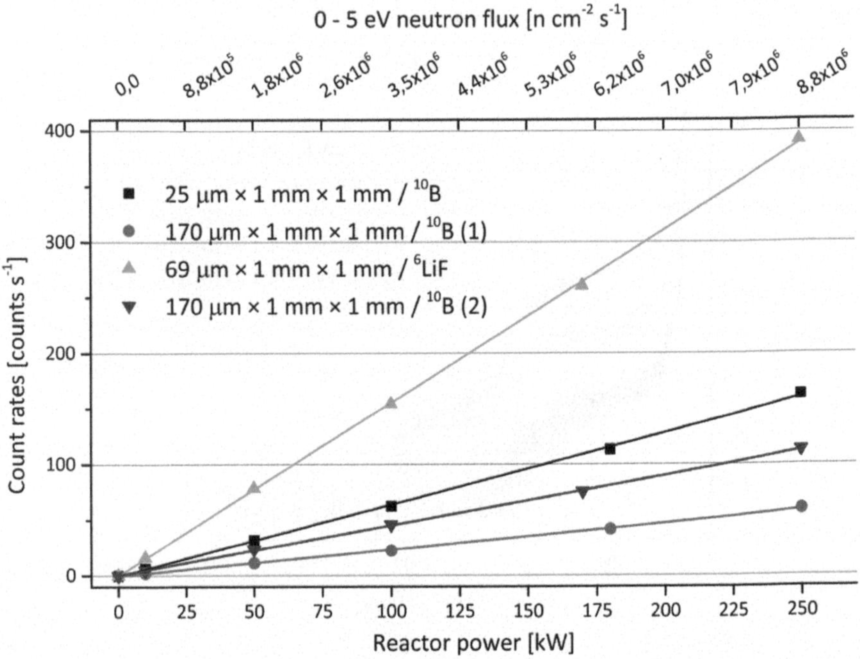

Fig. 8.4 Total detected count rates (above channel no. 500, i.e. around 600 keV as a function of reactor power/neutron flux. The lines represent fitted linear relationships to the data

neutron flux as depicted in Fig. 8.4. The figure shows excellent response linearity, irrespectively of the converter layer employed or the thickness of the epitaxial 4H-SiC front layer. The sensitivity of the detector per unit of neutron flux is given by the slope in the graphs. On average, these were found to be over 10^{-5} (counts s^{-1})/(n $cm^{-2}s^{-1}$) [9].

8.4 Conclusions

Here we present the development and testing of SiC SBD neutron detector prototypes and acquisition systems performed in the NATO SPS-funded project e-SiCure, with the motivation to pave the way for the application of SiC detector technology for enhanced border and port security barriers. It is demonstrated that simple detection devices based on 4H-SiC SBDs, equipped with neutron to charged particle converter films, based on the ^{10}B and ^{6}LiF isotopes, show an excellent neutron response, which varies linearly with the incident neutron flux.

References

1. Ito M, Storasta L, Tsuchida H (2008) Development of 4H–SiC epitaxial growth technique achieving high growth rate and large-area uniformity. Appl Phys Express 1:015001. https://doi.org/10.1143/APEX.1.015001
2. Han SY, Kim KH, Kim JK et al (2001) Ohmic contact formation mechanism of Ni on n-type 4H–SiC. Appl Phys Lett 79:1816–1818. https://doi.org/10.1063/1.1404998
3. Brodar T et al (2018), Laplace DLTS study of deep defects created in neutron-irradiated n-type 4H-SiC, Nuclear Inst. and Methods in Physics Research B, 437, pp. 27–31. https://doi.org/10.1016/j.nimb.2018.10.030
4. Capan I, Brodar T, Coutinho J et al (2018) Acceptor levels of the carbon vacancy in 4H-SiC: combining Laplace deep level transient spectroscopy with density functional modeling. 245701. https://doi.org/10.1063/1.5063773
5. Heyd J, Scuseria GE (2004) Efficient hybrid density functional calculations in solids: assessment of the Heyd-Scuseria-Ernzerhof screened Coulomb hybrid functional. J Chem Phys 121:1187–1192. https://doi.org/10.1063/1.1760074
6. Radulović V et al (2012) Characterization of ex-core irradiation facilities of the JSI TRIGA Mark II reactor. In: Proceedings, 21st International Conference Nuclear Energy for New Europe, September 5–7, 2012, Ljubljana, Slovenia
7. Mandić et al (2004) Bulk damage in DMILL npn bipolar transistors caused by thermal neutrons versus protons and fast neutrons. IEEE Trans Nucl Sci 51:1752. https://doi.org/10.1109/TNS.2004.832927
8. Mandic et al (2007) Online integrating radiation monitoring system for the ATLAS Detector at the Large Hadron Collider. IEEE Trans Nucl Sci 54:1143. https://doi.org/10.1109/TNS.2007.895120
9. Radulović V et al (2020) Silicon carbide neutron detector testing at the JSI TRIGA reactor for enhanced border and port security. Accepted for publication in Nuclear Instruments and Methods in Physics Research, A

Chapter 9
Infrared Transparent Ceramic Windows 2 for High-Speed Vehicles

Andrey Ragulya, V. Kolesnichenko, and M. Herrmann

Abstract The project has created new technical approaches to manufacture large size transparent ceramic windows (transparency 85% and higher in the IR part of spectrum) using advanced consolidation techniques such as 3D printing (binder jet printing of green prototypes up to 120 mm in diameter) and spark plasma sintering (enable to form domes of 70 mm in diameter).

The proposed approach has required the development of ceramic nanopowders specifically formulated by internal structure to be applied in 3D binder-jet technology of meniscus-shape domes. The technologies for the synthesis of nanopowders of magnesium fluoride (MgF_2) and magnesium-aluminum spinel ($MgAl_2O_4$) have been developed, and batches of 3 kg of the both powders have been manufactured. According to the results of the analysis, it was found that the powders are thermally, mechanically and optically-suitable for the manufacture of IR-transparent windows. The specific surface of MgF_2 powder is 38–40 m^2/g, for $MgAl_2O_4$ powder it is 27 m^2/g.

Shaping technologies by slip casting and 3D-printing suitable for sintering under pressure have been developed. Granules from KPI powders and Baikovsky powders have been prepared. Based on the granules a printing process has been developed using CJP 360 printer.

Before sintering stage, the FEM calculations of the press instrument design and shape were carried out and optimal construction providing uniform temperature and strain rate distributions were defined. FEM simulation of the sample under pressure consolidation conditions (spark-plasma sintering and hot pressing) resulted in manufacturing of molds valid for the pilot production of dome-shaped windows. Optimization of the graphite press mold using FEM allows five times shortening of operation regime.

A. Ragulya (✉) · V. Kolesnichenko
Igor Sikorsky Kiev Polytechnic Institute, Kyiv, Ukraine

Frantsevich Institute for Problems in Materials Science NAS of Ukraine, Kyiv, Ukraine

M. Herrmann
Institute for Ceramic Technologies and Systems, Dresden, Germany

C. Palestini (ed.), *Advanced Technologies for Security Applications*, NATO Science for Peace and Security Series B: Physics and Biophysics,
https://doi.org/10.1007/978-94-024-2021-0_9

The developed technologies of spark-plasma sintering (SPS) and hot pressing (HP) are key-enable technologies to produce transparent (88% and above) in the IR range. It has been shown that rapid sintering conditions (heating rates above 100 °C/min) and variable pressure application can achieve high values of density (above 99.95% of theoretically possible) and optical transparency (up to 91%) and avoid excessive grain growth. In this way, SPS-forging allows manufacturing of infrared lenses from both MgF_2 and $MgAl_2O_4$ for complex windows for sensors. For spark-plasma sintering in sinter- forging mode, it was first performed for optical materials. It was found that the strain rates at high temperatures correspond to the superplastic deformation rates $1*10^{-3}$ to $5*10^{-3}$ s^{-1}. The experimental batch of 10 pieces was used for aerokinetic tests. Studies of the properties of manufactured samples of IR-windows have demonstrated the feasibility of the proposed technological approaches (spark-plasma sintering or hot pressing) for the pilot production of dense homogeneous ceramics from magnesium fluoride (MgF_2) and magnesium-aluminum spinel ($MgAl_2O_4$) with a fairly high level of transparency (higher than 88–91%) in the infrared diapason. Specific design of this graphite instrument was helpful to manufacture optically transparent domes with uniform transparency up to 91%.

Keywords IR transparent ceramic domes · Spark plasma sintering · Sinter-forging · Nanopowders of MgF_2 and $MgAl_2O_4$

9.1 Synthesis and Characterization of MgF_2 and $MgAl_2O_4$ Nanopowders

9.1.1 Optimization and Upscaling of the MgF_2 Powder Synthesis

Optical materials are important objects in various applications. Transparent crystals are widely used in the optical devices. The so-called optical ceramics – polycrystalline transparent materials obtained by sintering of powders under pressure are considered one of the hot topics of modern ceramic science. The transparent ceramics is preferable for those materials, which are impossible (or difficult) to obtain in the form of single crystals or glasses. Ceramic technologies open the possibility of choosing the shape and size of optical parts, achieving a significant increase in mechanical strength of materials. Optical materials are usually fragile crystals, chemically cleaner, but more susceptible to cracking on the surfaces of cleavage. The manufacturing of transparent ceramics is widely described for various compositions: oxides, fluorides, nitrides. Nowadays, due to increasing requirements to the quality of optical materials the methods for their synthesis are being improved.

Chemistry of inorganic fluorides (MgF_2) differs considerably from the chemistry of simple and complex oxide compounds ($MgAl_2O_4$), which are mainly used as the basis for the creation of optical ceramics. In this work, synthesis of the MgF_2 powder was carried out by the reaction:

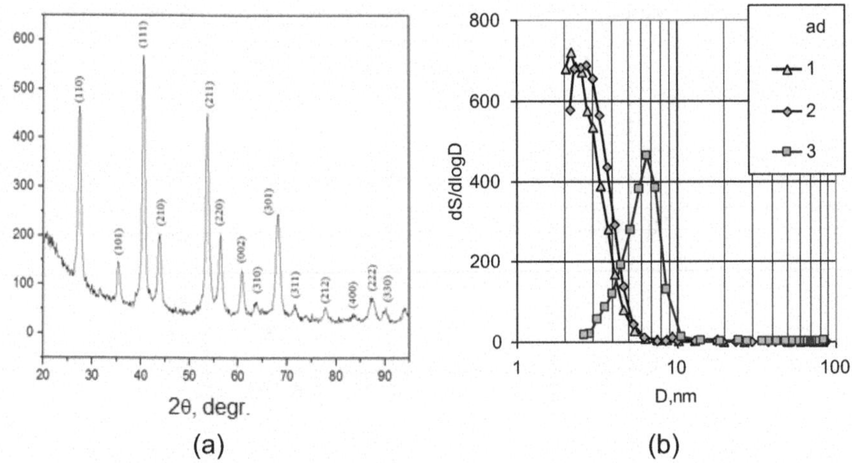

Fig. 9.1 Characterization of MgF_2 nanopowders by sol-gel technique: (**a**) diffraction pattern after annealing at 390 °C; (**b**) pore size distribution in the MgF_2 powders annealed at various temperatures: 1 – dried at 100 °C; 2 –at 200 °C; 3 – dried at 390 °C – final MgF_2 powder

$$Mg(NO_3)_2 + HF \rightarrow \downarrow MgF_2 + 2HNO_3$$

Mixing of the input solutions was carried out using stirrer to form a gel-like precipitate. After centrifugation, the resulting precursor was dried in an oven at 80–120 °C for 24 h to remove water. The cake was hot ground at 120 °C until a homogeneous powder. Then the powder has been annealed at temperatures of 120, 200, 300 and 390 °C. The XRD and adsorption spectroscopy data are presented in Fig. 9.1. Good quality crystalline magnesium fluorite nanosize powders have been prepared. After granulation, the powder was examined in slip casting, or 3D printing operations, and then, the technology was scaled up to several kilos yield for further manipulation.

9.1.2 Low-Temperature Synthesis and XRD Analysis of $MgAl_2O_4$ Spinel Nanopowder

The $MgAl_2O_4$ spinel nanopowder has been prepared from the complex precursor containing Mg and Al nitrates and glycine at temperature of 80 °C and pH = 5.5. The gel has been composed with the molar ratio of Mg^{2+}: Al^{3+}: glycine = 1: 2: 5. The gel was dried in air for 50 h, followed by pyrolysis of the xerogel at t = 350°C in oven for 1 h. A foamed precursor MgO-Al_2O_3-C was obtained. Thermal analysis of the obtained xerogel in the temperature range of 25–800 °C (Fig. 9.2) showed endothermal effects at temperatures of 105, 140, and 187 °C indicated the loss of xerogel' structural water.

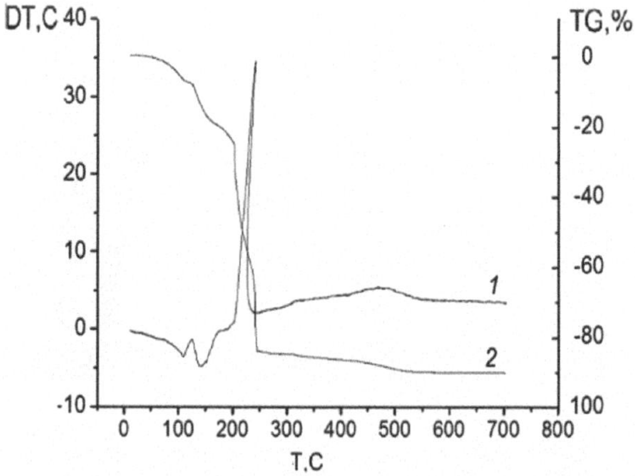

Fig. 9.2 Thermal analysis of the xerogel based on complex Mg:Al:glycine – (1) DT, (2) TG

Fig. 9.3 Rotary tube furnace for synthesis of nanopowders

A strong exothermal effect with a maximum at 240 °C refers to the pyrolysis of the precursor, which is accompanied by the weight losses of 83%. Any localized thermal effects were not found in the temperature range of 550–800 °C. Weight loss of the sample in the range of 25–800 °C was about 90%.

The XRD analysis of the as-prepared powders of the precursor MgO-Al_2O_3-C just after pyrolysis at 350 °C, its xerogel Mg^{2+}:Al^{3+}: glycine (1) and its intermediate products after further decomposition in the rotary kiln (Fig. 9.3) in air at temperatures 600 (2), 650 (3) and 700 °C (4) for 3 h (Fig. 9.4). As a result, the cubic phase of spinel only (*Fd-3m*) in the nanocrystalline state has been found in the samples.

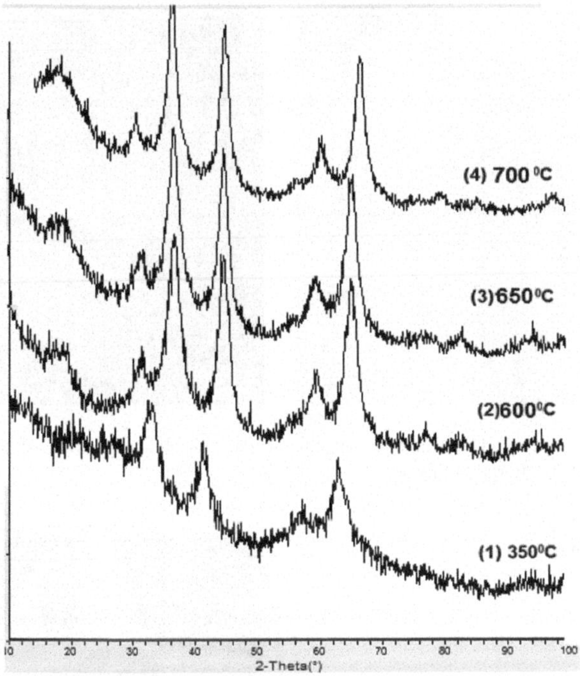

Fig. 9.4 The X-ray diffraction patterns of the MgO-Al_2O_3-C precursor powder after pyrolysis (350 °C) of xerogel of Mg^{2+}: Al^{3+}: glycine (1) complex and its further heat treatment in air at temperatures of 600 (2), 650 (3), 700 °C (4) – exposure time of 3 h.

The specific surface area of the powder sample was calculated using static adsorption-structural method based on the interpretation of adsorption isotherms: 1-precursor MgO-Al_2O_3-C after pyrolysis (350 °C) of xerogel Mg^{2+}:Al^{3+}:glycine – $S_{BET} = 16{,}5\ m^2/g$; 2 – Spinel (as prepared at 600 °C, milled) – $S_{BET} = 12{,}3\ m^2/g$; 3 – Spinel (as prepared at 600 °C, no milling) – $S_{BET} = 14{,}7\ m^2/g$; 4 – Spinel (as prepared at 700 °C, preliminary milled) – $S_{BET} = 13{,}7\ m^2/g$. Large surface area and the smallest particle size allowed spinel powder to be obtained at optimal temperature of 700 °C without preliminary milling of the precursor MgO-Al_2O_3-C.

From the TEM observation of nanopowders, it follows that the samples are strongly agglomerated. In general, the particles form porous structure with an average pore diameter of 400–500 nm. Switching from 600 °C to 700 °C in the synthesis temperature (Fig. 9.5) does not lead to substantial changes in the morphology of nanopowders or pore size distribution.

At the same time, in the analysis of electron diffraction patterns (SAED – selected region electron diffraction), the crystallinity of the structure increases with increasing temperature of synthesis. Namely, when the temperature varies from 600 to 700 °C, the rings of the diffraction patterns become more distinct, and the

Fig. 9.5 TEM-images and diffraction patterns of the samples $MgAl_2O_4$ after calcination at temperatures 700 °C after grinding

Fig. 9.6 General view of CJP 360 3D Printer

appearance of bright dots indicates the transition of the material from amorphous to polycrystalline state.

The self-made powders (I) and standard spinel $MgAl_2O_4$ nanopowders (Baikalox, II) have been used to consolidate the dense ceramic pieces under spark-plasma sintering conditions.

By the BET data (ASAP-2000) the specific surface area of the powder I is equal to 39,8 m^2/g, and of the powder II – 29,5 m^2/g. X-ray fluorescent analysis (Rigaku Primini) of the original powders against reference samples was used to discriminate excess of oxides over stoichiometric spinel compound (Fig. 9.5).

The granulated MgF_2 and $MgAl_2O_4$ powders for binder jet CJP360 3D printer (Fig. 9.6) were prepared from both as-synthesised nanopowder and Baikalox powder by their mixing with hydrophilic binder solution and then dried in air for several hours to eliminate solvent and remain mesoporous cake followed by gentle grinding of the cake in the plastic drum until granules of 20–80 μm in size. The pore size distribution in dry samples was thoroughly controlled using ASAP2000 device.

9.2 Manufacturing of 3D Printed Green Domes

The 3D model (computer-aided design or computer modelling) of domes followed by layer-by-layer printing out of the model has been elaborated.

The printer' box was filled with granulated powders of MgF_2 and $MgAl_2O_4$. Important characteristics for granules were achieved: definite porosity, pore size distribution and mesoporous structure. The adsorption spectroscopy showed us the pore size distribution in the initial MgF_2 powder annealed at 200 °C (Fig. 9.7a) as well as in the granules. The as-printed domes are shown in Fig. 9.7b. After slow de-binding, the printed green domes had porosity of 52 ± 0.4%. The slip casted analogues were much denser (porosity of 36 ± 0.9%). Further operations were carried out using both as-printed and as-casted green domes. The printing operation represents simultaneous manufacturing of 20–24 very similar green domes from one powder charge for 6 h printing.

9.3 Spark Plasma Sintering of Domes

The HPD-25 FCT Systems GmbH SPS machine was used (Fig. 9.8). The spark plasma sintering of transparent MgF_2, $MgAl_2O_4$ ceramics is not industrial process yet and considered here an innovation. For both sorts of ceramics, the following three challenges have been met: (i) minimal grain growth; (ii) minimal carbon

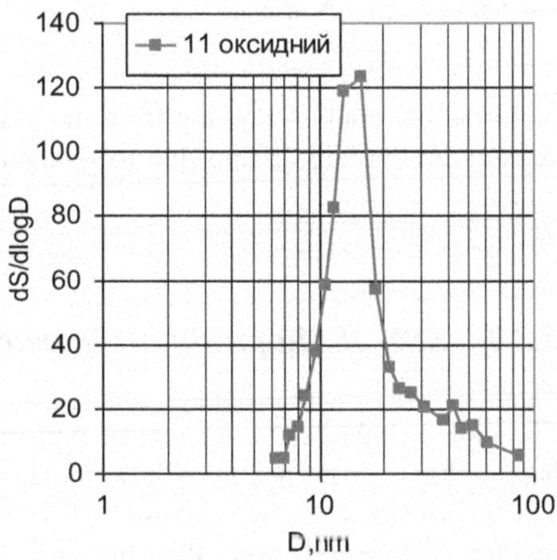

Fig. 9.7a Pore size distribution in the as-prepared MgF_2 granules

Fig. 9.7b As-printed prototypes of IR-domes (green samples)

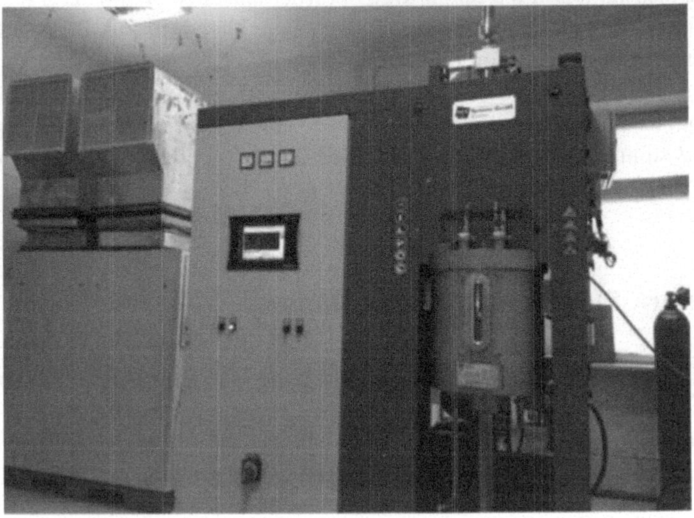

Fig. 9.8 HPD 25 SPS machine in the I. Sikorsky KPI laboratory

pollution of ceramics coming from the graphite press mold; (iii) minimal thermal stresses on cooling arising due to temperature gradient between top and bottom punches.

9.3.1 SPS of Magnesium Fluorite

The most suitable sintering regime performing highest density and optical transparency has been refined from the systematic study of densification in both as-casted and as-printed discs in vacuum (~ 0.1 mbar) under the numerous heating rate regimes. Figure 9.9 shows the kinetics of the shrinkage during SPS of the MgF_2

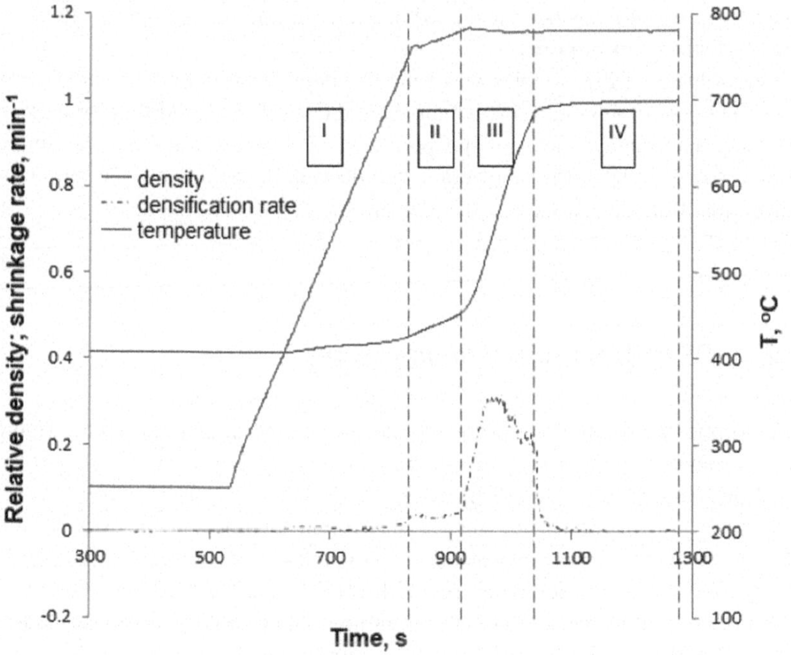

Fig. 9.9 SPS regimes for MgF_2 nanopowder

Table 9.1 Relative density, hardness and IR-transparency of the sintered tablet shape samples of MgF_2 of 20 mm in size ($T_{sint} = 780$ °C, P = 70 MPa)

(№ sample)	Relative density, %	Transparency[a], %	Hardness, HV, GPa (50 g)
1	99,62	88	4,26 ± 0,35
2	99,36	77	5,26 ± 0,56
3	99,63	84	4,94 ± 0,56
4	99,85	90	4,72 ± 0,7

[a]Wave length 4,6 μm (thickness 2,5 mm)

powder. As the best, the sintering mode provided several heating stages: I (fast heating) to 750 °C, 100 °C/min, 16 MPa; II (slow heating) to 780–800 °C, 20 °C/min, 16 MPa; III (increase of pressure) at 780 °C, 0 °C/min, 70 MPa; IV (isothermal hold) at 780–800 °C for 4 min under 70 MPa. The initial shrinkage has been observed at temperature of ~ 420 °C, and the main shrinkage (from 51% to 98% of theoretical density) occurs when higher pressure applied. The shrinkage ended (the material achieved the maximum density) after 1.5–2.0 min of exposure at maximal sintering temperature.

The relative density, transparency and hardness of the sintered magnesium fluoride are given in Table 9.1. All samples achieved high relative density of 99.4–99.8% of theoretically possible, but the highest relative density of 99.85% at top transparency of 90% has been achieved due to sintering regime optimization at the stage of pinching off of pores.

The hardness of the samples was found to depend on grain size and measured within the range of 4.3–5.3 GPa, which is good enough pre-condition for high wear resistance. Thorough control of the pore size evolution throughout the sintering regime resulted in late closure of pores and maximal pore numbers located mostly in triple junctions. In this case, full densification and minimal grain growth was observed [1].

9.3.2 SPS of Magnesia–Alumina Spinel

To determine the optimal sintering regime for magnesia-alumina spinel (temperature, pressure, heating rate and duration of heating hold) on densification and optical-mechanical properties of Baikovsky S25CRX, the experiments were carried out according to the following regimes: (i) rapid heating rate of 200 °C/min to 1100 ° C to avoid particle coalescence; (ii) heating with different moderate rates (100, 50, 20 °C/min) to temperatures of 1400 °C as a minimum or 1600 °C as a maximum without isothermal holds or durable hold up to 20 min; (iii) pressures of 20, 35, 50 MPa have been applied at the beginning of heating cycle. Then it was maintained constant to the maximum sintering temperatures of 1400 or 1600 °C. When cooling, the pressure was smoothly decreased to zero at temperature of 1100 °C. The sample was cooled down till room temperature without pressure and without control of the cooling rate. ASAP 2000 was used to determine porosity and pore's size distribution in the temperature interval 900–1600 °C, when porosity allowed (Table 9.2).

The grain size (GS) of the sintered spinel changes with temperature and duration of hold: at the temperature of 1200 °C GS is about 0.5 μm, but after sintering at 1400 °C without hold, GS is 0.8–1 μm. When the sample was hold at 1450 °C for 20 min, the grain size increased to 1.5–1.7 μm. But the best transparency of 85% has been achieved at density of ~99.9%, and GS less than 1 μm, which corresponds to regime of 0–10 min hold at 1450 °C. Further grain growth leads to substantial decrease of transparency due to increasing porosity and larger grains.

When the sintering temperature increased to 1600 °C, the grain size is growing up to ~ 2–3 μm and transparency decreased to 75% especially in the short-

Table 9.2 Best results of spark plasma sintering for spinel powder

## of samples	T, °C/ hold, min	Open porosity, %	Relative density, %	HV100, GPa	Transparency, %
FCT-1	1600/0	0,04	99,86	10.69 ± 1.09	75
FCT-2	1450/0	0,01	99,87	11,18 ± 1,25	84
FCT-3	1450/10	0,02	99,89	11,79 ± 1,0	85
FCT-4	1450/20	0,08	99,77	13,67 ± 0,79	81,5

wave area. Thus, the optimization concerned temperature and isothermal hold. For manufacturing of domes, the temperature of 1450 °C has been considered a maximum safe temperature for SPS regimes.

The adsorption spectroscopy analysis of the porous structure in the particulate samples was carried out for: mesoporous nanosize powder and granules, slip casted domes, 3D printed domes, sintered tablets and domes. Application of pressure resulted in destruction of the multi-level aggregates in the powder. The dependence of the total and open porosity on sintering temperature (force of 11 kN) has non-linear changes in macroporosity at low heating rates (less than 10 °C/min) – pore coarsening and shrinkage, while the rapid heating rates (above 200 °C/min) demonstrate linear macroporosity behaviour vs. temperature – pore shrinkage only.

9.4 Characterization of Transparent Domes

Extensive optical characterization of domes has been carried out using IR spectrophotometer. Figure 9.10a, b demonstrate best performance of samples made from MgF_2 and $MgAl_2O_4$. Using the optimal batch of MgF_2 nanopowder, its 3D printing to green domes followed by SPS at optimal pressure and hot de-loading we succeeded in 90 ± 1% transparent ceramic pieces (theoretical value is 93%). To meet technical requirements of optical characteristics of the domes, the Finite Element Modelling was applied to get optimal design of graphite matrix providing uniform distribution of current and temperature in the area of the dome. In the best $MgAl_2O_4$ samples, it was found an appropriate regime to manufacture green domes (3D printing) followed by optimal sintering regime. The transmittance plotted against wavelength for the sintered magnesium fluorite samples achieve d ~

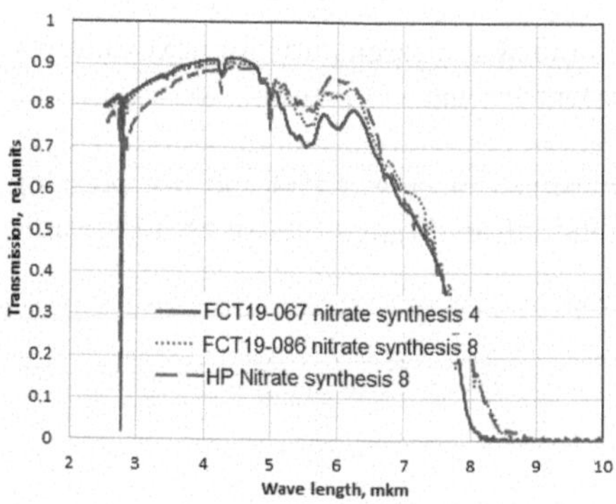

(a)

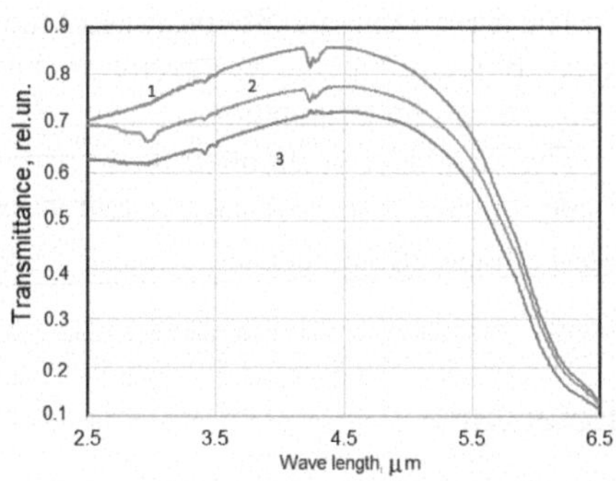

(b)

Fig. 9.10 Infrared transmission curves versus wave length for (**a**) MgF2 made form self-prepared MgF_2 nanopowder, (**b**) $MgAl_2O_4$ (Baikovsky Corp.), where #1 is identical sample annealed at 1200 °C in air for 3 h until removing blackness. #2 and #3 are non-annealed samples

Fig. 9.11 Polished dome, prepared from magnesium fluoride by 3D printing combined with spark plasma sintering

80–85% of theoretical value in the range of 2.5–5.0 μm (Fig. 9.10a). The curve #1 in the Fig. 9.10b corresponds to SPSed sample of $MgAl_2O_4$ ceramics annealed in air for several hours to remove blackness existing in the sample #2, which has been SPSed as well.

Finally, the as-sintered fully dense and 90% transparent in the IR diapason of 3–6 μm dome is shown in Fig. 9.11.

9.5 Conclusions

Technologies for the synthesis of nanopowders of MgF_2 and $MgAl_2O_4$ have been developed. The specific surface of MgF_2 powder is 38–40 m^2/g, $MgAl_2O_4$ powder was 37.2 m^2/g.

The slip casting and 3D-printing of the dome-shape green parts for next sintering under pressure have been developed. Powders for the 3D printing were prepared from slurries by granulation.

The developed technologies of spark-plasma sintering (SPS) and hot pressing made it possible to produce transparent in the infrared range (above 88%) ceramic parts with complex shapes.

Reference

1. Ragulya AV (2008) Consolidation of ceramic nanopowders. Adv Appl Ceram 107(3):118–134

Chapter 10
Advanced Nanotechnologies for Multivariate Sensor Fabrication

Iryna Krishchenko, Eduard Manoilov, Borys Snopok, Alberto Verdini, and Andrea Goldoni

Abstract The interaction between plasmonic nanoparticles (PNPs) is of the utmost importance for the optimal design of surface enhanced Raman spectroscopy (SERS) substrates. This interaction can either quench or enhance the strength of the localized plasmon resonance of nanoparticles depending on the size of the PNPs themselves, the pores between them, their size distribution and spatial locations. Variating the degree of porosity opened wide prospects to realize both localized surface plasmon resonance and plasmon polariton resonance on the same material. All these features make it possible to create a new type of tunable SERS substrates using wedge-shaped porous gold structures with variable properties. This paper conceptually describes the technology for obtaining such structures, developed in the frame of the NANO SPS G5140 project (ANT-MUSE).

Keywords Porous materials · Porous gold · Raman scattering · SERS substrates · Pulced laser deposition · PLD · Backward particle flow · Local plasmon resonance · Hot spots · Plasmonic nanostructures · Nanophotonics

10.1 Introduction

A xenobiotic "attack" of any sort (e.g. terrorist act, technological accident, natural disaster, traffic accident etc.) is typically a very low-probability event. Therefore, it is fundamentally impossible to predict which potentially toxic chemicals must be

I. Krishchenko · E. Manoilov · B. Snopok (✉)
V.E. Lashkarev Institute of Semiconductor Physics, National Academy of Sciences, Kyiv, Ukraine
e-mail: snopok@isp.kiev.ua

A. Verdini
Institute of Materials of the National Research Council, Trieste, Italy

A. Goldoni
Elettra – Sincrotrone Trieste, Trieste, Italy

C. Palestini (ed.), *Advanced Technologies for Security Applications*, NATO Science for Peace and Security Series B: Physics and Biophysics,
https://doi.org/10.1007/978-94-024-2021-0_10

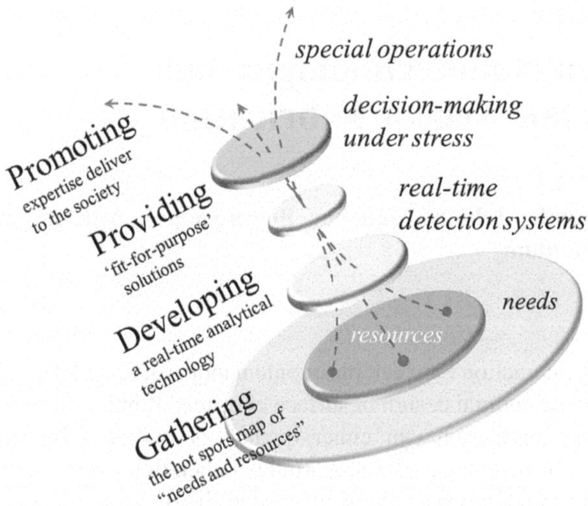

Fig. 10.1 Overview of the analytical workflow specific for the development of special operations for the effective treatment of casualties

detected, tracked and identified for optimal source assessment, especially when the situation diagnosis will be performed in the field or in an industrial environment. Traditional military approaches to the battlefield detection of chemical and biological hazards are not necessarily suitable or easily adapted for use by civil defense providers dealing with heterogeneous population in a peacetime civilian setting [1]. Moreover, a potential complication is that an accident may involve the use of a priori unknown or multiple toxic agents; therefore, simultaneous processing of multiple, different analytes is necessary. The situation diagnosis is aimed to decrease as much as possible the uncertainties concerning pollution content arising after the accident; it must be processed in real time. This overriding task will be realized through the development of multilevel strategies (see Fig. 10.1) utilizing a specially designed high-throughput intelligent screening platform for multi-scale environmental characterization.

"What-if" evaluation for the effective treatment of casualties requires *objective scoring of environment pollution* using mobile detectors or monitors. The ultimate goal of such mobile system is to rapidly and inexpensively detect the presence of potent xenobiotics in the field using portable instruments and transfer the sampling estimate over standard communication channels for data analysis and a decision on the "fit-to-problem" solutions. Although there is enormous interest in scientific and engineering communities to find the optimal construction of such alarm-type xenobiotic monitors, no complete method exists yet for fast and simple detection of a priori unknown hazardous materials in the field.

A typical solution used to improve the performance of post-accident assessment is the integration of two or more sensors in a single platform. The conventional concept is based on the model of "univariate" arrays with completely independent channels: one agent - one sensor. The number of objects that can be identified by such a system directly depends on the number of integrated sensors. It is reasonable to conclude that this concept implies the "arithmetic addition" of different sensors which are not sufficiently reliable for the identification of unexpected situations being the most dangerous in a civil or industrial environment; a more sophisticated analytical concept must be developed, practically realized and validated. Moreover, the development of equipment for each individual dangerous substance is economically inexpedient and senseless for real-time analysis, taking into account the limitless diversity of potent xenobiotics. The need to develop new interpretative descriptors for potentially hazardous materials in terms of their "class affiliation" or even "personal identity" is a very pressing challenge.

Identification of unknown or well-known, but very distinct in their properties chemicals, is only possible with the greatest efficiency when employing the concept of multifactor fingerprinting, when the image of the object (a "fingerprint") in the response space depends on different characteristics of the classified object in a non-linear manner [2]. All "quick response" systems of living organisms, like the sense of smell or taste, are built precisely according to this principle. The same technology is used for goods labeling by bar- or QR-codes. In this case, the chemicals are represented by some abstract image - a virtual fingerprint, whose specific form depends on the desired visualization system (like human fingerprints or iris scanners). These virtual fingerprints can be used for further classification (multivariate "calibration" procedure) and following identification (e.g. screening during the analysis) based on image recognition techniques [3].

Practical implementation of the multivariate analysis concept involves a reasonable choice of key identification descriptors, and, accordingly, analytical methods which are intended to identify the set of fingerprint features that can segregate identifiable objects into separate classes of potentially dangerous, not dangerous or unknown environments. The concept may be realized using multichannel cross-reactive transducer-based sensing elements (e-nose [4, 5]), as well as multichannel virtual sensor approaches, i.e. chromatography (GS) or mass-spectroscopy (MS) [6]. Raman spectroscopy has also emerged as a useful technique, which can produce rich chemical information using visible light and characterize compounds by vibrational fingerprints (virtual chemical images) both in water and air environments [7, 8]. With regard to the specific case of Raman scattering, the signal is relatively weak; so optical energy concentration and additional scattering enhancement is necessary for a sensitive enough optical readout in the field. An enhancement of Raman signals can be achieved by the magnification of the incident electromagnetic field, the amplification of Raman scattering emission and the direct modification of the scattering process itself [9]. An enhancement of the analyte-specific response is particularly strong when both incident and scattered fields are in resonance with the electromagnetic excitations in a nanosized "antenna" and is frequently observed on the rough surface of noble metals; this is surface enhanced Raman scattering, or SERS. Therefore, for the practical applications of Raman spectroscopy, spatially organized composite nanostructured materials with embedded "antennas" enabling an intimate contact with the analyte are required. In this case, the potential xenobiotic or class of xenobiotics can be easily identified and the following analytical validation and effective deactivation timely performed.

10.2 Advanced Technology for the Low-Cost Production of Tunable SERS Substrates Based on Porous Solids with a Wedge-Like Profile

Seeking efficient ways to diagnose the situation after a xenobiotic attack would be impractical without a simple and scalable production technology of highly efficient and stable sensors that will *in-situ* detect, identify and characterize chemical compounds embedded in complex matrices in a rapid, accurate and reproducible way. It is necessary to highlight that the identification of specific chemical functionality (*e.g.* a molecule-specific fingerprint) may be realized only within a certain scale length of structured matter - at the nanoscale, where adequate approaches have only become available in recent years. Surface-enhanced Raman scattering is the best known example of near-field (nano-scale) optical effects being detectable in the far-field (macro-scale).

Metal surfaces with natural roughness do not provide the highest SERS enhancement, and special attention is paid to the formation of 3D nanostructured surfaces. Indeed, a nanoscale porous structure is optimal for SERS substrates: the highest Raman enhancement occurs within the "hot spots", which are typically between the nanoparticles. There are a few technological strategies to create nano-porous materials, but very often they are costly due to the use of various expensive fabrication processes and techniques, such as focused ion beam and electron beam lithography, as well as various etching protocols. Moreover, for the expensive substrates fabricated using these processes, a conflict between detection limit and stability is typical: the more sensitive (detection limit) the structure, the less stable it is during long-time analytical procedures. To overcome this contradiction, we developed a new technology for the fabrication of nanocomposite porous plasmonic structures, where functional properties are realized through the spatial organization of the material at the nanoscale, whereas the stability is determined by the nature of the material itself (*e.g.* gold).

Among the variety of methods for the fabrication of porous gold layers, pulsed laser deposition (PLD) of metals in vacuum is of great importance, as it is simple and cheap. PLD is a coating technique carried out under specific atmosphere and typically involves three stages: the laser evaporation of the coating material, the transportation of the evaporated material to the surface and its condensation over the surface as a thin film. As a result of laser irradiation, the target is now plasma consisting of atoms, ions and clusters of target material. Further adiabatic expansion of this plasma in the form of an erosion torch results in the interaction of torch particles with the atoms of a working (inert) gas and the respective dissipation

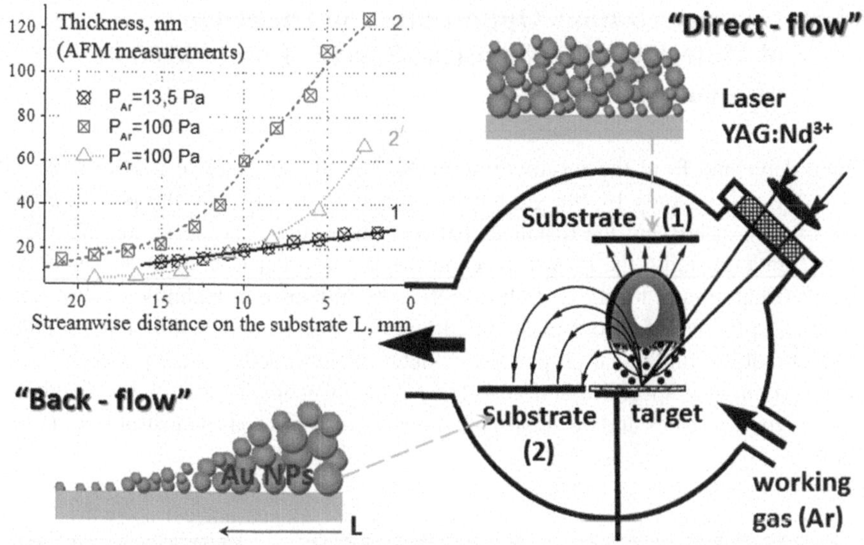

Fig. 10.2 The schematic of typical metal films and the view of the bw-PLD setup (argon pressure P_{Ar} (10–100 Pa), laser pulse number YAG:Nd3+ N (3–45)·10^3 and energy density per pulse j (5-J/cm^2), pulse duration 10 ns, pulse repetition rate 25 Hz). Insert: the thickness of surface layers extracted from Atom Force Microscopy (AFM, Nanoscope IIIa, Digital Instrument; tapping mode, silicone nitride tip) images (Deposition mode: j = 20 J/cm^2, N = 30·10^3, p_{Ar}, Pa: 13,5 (curve "1"), 100 (curve "2"). Curve "2/"– at j = 5 J/cm^2, N = 30·10^3, p_{Ar} = 100 Pa).

of their energy. The deposition may be performed under different atmospheres and pressures. Energy losses during the transport to the substrate are lower when depositing from forward particle flow ("Direct flow", Fig. 10.2), and higher during the deposition from backward particle flow ("Back flow", Fig. 10.2). Thin non-porous metal films are usually prepared using the classical PLD method with forward high-energy flow of erosion torch particles in high vacuum conditions: the substrate is placed perpendicular to the torch axis at a distance of 20–25 mm from the target.

Contrary to the classical PLD process, the film grows on the substrate located in the target plane from the backward flow (bw-PLD) of low-energy erosion torch particles, which causes the formation of porous films with a wedge-like profile. In the "back flow" mode, the separation of Au NP by their sizes occurs: large particles are deposited closer to the torch axis, while smaller ones - wide apart. The structure of the films depends on gas (*e.g.* argon) pressure, energy density of laser pulses, their number and substrate position in relation to the torch axis.

10.3 Characterization, Optimization and Selection of Materials with Nanoscale-Specific Local Plasmon resonance Effect

When deposited from the reverse flow of Au NP, the structure of 2D/3D films is characterized not only by the gradient of thickness, but also by the porosity, the size of the Au NP and the distances between them along the substrate. With the aim of understanding the growth mechanism and in order to control the quantity, size distribution and location of "hot spots", the influence of technological modes on the morphology of porous plasmonic films was investigated. Taking into account the fact that the functional properties are determined directly by the parameters of the growth modes, special attention will be paid to investigating the relation between optical characteristics and the nanoscale organization of porous nanostructures (Fig. 10.3).

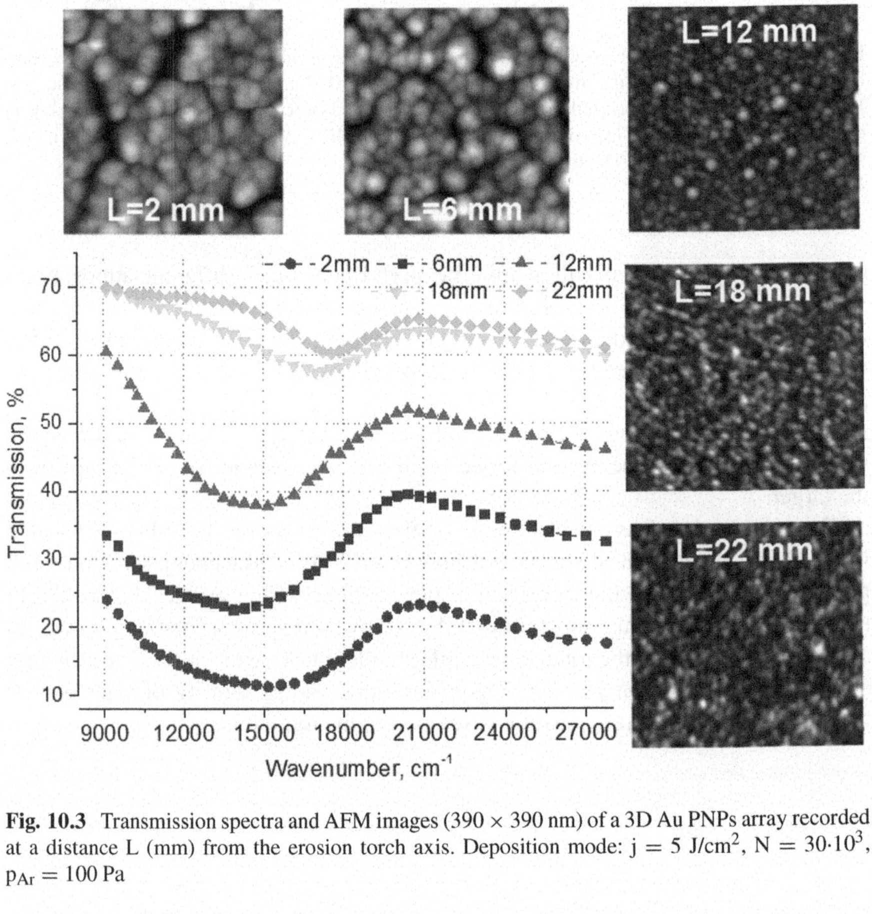

Fig. 10.3 Transmission spectra and AFM images (390 × 390 nm) of a 3D Au PNPs array recorded at a distance L (mm) from the erosion torch axis. Deposition mode: $j = 5$ J/cm^2, $N = 30{\cdot}10^3$, $p_{Ar} = 100$ Pa

It has been established that the method of deposition from the backward flow of particles in an argon atmosphere enables the formation of films with gradients of their thickness, porosity, PNPs size and distances between them. Changing the PNPs array structure from 3D to 2D along the substrate makes it possible to control their optical spectral characteristics – the spectral position of the adsorption bands, which are due to the effects of local plasmon resonance in the nanostructures.

The optimal pulsed laser deposition conditions for the formation of films with Au PNP arrays with the manifestation of localized surface plasmon resonance in the wavelength range of 535–740 nm for Au nanoparticle arrays were determined [10]. It was established that the variation of the argon pressure, the number of laser pulses and the energy density in a pulse allows to form 2D and 3D arrays of Au PNPs and to manage their structural parameters – PNP size, distances between them, thickness and porosity of films, which enables the obtaining of nanostructures with given resonance-optical properties. The porosity of the films lies in the range of 15–70% and increases with the increase of argon pressure, the decrease of the number of pulses and the energy density in a pulse. Increasing the porosity value resulted in a shift of the transmission minimum into the shortwave region and has a linear dependence. For 3D layers, the structure is characterized not only by a wide distribution of the PNP sizes, but also by a wide distribution of distances between them and the presence of chains of tangent PNPs. In such 3D nanosystems, new plasmon modes may emerge as a result of the interaction of PNPs plasmons - both laterally and normal to the film surface. It is in such structures that a large variety of "hot spots" of different sizes and structures can be formed.

10.4 Tunable Substrates Based on Wedge-Like Porous Metal Layers for Surface Enhanced Raman Spectrocopy

A porous nanoscale structure with tunable local SPR excitations is optimal for substrates used in Surface Enhanced Raman Spectroscopy – a powerful vibrational spectroscopy technique that allows a highly sensitive detection of low-concentration analytes through the amplification of electromagnetic fields near nanostructured surfaces. The highest Raman enhancement occurs within the "hot spots", which are between the nanoparticles. These are the areas of the largest field density, which is a result of the interference of all fields scattered by nanoparticles. Since the resonance-optical properties of plasmonic structures discussed above are also determined by their morphology, it is reasonable to suggest that the efficiency of SERS analysis may be tuned by shifting across the porous film.

Nanostructures with Au PNPs arrays optimal for recording the spectra with excitation wavelength of 514.5 nm were obtained in an argon atmosphere with a pressure of 100 Pa, an energy density in a pulse of 5 J/cm^2, and a pulse number of $30 \cdot 10^3$. Such deposition conditions made it possible to form films with the highest porosity values from ~40% to 65%, film thickness from 67 nm to 6 nm, distribution

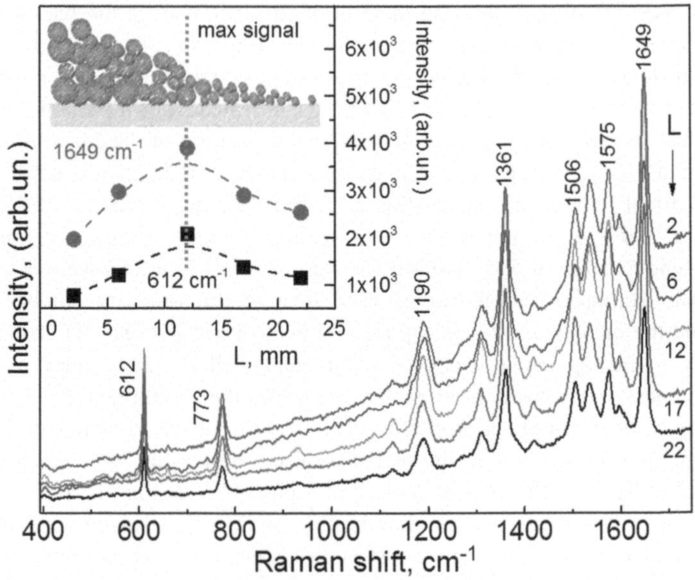

Fig. 10.4 Typical SERS spectra of rhodamine 6G with a concentration of 10^{-7} M/L on Au-por SERS substrates at different distances from the erosion torch

of Au PNP diameters from 5–60 nm to 2–14 nm with the distance from the torch axis of 2–22 mm. Fig. 10.4 shows the results of measurements of SERS spectra of rhodamine 6G, which served as a standard marker, with a concentration of 10^{-7} M/L [11].

The optimal distance to obtain the maximum SERS signal is L = 12 mm, which corresponds to average Au PNP size of 8 nm (with the range of sizes 2–20 nm), average pore diameter of 5 nm (with the range of sizes 2–10 nm) and thickness of c.a. 20 nm. The signal intensity in this area is more than two times higher than at other points of the film. The values of the enhancement factor at rhodamine concentrations of 10^{-5}, 10^{-7}, 10^{-9} and 10^{-10} M were 3.1×10^3, 7.2×10^4, 5.2×10^6 and 3.9×10^7 respectively, for the optimal position on the por-Au film. The high values of the enhancement parameters are due to the large internal surface for the analyte and the presence of "hot spots" in the pores.

It is also reasonable to highlight a principal advantage of the developed materials based on porous gold films, which enable enhanced Raman scattering and simultaneously reduce the analyte-specific luminescence to zero. The latter in combination with the possibility to tune the sensitivity of the substrate through the variation of the measurement position on the chip (with different nanostructured organizations) clearly demonstrates the advanced functional properties of the materials under consideration.

10.5 Hot-Spots Examination Method for Porous Gold Nanoparticle Arrays for the Purposeful Development of SERS Substrates

Both theoretical calculations and experimental results unambigously indicate that the highest Raman enhancement occurs within the "hot spots", which are typically between the nanoparticles. These are the areas of the highest field density, which is a result of the interference of all fields scattered by nanoparticles. Therefore, in order to confirm that this mechanism also occurs in the materials considered in this case, it is necessary to show the correlation between the number of such potentially high energy regions and the SERS enhancement value (Fig. 10.4, Insert). Using self-developed software for the layer-by-layer (sections in thickness) analysis of AFM images, it was confirmed that the greatest amplification of the SERS signal corresponds to the largest number of "hot spots" within the film [12].

From the dataset presented in Fig. 10.5, one can see that the dimensions of single crystallites and pores and the width of the respective distributions decrease when

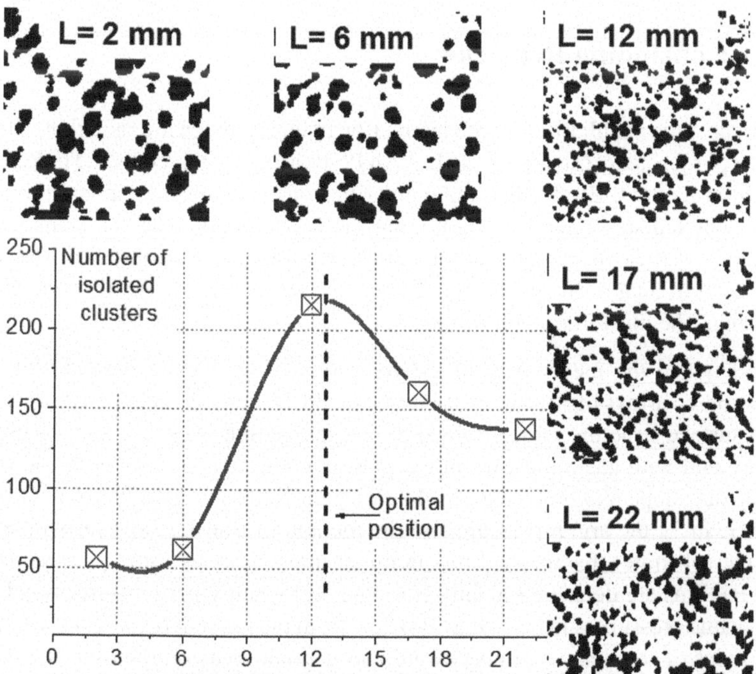

Fig. 10.5 The dependence of the number of isolated clusters in the cross section of a por-Au film, calculated from AFM images (390 × 390 nm), on the distance from the erosion torch axis. The figure shows images of sections parallel to the plane of the substrate, located at a distance of 31.5% of the film thickness from above

the distance from the erosion torch axis increases. If the maximum of the Au PNP distribution near the erosion torch axis is 25 nm, then at the farthest surface point it is 7 nm. The sizes of Au PNPs and pores are approximately the same at distances greater than 12 mm; near the erosion torch (L = 2 mm), the pores are 2 times smaller than Au NPs. It should be noted that there is a correlation between the film thickness and the sizes of Au PNPs and pores. Moreover, at thicknesses in tens (units) of nanometers, the Au PNPs sizes and pores are of the same order of magnitude. These data indicate significant porosity of por-Au films.

For all layers, with the depth increase, the number of isolated clusters increases, reaches maximum in the range of 20–30%, and then decreases sharply. This indicates that the maximum efficiency of such SERS substrates is achieved not at the film surface itself, but inside the film at a certain distance from the surface. In this case, the maximum value of the number of clusters is observed for the sample obtained at a distance of 12 mm from the torch axis, which corresponds to the point with the maximum increase of the SERS signal (Fig. 10.4). At the same time, the efficiency of "hot spots" can also be affected by the development of the boundaries of individual grains. Here, we also have a correlation with the experimental result.

10.6 Concluding Remarks

Investigated the peculiarities of the interaction of electromagnetic radiation with disordered porous nanostructures based on Au PNPs arrays obtained by the pulsed laser deposition method. Established the regularities of the resonance-optical properties of the obtained structures by determining the relationships with the technological conditions of their formation and the structural properties, for purposeful practical application in analytical systems based on the surface enhancement of the Raman scattering. The most important results are:

1. Investigated the mechanism of the formation of composite porous films with disordered arrays of Au nanoparticles by the PLD method: the process of forming the surface structure is determined by the interaction of metal atoms with each other and with gas molecules in the plasma torch, outside the torch and on the substrate, depending on the basic ablation parameters (the number of laser pulses YAG:Nd^{3+}, the argon pressure and the method of transferring plasma particles to the substrate). Established that in direct transfer, at an argon pressure of 10–20 Pa films are deposited mainly from the high-energy particle stream, whereas when the pressure is increased to 100 Pa – from the low-energy stream with fewer particles of smaller size. In backward nanoparticle transfer, the films are formed from low-energy streams with nanoparticles size distribution along the substrate.
2. Determined that the peculiarities of the formation of the structural properties of films correspond to the processes of their obtainment and are characterized by composite composition (metal nanoparticles and pores), disordered nanostructures from 2D with isolated nanoparticles to 3D with interparticle interactions.

When depositing from the direct particle flow, with the increase of the argon pressure the porosity of the films increases to 50–60%, while the PNP sizes and distances between them decrease. 2D nanostructures with isolated Au PNPs with average sizes of 7–10 nm are formed when the number of pulses is lower than $N = 3 \times 10^4$; when the number is increased to $N = 15(45) \times 10^4$, 3D porous nanostructures with a large dispersion of nanoparticle sizes are formed. As the number of pulses increases, the effective film thickness increases from 5 to 50–120 nm, and the Au PNP sizes also increase. When deposited from the backward PNP flow, the films structure is characterized by the gradients of their thickness, porosity, PNP sizes and distances between them; the nanostructure changes from 3D to 2D along the substrate.

3. Established that the resonance-optical properties of the obtained composite porous films with Au PNP arrays are due to the excitation of the local SPR. Showed that controlling the technological parameters (deposition mode, argon pressure, number of laser pulses, energy density in a pulse) forms composite films with Au PNP arrays with controlled resonance-optical properties in the 535–740 nm range of local SPR wavelengths for Au PNP arrays.
4. The determined connections between conditions, formation processes, structural and resonance-optical properties of films with Au PNP arrays allowed to obtain SERS structures with high values of local electric fields in "hot spots" with the increase of the Raman signal to $3\text{–}4 \times 10^7$.

Finally, determined the optimal technological parameters ensuring the highest efficiency of SERS substrates based on porous Au films. Fabricated stable, low-cost and tunable por-Au SERS substrates with a large surface area and a high density of "hot spots". SERS substrates based on 2D or 3D NPs random arrays allow tuning the enhancement factor by changing the analysis position along the substrate: the change of the porous structure with the distance from the erosion torch allows to maximize the number of "hot spots" for an analyte with a certain size and chemical functionality. Demonstrated the possibility of express registration of organic molecules with a local enhancement factor of $3\text{–}4 \times 10^7$, which allows to identify the analyte under real conditions with concentrations up to 10^{-10} M/L.

Acknowledgment This research was funded by the NATO SFP Program (G5140 Grant). The authors expresses their sincere gratitude to E. B. Kaganovich for the initiation of this technology, S. Kravchenko, O. Kolomys, A. Nikolenko and P. Lytvyn for help with Raman and AFM measurements and Oleksii Snopok for the assistance in the preparation of the present manuscript.

References

1. Snopok B (2008) In: Bonca J, Kruchinin S (eds) Electron transport in nanosystem. Springer, New York, pp 331–339
2. Burlachenko J, Kruglenko I, Snopok B, Persaud K (2016) Sample handling for electronic nose technology: state of the art and future trends. Trends Anal Chem 82:222–236

3. Snopok BA, Kruglenko IV (2002) Multisensor systems for chemical analysis: state-of-the-art in electronic nose technology and new trends in machine olfaction. Thin Solid Films 418:21–41
4. Kruglenko IV, Snopok BA, Shirshov YM, Venger EF (2000) Digital aroma technology for chemical sensing: temporal chemical images of complex mixtures. Semiconductor Phys Quantum Electronics Optoelectronics 3(4):529–541
5. Burlachenko JV, Snopok BA (2008) Multisensor arrays for gas analysis based on photosensitive organic materials: an increase in the discrimination capacity under selective illumination conditions. J Anal Chem 63:610–619
6. Burlachenko J, Kruglenko I, Manoylov E, Kravchenko S, Krishchenko I, Snopok B (2019) Virtual sensors for electronic nose devises. 2019 IEEE International Symposium on Olfaction and Electronic Nose (ISOEN), Symposium proceedings, pp 77–79
7. Snopok B, Naumenko D, Serviene E, Bruzaitc I, Stogrin A, Kulys J, Snitka V (2014) Evanescent-field-induced Raman scattering for bio-friendly fingerprinting at sub-cellular dimension. Talanta 128:414–421
8. Naumenko D, Snitka V, Serviene E, Bruzaite I, Snopok B (2013) In-vivo characterization of protein uptake by yeast cell envelope: single cell AFM imaging and μ-tip-enhanced Raman scattering study. Analyst 138:5371–5383
9. Naumenko D, Snitka V, Snopok B, Arpiainen S, Lipsanen HI (2012) Graphene-enhanced Raman imaging of TiO_2 nanoparticles. Nanotechnology 23:465703
10. Krishchenko I (2018) Resonance-optical properties composite porous films with Au(Ag) nanoparticles arrays obtained by pulsed laser deposition. PhD thesis, Kyiv
11. Kolomys O, Nikolenko AS, Lytvyn PM, Manoilov EG, Krishchenko IM, Strelchuk VV, Snopok BA (2019) SERS substrates with porous gold films obtained by pulsed laser deposition for bio-imaging. In: 16th International Conference on Nanoscience and Nanotechnologies, 2–5 July 2019, Thessaloniki, Greece, Abstract Book, p 157
12. Kravchenko S, Krishchenko I, Manoilov E, Snopok B (2019) Hot-spots examination method for porous gold nanoparticle arrays in SERS application. In: 13th International Summer Schools on Nanosciences and Nanotechnologies, 29-6 July 2019, Thessaloniki, Greece, Abstract book, p 150

Chapter 11
Continuous and Time-Resolved Cathodoluminescence Studies of Electron Injection Induced Effects in Gallium Nitride

Sushrut Modak, Leonid Chernyak, Igor Lubomirsky, and Sergey Khodorov

Abstract The purpose of this study is to experimentally determine the effect of electron injection on the minority carrier lifetime in Gallium Nitride. Earlier studies of electron injection in GaN have provided an indirect proof of lifetime enhancement through the increase of minority carrier diffusion length and the decrease in the cathodoluminescence intensity. These changes in the minority transport properties, caused by electron injection, are brought forth through defect and trap levels in the bandgap. Furthermore, a thorough discussion of the electron injection model and role of these trap levels is presented.

Keywords Electron injection · Lifetime · Minority carrier transport · Gallium nitride

11.1 Introduction

Over the last decade, technological advances in the semiconductor processes have enabled a new trend. Wide-bandgap semiconductors, which were previously thought of to be insulators, have become more accessible. Extensive research on these materials have enabled many of them to find applications in industry such as blue/UV LEDs and use in high-power electronics and radiation prone zones.

S. Modak
University of Central Florida, Orlando, FL, USA

L. Chernyak (✉)
Department of Physics, University of Central Florida, Orlando, FL, USA
e-mail: Leonid.Chernyak@ucf.edu

I. Lubomirsky · S. Khodorov
Weizmann Institute of Science, Rehovot, Israel

C. Palestini (ed.), *Advanced Technologies for Security Applications*, NATO Science for Peace and Security Series B: Physics and Biophysics,
https://doi.org/10.1007/978-94-024-2021-0_11

Gallium nitride and zinc oxide are examples of such materials which have matured over the last two decades [1–4] with gallium oxide on the horizon [5].

But due to the large bandgap that these materials possess and technological challenges, they suffer from an array of point and extended defects [3, 6–11]. These defects not only cause carrier scattering, but also introduce trap levels in the bandgap, which degrade the performance of the material through the Shockley-Read-Hall (SRH) recombination. Apart from the heavy nature of holes in wide bandgap semiconductors, the above-mentioned defects also play a role in the reduction of the carrier diffusion length and lifetime [2, 12]. Therefore, there is still room for more work to be done for exploiting GaN-based devices to their full potential.

It has been previously shown that electron injection provides a way for control over SRH recombination process in GaN and ZnO [13–20]. One of the ways for realization of electron injection, is irradiation of the material with low energy electrons (10–30 keV) for extended durations. Material's excitation with these electrons create non-equilibrium electron-holes pairs, which locally saturate the trap levels, causing the lifetime of non-equilibrium carriers to increase. Previous studies of electron injection in p-type Zinc Oxide and Gallium Nitride showed an improvement in the carrier transport properties, such as minority carrier diffusion length, and a decrease in the cathodoluminescence intensity in the material from the region of excitation, thus indicating improved carrier lifetime [13, 14, 18]. In this study, we provide a direct proof for the impact of electron injection on non-equilibrium carrier lifetime at room temperature and further develop the underlying mechanism responsible for this effect.

11.2 Experimental

In this study, we employ continuous wave and time-resolved cathodoluminescence (CL and TRCL, respectively) to study the impact of electron injection on non-equilibrium carrier lifetime in Gallium Nitride. The above techniques are available in Attolight Allallin Chronos Scanning Electron Microscope (SEM). An electron beam of 10 keV was used to irradiate an area of constant size (500 × 500 nm) for all measurements. Electron injection is performed by continually scanning the beam within the selected area with a resolution of 128 × 128 points for 0, 420, 840, 1260, 1680, 2100 and 2520 s. Since the area exposed to beam is comparable with the range of electrons in the sample under test, the used method of irradiation is equivalent to a continuous exposure of the material with an electron beam at a single point. Emitted cathodoluminescence from the material is collected by parabolic mirror and passed through a UV-visible spectrometer (Horiba iHR320). The spectrum is recorded with a thermoelectrically cooled CCD array (Andor Newton 920P) kept at −70 °C. Hyperspectral cathodoluminescence maps are recorded by repeating the above process of synchronously irradiating with an electron beam (dwell times of 2 ms) in an grid of 128 × 128 points over the area of interest, and recording the CL spectrum with the CCD camera at each point. Electron pulses in the time-resolved mode of operation of the SEM are attained by focusing a laser (Onefive Genka HP-03, 80 MHz) on the tip of the gun after the continuous electron beam is turned off. The resulting electron pulse is guided towards the sample with a magnetic lens assembly of the microscope. Laser pulses are synchronized with a streak camera (Optronis Optoscope) attached to the spectrometer to collect streaks. All recorded streaks are accumulated for 400 collection cycles. After every cycle of electron injection, a hyperspectral map in CL mode and a minimum of ten streaks in TRCL mode within the area of electron injection are collected.

The samples used in this study consist of a crack-free, free standing 400 μm thick GaN layers. These layers were grown by Halide Vapor Phase Epitaxy (HVPE) on a two-step epitaxial lateral over-grown GaN template on sapphire. The 'as-grown' GaN epitaxial layer is n-doped with carrier concentration of ~ 1×10^{17} cm^{-3} and mobility ~ 750 cm^2/Vs at room temperature. Threading dislocation density of 10^6 cm^{-2} was discovered by polychromatic CL mapping.

11.3 Results

The effect of electron injection on minority carrier transport in GaN has been extensively studied in the past. Diffusion length (*L*), which is one of the most important parameters in minority carrier transport, shows the ability of the carrier, generated due to external mechanism, to diffuse without recombination or scattering [20–23]. *L* is defined as

$$L = \sqrt{D\tau}, \tag{11.1}$$

where D is the diffusion coefficient and τ is the minority carrier lifetime. Previous studies have shown that electron injection can cause L to increase linearly from 0.5 to 2 μm (a quadruple increase) as a function of charge injection density from 0 to 2.0×10^{-7} C/μm^3 [18]. This has a significant positive impact on the quantum efficiency of photodetectors, η, which is defined as

$$\eta = (1 - R)\left(1 - e^{-\alpha W}/(1 + \alpha L)\right) \tag{11.2}$$

where R and α are the reflectance and absorption coefficient respectively; W is the depletion layer width; and L is the minority carrier diffusion length. It can be seen from Eq. (11.2), that greater L leads to a more efficient device, especially in III-Nitrides, which have a small L (< 1 μm). L may change either due to the rise in lifetime or diffusivity (which is also directly proportional to mobility). Earlier studies [24] reported an electron beam-induced decrease of resistivity by three orders of magnitude in GaN. This was due to an increase in the majority carrier concentration with unchanged mobility. The latter result favors the independence of mobility (diffusivity) for minority carriers on electron injection [18]. Separately, it was also shown that electron injection causes deep trap levels, embedded into the bandgap, to capture non-equilibrium charges, thus making these levels unavailable for SRH recombination events [15, 16, 19, 20, 25]. The above-reported effects suggest that longer L in GaN is caused by increased lifetime of minority carriers.

The change in lifetime of a material can be probed with CL measurements [15]. If the rise in L is due to increased lifetime, the number of recombination (radiative) processes should decrease (carriers stay active longer) together with the intensity of CL signal. The reduction in the CL intensity due to electron injection in GaN is reported previously and conforms to the analysis above [18]. The final piece of the puzzle towards completion of this analysis is the direct measurement of minority carrier lifetime as a function of electron injection duration, which is accomplished in this study.

The experiments were started with continuous cathodoluminescence study of GaN sample. An area of 500 × 500 nm was irradiated with 10 keV electron beam for 0, 420, 840, 1260, 1680, 2100 and 2520 s, as mentioned above. CL spectra were periodically collected (Fig. 11.1).

Figure 11.1 showed a decrease in the CL amplitude by ~ 20% as a result of electron injection. As is previously noted, CL intensity is directly proportional to the inverse lifetime of non-equilibrium carriers: the longer the carriers are alive, the fewer is the number of recombination events, and vice versa. This information, coupled with Eq. (11.1), implies a quadratic dependence of I^{-1} on electron injection duration. This relationship is shown with the acquired data in Fig. 11.1, inset. The linear dependence between $\sqrt{I^{-1}}$ and electron injection duration is clearly

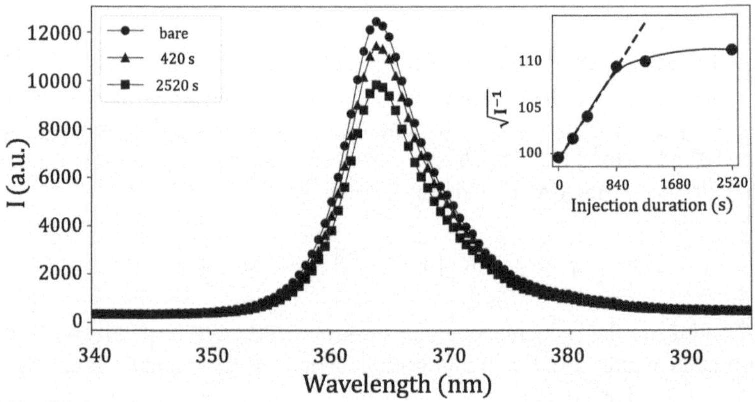

Fig. 11.1 Cathodoluminescence spectra collected at 0, 420 and 2520 s of continuous GaN exposure to a 10 keV electron beam. Inset shows the linear dependence of $\sqrt{I^{-1}}$ on electron injection duration. The saturation of this effect occurs after 840 s of electron beam exposure, as seen from the inset

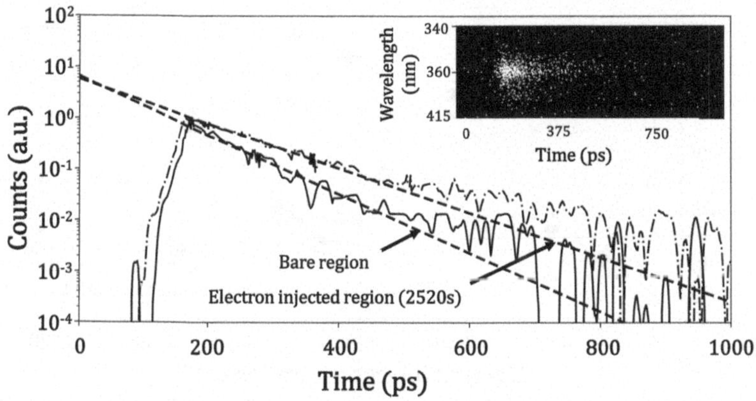

Fig. 11.2 Streak signal for the bare region and 2520 s electron injection region. Lifetime, which is inverse of the slope of the streak decay, is higher for the electron injection region with lower slope. Inset shows all photon counts in a raw streak image

portrayed. The saturation of decrease in the intensity occurs approximately after 840 s of exposure to electron beam.

After every electron beam injection cycle, several TRCL streaks were taken within the entire exposed area. An electron pulse of 10 ps duration is used to collect CL signal with a streak camera, to measure the radiative recombination lifetime. An integral accumulation of 400 spectra (with photon thresholding and counting) was used to generate a streak. An example of collected streaks, corresponding to 0 and 2520 s of electron injection, is shown in Fig. 11.2. Lower slope of the electron injected streak, compared to the streak from the bare region, is indicative of longer minority carrier lifetime. The inset of Fig. 11.2 shows a raw image of the streak displaying the photon counts over 400 collection cycles. The lifetimes for 0 and 2520 s of electron injection were found to be 77 ± 7 and 101 ± 6 ps. These values fall within an acceptable range reported by an independent study [26] using a TRCL technique.

Similar to the saturation in decrease of CL intensity I, with electron injection, a saturation in τ is also observed at 840 s mark, with a lifetime of 95 ± 8 ps. For these measurements, a minimum of ten streaks were collected over the electron injected region, which contains places with threading dislocation. It has been found that the dislocations have a slightly lower time response, leading to slightly higher values of standard deviation in these measurements [26, 27]. Increase of lifetime with respect to the injection duration is shown in Fig. 11.3.

The trend of lifetime extension saturates at about 840 s of electron injection, same as the saturation time for CL intensity decrease (cf. Fig. 11.1). Inset of Fig. 11.3 shows the linear increase of $\sqrt{\tau}$ with respect to electron injection duration, as predicted by Eq. (11.1). Moreover, the linear dependence of I^{-1} on τ, is depicted in Fig. 11.4 by the dashed line. This linear relation between I^{-1} and τ is fundamental to the behavior of non-equilibrium carrier recombination and occurs irrespective of electron injection duration.

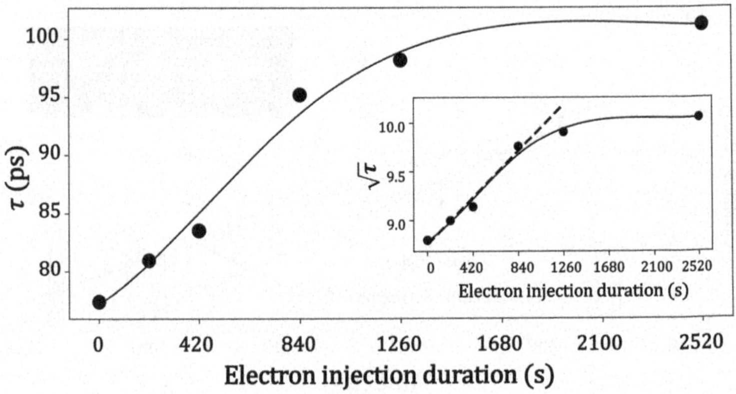

Fig. 11.3 The measured lifetime of minority carriers as a function of electron injection duration. The saturation of the lifetime increase occurs at around 840 s (matches with Fig. 11.1), after which higher injection duration produces very little effect on it. Inset shows the linear trend of $\sqrt{\tau}$ with injection duration, as indicated by Eq. (11.1)

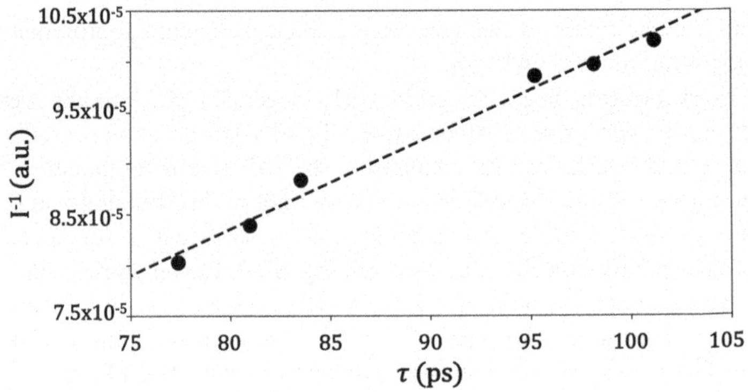

Fig. 11.4 Plot depicting inverse of the cathodoluminescence intensity as a function of lifetime of non-equilibrium carriers. The rise in τ is accompanied with decrease in I

All the results listed above are in congruence with the previously performed studies on electron injection in GaN [15, 16, 18–20, 28]. Figure 11.5 summarizes the model for effects of electron injection on minority carrier transport in GaN:

- Electron beam irradiation creates non-equilibrium electron-hole pairs in GaN, which can consequently recombine either band-to-band or through a trap level in the forbidden gap (cf. Fig. 11.5, I).
- As electron injection proceeds, a portion of non-equilibrium electrons is captured by trap levels in the forbidden gap (cf. Fig. 11.5, II) [15, 16]. As a result, these trap levels stop playing a role in the recombination process, which is mostly radiative in nature [18]. This explains a decrease of CL intensity with duration of electron injection as seen in Fig. 11.1.

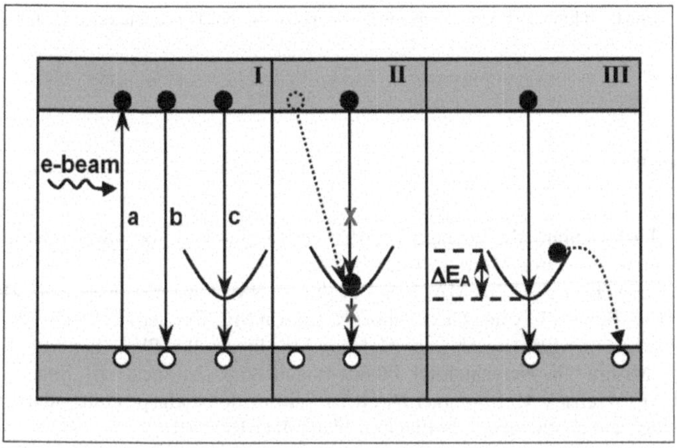

Fig. 11.5 The model for impact of electron injection on the minority carrier transport in GaN. (*I a*) Electron-hole pair generation, (*I b*) Band-to-band recombination, (*I c*) Recombination through a trap level; (*II*) Effect of electron injection in filling up the trap level with a non-equilibrium electron and making the trap unavailable for further recombination events; (*III*) The trap level, when supplied with the activation energy corresponding to the depth of the trap, releases the trapped electron (hole capture) and becomes available again for SRH recombination

- Electron blockade of recombination levels in the forbidden gap (cf. Fig. 11.5, II) causes the non-equilibrium electrons and holes to live longer in the conduction and valence bands, respectively. This effect manifests in the form of increase in the lifetime (Fig. 11.3) in the electron injected region and, hence, longer diffusion length of the minority carriers as was previously reported.
- As electron injection continues, a saturation in the number of traps, available for non-equilibrium electron capture, occurs. This is manifested in the saturation of CL intensity decrease (Fig. 11.1) and slower growth in the lifetime (cf. Fig. 11.3).
- The trap levels in the bandgap have their own specific activation energies and can be activated optically or thermally [18]. This causes an escape of the trapped non-equilibrium electron from the local potential barrier, created by a trap (cf. Fig. 11.5, III). The latter becomes again available for recombination.
- The effect of electron injection on minority carrier transport properties has been shown to last for several days (due to the meta-stability of the trap levels) before the deterioration in these properties becomes significant [18].

In conclusion, we experimentally measured the changes in the lifetime of non-equilibrium carriers with TRCL due to electron injection. The measured values of non-equilibrium carrier lifetime are in agreement with independent studies on GaN [26]. The present study served as the missing link in the proposed model of electron injection, attributing the increase of diffusion length and decrease in cathodoluminescence intensity to the increase in lifetime due to electron injection, rather than changes in the carrier mobility.

Acknowledgments This work was supported in part by the NATO Science for Peace and security award G5453.

References

1. Flack TJ, Pushpakaran BN, Bayne SB (2016) GaN technology for power electronic applications: a review. J Electron Mater 45(6):2673–2682
2. Tsao JY, Chowdhury S, Hollis MA, Jena D, Johnson NM, Jones KA, Kaplar RJ, Rajan S, Van de Walle CG, Bellotti E, Chua CL, Collazo R, Coltrin ME, Cooper JA, Evans KR, Graham S, Grotjohn TA, Heller ER, Higashiwaki M, Islam MS, Juodawlkis PW, Khan MA, Koehler AD, Leach JH, Mishra UK, Nemanich RJ, Pilawa-Podgurski RCN, Shealy JB, Sitar Z, Tadjer MJ, Witulski AF, Wraback M, Simmons JA (2018) Ultrawide-bandgap semiconductors: research opportunities and challenges. Adv Electron Mater 4(1):1600501
3. Wang J, Mulligan P, Brillson L, Cao LR (2015) Review of using gallium nitride for ionizing radiation detection. Appl Phys Rev 2(3):031102
4. Kolodziejczak-Radzimska A, Jesionowski T (2014) Zinc oxide-from synthesis to application: a review. Materials (Basel) 7(4):2833–2881
5. Pearton SJ, Yang J, Cary PH, Ren F, Kim J, Tadjer MJ, Mastro MA (2018) A review of Ga2O3 materials, processing, and devices. Appl Phys Rev 5(1):011301
6. Reshchikov MA, Morkoç H (2005) Luminescence properties of defects in GaN. J Appl Phys 97(6):061301
7. Reshchikov MA, Usikov A, Helava H, Makarov Y, Prozheeva V, Makkonen I, Tuomisto F, Leach JH, Udwary K (2017) Evaluation of the concentration of point defects in GaN. Sci Rep 7(1):9297
8. Kamyczek P, Placzek-Popko E, Zielony E, Zytkiewicz Z (2013) Deep levels in GaN studied by deep level transient spectroscopy and Laplace transform deep-level spectroscopy. Mater Sci-Pol 31(4):572–576
9. Monemar B (2001) Bound excitons in GaN. J Phys Condens Matter 13(32):7011–7026
10. Look DC, Reynolds DC, Hemsky JW, Sizelove JR, Jones RL, Molnar RJ (1997) Defect donor and acceptor in GaN. Phys Rev Lett 79(12):2273–2276
11. Boguslawski P, Briggs EL, Bernholc J (1995) Native defects in GaN. Phys Rev B 51(23):17255–17259
12. Park JH, Kim DY, Schubert EF, Cho J, Kim JK (2018) Fundamental limitations of wide-bandgap semiconductors for light-emitting diodes. ACS Energy Lett 3(3):655–662
13. Lopatiuk-Tirpak O, Chernyak L, Mandalapu LJ, Yang Z, Liu JL, Gartsman K, Feldman Y, Dashevsky Z (2006) Influence of electron injection on the photoresponse of ZnO homojunction diodes. Appl Phys Lett 89(14):142114
14. Lopatiuk-Tirpak O, Chernyak L, Xiu FX, Liu JL, Jang S, Ren F, Pearton SJ, Gartsman K, Feldman Y, Osinsky A, Chow P (2006) Studies of minority carrier diffusion length increase in p-type ZnO:Sb. J Appl Phys 100(8):086101
15. Burdett WC, Lopatiuk O, Osinsky A, Pearton SJ, Chernyak L (2004) The optical signature of electron injection in p-(Al)GaN. Superlattice Microst 34(1–2):55–62
16. Chernyak L, Burdett W, Klimov M, Osinsky A (2003) Cathodoluminescence studies of the electron injection-induced effects in GaN. Appl Phys Lett 82(21):3680–3682
17. Chernyak L, Schulte A, Osinsky A, Graff J, Schubert EF (2002) Influence of electron injection on performance of GaN photodetectors. Appl Phys Lett 80(6):926–928
18. Chernyak L, Osinsky A, Schulte A (2001) Minority carrier transport in GaN and related materials. Solid State Electron 45(9):1687–1702
19. Chernyak L, Nootz G, Osinsky A (2001) Enhancement of minority carrier transport in forward biased GaN p-n junction. Electron Lett 37(14):922–923

20. Chernyak L, Osinsky A, Fuflyigin V, Schubert EF (2000) Electron beam-induced increase of electron diffusion length in p-type GaN and AlGaN/GaN superlattices. Appl Phys Lett 77(6):875–877
21. Leamy HJ (1982) Charge collection scanning electron microscopy. J Appl Phys 53(6):R51–R80
22. Ioannou DE, Dimitriadis CA (1982) A SEM-EBIC minority-carrier diffusion-length measurement technique. IEEE Trans Electron Devices 29(3):445–450
23. Berz F, Kuiken HK (1976) Theory of life time measurements with the scanning electron microscope: steady state. Solid State Electron 19(6):437–455
24. Nakamura S, Senoh M, Mukai T (1991) Highly p-typed Mg-doped GaN films grown with GaN buffer layers. Jpn J Appl Phys 30(Part 2, 10A):L1708–L1711
25. Burdett W, Osinsky A, Kotlyarov V, Chow P, Dabiran A, Chernyak L (2003) Impact of aluminum concentration and magnesium doping on the effect of electron injection in p-$Al_xGa_{1-x}N$. Solid State Electron 47(5):931–935
26. Liu W, Carlin JF, Grandjean N, Deveaud B, Jacopin G (2016) Exciton dynamics at a single dislocation in GaN probed by picosecond time-resolved cathodoluminescence. Appl Phys Lett 109(4):042101
27. Sonderegger S (2007) Picosecond time resolved cathodoluminescence to study semiconductor materials and heterostructures (École polytechnique fédérale de Lausanne)
28. Chernyak L, Osinsky A, Temkin H, Yang JW, Chen Q, Asif Khan M (1996) Electron beam induced current measurements of minority carrier diffusion length in gallium nitride. Appl Phys Lett 69(17):2531–2533

Chapter 12
Creation of New Generation Titanium Diboride Composite Armour Material

Gabriel Benga, Nikoloz Iakobidze, Danut Savu, Sorin Savu, and Iulian Stefan

Abstract The next generation Armour Piercing Bullet with Tungsten Carbide core has been introduced into the world's leading military forces by the beginning of the twenty-first century, and later, into the Russian army, which has resulted in a sharp increase in penetrability of the existing standard composite armour plates. The only armour system that effectively stops these bullets is a combined structure obtained through a hot-pressing technology and composed of multilayer fabric made of high-molecular organic fibres. Despite the high armour properties of the above-mentioned structure, its components (separately and together) have some negative sides as well: high cost, low production capacity and low firmness of armour ceramics, which hinders its widespread use in personal protective equipment and in armoured combat vehicles as well. Therefore, there was a need to find new approaches in processing technologies of protective structure and its new components in order to create a new generation armour material for ceramic armour characterized by high hardness and strength (not less than boron carbide), low cost and high capacity and with the following physical-mechanical properties: micro hardness – 3600–3800 kg/mm^2; bending strength 70–80 kg/mm^2; density – 4.0–4.5 g/cm^3. The obtained results will allow us to replace the boron carbide ceramics with a relatively inexpensive and perspective armour material. With regard to the second element of armour structure, a new approach has also been found to composition of multilayer fabric and manufacturing technology, which involves the production of a high performance industrial thermo-hydro forming unit, which in liquid, under high pressure and temperature, would thermoform composite armour products (ballistic ceramic plate) of complex configurations. This method gives us the opportunity to

---

G. Benga (✉) · D. Savu · S. Savu · I. Stefan
Faculty of Mechanics, Department of IMST, University of Craiova, Drobeta Turnu Severin, Romania
e-mail: benga.gabriel@ucv.ro

N. Iakobidze
Georgian Technical University, Tbilisi, Georgia

C. Palestini (ed.), *Advanced Technologies for Security Applications*, NATO Science for Peace and Security Series B: Physics and Biophysics,
https://doi.org/10.1007/978-94-024-2021-0_12

get a composition of thermally pressed multilayer fabric (ceramic plate) composed of relatively cheap, impact-resistant and puncture-resistant super-hard materials.

Keywords Armour plates · TiB_2 · Nanosized composite powders · Microwave sintering · Thermoforming

12.1 Introduction

In modern Armour Piercing Bullets hardened steel core is replaced by hard alloy of tungsten carbide and cobalt, the hardness of which 91–92 HRA is much higher than the hardness of hardened steel core. This makes it possible to increase the penetrability of armour plates by 60–70% compared to Armour Piercing Bullet with hardened steel core and with the same calibre and energy. The use of such bullets poses serious problems for personal protective equipment, such as armour vests and combat helmets, and for armoured vehicles as well. Among the most common Armour Piercing Bullets with Tungsten Carbide core in NATO subdivisions are: small calibre -5.56 × 45 mm AP M995, 7.62 × 51 mm AP M993, 7.62 × 51 SWISS P AP, 8,6 × 70 mm AP485 and AP529, 12,7 × 99 mm AP-S NM185 and 12,7 × 99 mm API-S NM173. Large calibre automatic guns -20 × 102 mm FAP and 25 × 137 mm M791 APDS-T; while in Russian Army – 5,45 × 39 mm AP 7N24, 12,7 × 108 mm API 7-BZ-1 and 14,5 × 114 mm AP BS-41 (special).

Many researchers developed tests in order to determine the behaviour of different ceramic composites in ballistic tests [1–3] revealing a positive effect in terms of ballistic performance.

Currently, the most efficient armoured system that stops these bullets is a combined structure obtained through a hot-pressing technology and composed of multilayer fabric made of high-molecular organic fibres. Despite the high armour properties of the above-mentioned structure, its components (separately and together) have some negative sides as well: high cost, low production capacity and low firmness of armour ceramics, which hinders its widespread use in personal protective equipment and in armoured combat vehicles as well. Therefore, there was a need to find new approaches in processing technologies of protective structure and its new components in order to: create a new generation armour material for ceramic armour based on TiB_2, characterized by high hardness and strength (not less than boron carbide), low cost and high capacity and with the following physical-mechanical properties: micro hardness – 3600–3800 kg/mm^2; bending strength 70–80 kg/mm^2; density – 4.0–4.5 g/cm^3 [4]. TiB_2 has been used on a large scale in various industrial and military applications due to its high melting temperature, hardness and thermal diffusivity [5–7].

12.2 Methods, Equipments and Workplan

The goal of the research is that the grain size of sintered composite material shall not exceed the micron size. For this purpose, it is necessary to develop the grinding and sintering modes, as well as to select special alloy components that will hinder the growth of grain size during sintering. The obtained results will allow us to replace boron carbide ceramic with a relatively prospective, cheap armour material with high physic – mechanical and ballistic properties.

The University of Craiova (UCV) will perform research activities oriented towards elaboration of nano sized composite powders powder metallurgy technologies based on wet grinding in an existing vario-planetary mill Pulverisette 4-FRITSCH followed by cold pressing using electro-hydraulic press OMCT 100 of the composite powder and sintering using microwave heating in a upgraded from 1.25 to 6 kW MUEGGE microwave sintering equipment with automatic matching load impedance TRISTAN. The characterization of the composite material will include grain distribution analysis using an existing BROOKHAVEN 90 PLUS equipment as well as microhardness testing with different dwell times using an existing microhardness tester NAMICON (Figs. 12.1, 12.2, and 12.3).

The proposed sintering technology in vacuum will ensure a sintering temperature up to 1700 °C which is far below the temperature obtained through a classical

Fig. 12.1 Fritsch Pulverisette 4 vario-planetary mill (UCV laboratory)

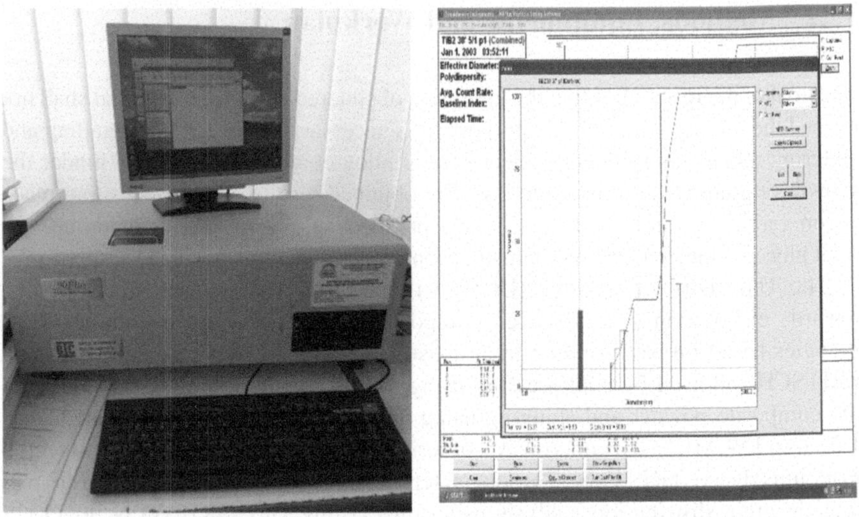

Fig. 12.2 Particle size analyzer Brookhaven 90PLUS (UCV laboratory)

Fig. 12.3 Microhardness tester NAMICON (UCV laboratory)

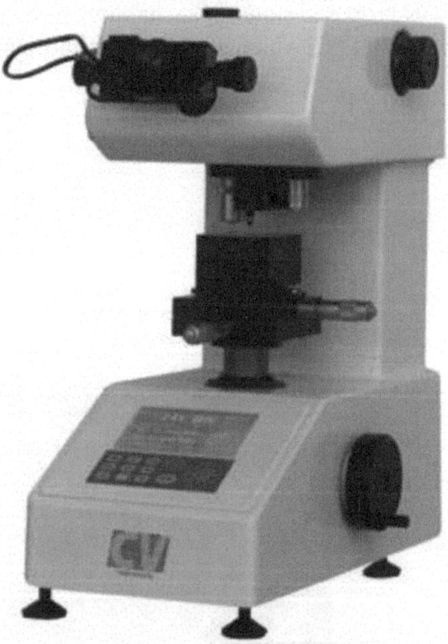

sintering procedure involving a high-temperature furnace. The lower sintering temperature will not allow an increase of the grain size and therefore the grain size can be maintained in a nano-size range (Fig. 12.4).

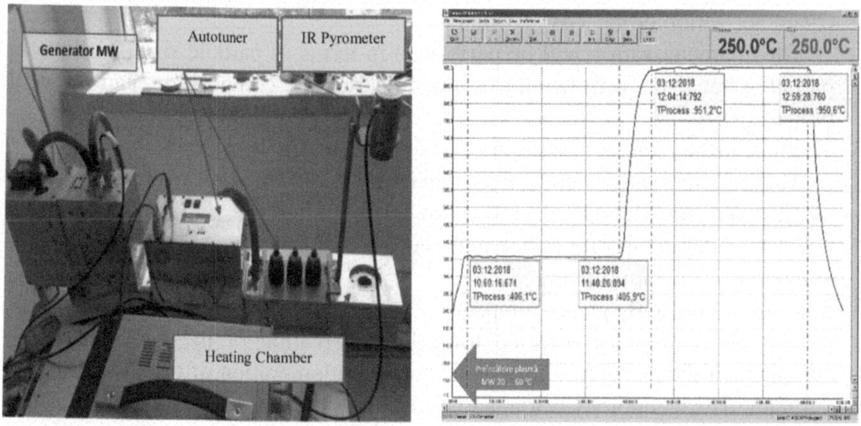

Fig. 12.4 Microwave sintering equipment MUEGGE (UCV laboratory)

As for the final direction, results obtained from direction through using the developed industrial technology will be gathered in solid structure of a single monolithic block with a thermal hydro forming method and received armour plates will be prepared for the ballistic tests.

Georgian Technical University's (GTU) laboratory for hard alloys shall start working on the creation of the new generation composite armour material through vacuum agglomeration of titanium diboride particles. The technology belongs to the liquid-phase agglomeration. Composite armour material consists of three components –base, bonding agent and alloying components. Bonding agent ensures high strength of armour material in bending, while alloying substances will increase the hardness and strength of the material. DINO-MILL Research Lab laboratory grinding mill is necessary for the creation of a new generation titanium diboride composition armour, which ensures fine (without metallic impurities) wet grinding of initial alloy powders up to nano size, as well as fine-grained particles and mixing. Material obtained after wet grinding should be dried in vacuum furnace, which will allow us to avoid the oxidation of metallic constituents of composite powders, which is undesirable due to the worsening of physic-mechanical properties in hard alloy. After thermoforming the obtained composite powder, it will be cold-pressed-formed and then sintered in a high-temperature furnace – CARBOLITE GERO-Laboratory High Temperature Graphite furnace top loader up to 3000 °C, Type LHTG 200-300/30-1G semi-automatic.

The workplan will be developed on several phases:

1. Elaboration of the nano-sized TiB_2 powders

   Within this activity the efforts will be focused on elaboration of the TiB_2 powders through different wet milling regimes of initial alloy powders up to nano size, as well as fine-grained particles and mixing. The wet milling regimes will take into account parameters such as: milling time, rotational speed, diameter and material of the balls.

2. Characterization of the TiB_2 powders
   The characterization will be performed in terms of particle size distribution, particle size, X-ray diffraction to study the structure, composition, and physical properties.
3. Thermoforming and cold-pressing-forming of the TiB_2 powders
   This activity consists in the determination of cold-pressing-forming regimens of plastered materials dried in vacuum furnace after wet milling of the initial powder of the composite armor material for its further agglomeration in high-temperature vacuum furnace.
4. Sintering of the TiB_2 green-compacts
   The efforts will be directed towards the creation of the new generation composite armour material through vacuum agglomeration of titanium diboride particles. The technology belongs to the liquid-phase agglomeration. Two different sintering methods will be applied to the green compacts.
5. Packaging of new composite armour material and armour ceramics composed of high-strength antiballistic polyethylene fabrics of HB-212 mark manufactured by DYNEEMA Company into a single monolithic structure through a new thermo-hydroforming method and development of thermoplasting method and preparation of trial samples for ballistic tests.
6. Development of a new thermoplasting method for packaging
   The new proposed technology will be divided into three successive basic stages: (1) Multilayer composite billets are placed in the molds of the configuration identical to the one of the articles made of thin metal plates (1.0–1.5 mm thick). (2) Mounting of the molds with billets in multi-cell cassettes. (3) The cassettes are placed in the special thermo-hydroforming high-pressure cylindric vessel, and the billets are successively subjected to heating – annealing and cooling – annealing – stabilization.
7. Ballistic tests according to International Ballistic Standards

12.3 Expected Results

The following results are expected:

1. Obtaining a nano-sized TiB_2 powder
2. Obtaining of a composite armour material which consists of three components – base, bonding agent and alloying components. Bonding agent ensures high strength of armour material in bending, while alloying substances will increase the hardness and strength of the material.
3. Developing an innovative hydro-thermoforming device which will be employed for the for the new thermoplasting packaging method.
4. Obtaining of a new generation of composite armour material based on $Ti\ B_2$ which will be further tested using ballistic tests. The ballistic tests will be performed in accordance to the NIJ standards. Particular tests NIJ IV – Car-

tridge/Projectile Type: 7.62 × 63 AP M2 and NIJ Special – Cartridge/Projectile Type: 7.62 × 51 SWISS P AP will be considered during the ballistic tests.

12.4 Conclusions

The research work which will be developed presents two major difference from any other research performed in the field of composite material for armour protective:

1. New sintering technology using pure microwave heating

Most of the researches performed in the field of sintering of TiB_2 composites materials using microwave heating are using microwave as thermal source for hybrid heating not pure microwave heating. The project partners will develop a special chamber for pure microwave heating and the results foreseen will ensure a lower sintering temperature as well as the inhibition of grain growth.

2. New thermoplasting method

The new thermoplasting method addresses a different material – TiB_2 and is based on three successive stages: (1) Multilayer composite billets are placed in the molds of the configuration identical to the one of the articles made of thin metal plates (1.0–1.5 mm thick). (2) Mounting of the molds with billets in multi-cell cassettes. (3) The cassettes are placed in the special thermo-hydroforming high-pressure cylindric vessel, and the billets are successively subjected to heating – annealing and cooling – annealing – stabilization.

The ultimate beneficiary of the research work results will be the military and law enforcement personnel. NATO troops and their allies will wear lighter and protective clothes which will reduce significantly the danger of injuries/deaths. Also military vehicles will use the new protective armour. Other indirect beneficiary will be the military industry taking into account that by up taking and deployment of the research results new production facilities will be open and new jobs will be created.

After successful testing of the new composite armour material, the project partners will submit a patent request (according to technological readiness level TRL 5) and once the patent is issued the technical specifications of the product will be provided to the representatives of the end-user in order to develop a TRL 6 product.

References

1. Tang RT, Wen HM (2017) Predicting the perforation of ceramic-faced light armors subjected to projectile impact. Int J Impact Eng 102:55–61
2. Chi R, Serjouei A, Sridhar I, Tan GEB (2013) Ballistic impact on by-layer alumina/aluminium armor: a semi-analytical approach. Int J Impact Eng 52:37–46

3. Rahbek DB, Simons JW, Johnsen BB, Kobayashi T (2017) Effect of composite covering on ballistic fracture damage development in ceramic plates. Int J Impact Eng 99:58–68
4. King DS, Fahrenholtz WG, Hilmas GE (2013) Silicon carbide-titanium diboride ceramic composites. J Eur Ceram Soc 33:2943–2951
5. Tang W, Zheng Z, Wu Y, Wang J, Lu J, Liu J (2006) Synthesis of TiB_2 nanocrystalline powder by mechanical alloying. Trans Nonferrous Metals Soc China 16:613–617
6. Demirskyi D, Agrawal D, Ragulya A (2014) Tough ceramics by microwave sintering of nanocrystalline titanium diboride ceramics. Ceram Int 40:1303–1310
7. Moallem M, Abbasi MH, Zadeh FK (2012) Synthesis and characterization of TiB_2 nanocrystalline powder by mechanical alloying. Int J Mod Phys Conf Ser 5:204–211

Chapter 13
Titanium Armor with Gradient Structure: Advanced Technology for Fabrication

S. V. Prikhodko, O. M. Ivasishin, P. E. Markovsky, D. G. Savvakin, and O. O. Stasiuk

Abstract Developing advanced cost-effective manufacturing technology and production of high-strength titanium-based armor materials is extremely important for the protection of military vehicles and personnel. Titanium alloys are high-strength and lightweight structural materials that can be thermally hardened to the level similar to steels but at much higher compared to steels weight efficiency. Armor plates fabricated using a novel approach, Blended Elemental Powder Metallurgy (BEPM), demonstrate superior antiballistic performance compared to the currently used titanium-based armor materials and can be fabricated at lower cost.

13.1 Introduction

The anti-ballistic protection of land systems, mobility and protection of the fighting vehicles and military personnel is vital in success of defense and anti-terrorist operations. Traditional material for armor is rolled homogeneous or high strength steel [1, 2]. Use of steel armor, however, can increase the overall weight of the fighting vehicle on 15–20%, which change the vehicle mobility, maneuverability, fuel efficiency and requires stronger breaks and more powerful engines [3]. The Army is in search of lightweight substitute for steel armor. Due to the high specific strength of titanium, materials on its base are contemplated as a viable alternative in low-weight armor production [3, 4]. However, when the armor parts are fabricated using traditional cast and wrought technology the feasibility of implementation

S. V. Prikhodko (✉)
Department of Materials Science and Engineering, University of California Los Angeles, Los Angeles, CA, USA
e-mail: sergey@seas.ucla.edu

O. M. Ivasishin · P. E. Markovsky · D. G. Savvakin · O. O. Stasiuk
G.V. Kurdyumov Institute for Metal Physics, National Academy of Science of Ukraine, Kyiv, Ukraine

C. Palestini (ed.), *Advanced Technologies for Security Applications*, NATO Science for Peace and Security Series B: Physics and Biophysics,
https://doi.org/10.1007/978-94-024-2021-0_13

is questionable due to the high cost of the armor parts. In this regard the use of blended elemental powder metallurgy (BEPM) instead of ingot metallurgy of Ti offers an effective cost reduction due to its ability to produce near-net shape parts, while the waste is considerably reduced [5]. Additionally, the price of BEPM fabricated components can be reduced through the use of low-cost hydrogenated titanium powder instead of conventionally used in powder metallurgy (PM) high-cost titanium metal powder [6]. When the sintering is conducted under the vacuum hydrogen can be removed from the final structure, so such approach combines the sintering and dehydrogenation in one processing stage and it has found to be practical in a lot of applications. It was also shown that temporary presence of hydrogen in compact during sintering can have beneficial effect on the final structure since it reduces impurities (including oxygen & chlorine) as well as facilitate the densification of the compact [6]. One of the key benefit of BEPM is that such fabrication can easily facilitate reinforcement of Ti-alloys with hard and light particles, for instance of TiC and TiB making metal matrix composites (MMC) with enhanced elastic moduli without compromising the material's low-weight so desirable for armor. Reinforcement of the alloy with hard particles, however, most likely lowers the value of toughness and plasticity of material. Yet, for durable armor, simultaneous high hardness and high plasticity are not essential through the entire structure: armor parts require high hardness and strength at the surface, whereas their core rather necessitates high toughness and ductility [7]. Such combination of mechanical properties can be achieved on layered structures that combine two and more layers of different materials with different chemical composition and/or microstructure within each individual layer. The objective of this study was development of cost-efficient technology for fabrication of low-weight and exceptional anti-ballistic properties multilayered structures (ML) made using BEPM of Ti alloy and composites on its base with TiB and TiC.

13.2 Experimental Procedure

13.2.1 Conceptual Design of Multi-layered Plates for Anti-ballistic Application

Combinations of high strength and ductility can be attained in part for anti-ballistic applications by creation of layered structure, which combines at least two materials with different properties. In manufacturing of armor, it is common to use a layered structures made of a front-facing layer, whose purpose is to blunt and abrade the incoming projectile and a back-facing supporting layer absorbing energy by deformation [7]. In our armor design (Fig. 13.1) we choose to combine the layers of Ti-6Al-4V (wt.%) alloy (Ti-64), which provide sufficient ductility, along with metal matrix composites (MMC) on a Ti-64 base reinforced with TiC or TiB particles to increase hardness. Layers with higher concentrations of reinforcement particles

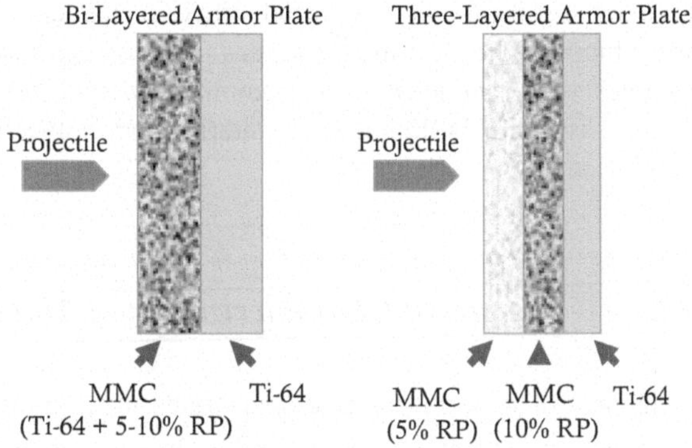

Fig. 13.1 Design concept of bi- and three-layered armor

have greater hardness. These high hardness layers build the front of the armor part, to enhance wear resistance and to deform and stop projectile upon ballistic impact. The Ti-64 alloy layer with higher ductility and toughness forms the backside of the plate to prevent crack propagation and plate destruction. For multi-layered (ML) structure with more than two layers, we considered different arrangements: (i) with having the hardest layer in the middle of the plate to prevent extensive spallation of this most brittle layer and (ii) in the front of the plate to maximize the abrasion of the projectile.

13.2.2 Fabrication of the Layered Structures Using BEPM

ML structures of Ti-64 alloy and its MMC with 5% and 10% (vol.) of TiC and TiB were fabricated using BEPM of hydrogenated Ti. PM processing utilizing TiH_2 is recognized as low-cost fabrication used in military, aerospace and other critical applications [6]. Four different powder sets of TiH_2 (3.5%H, wt.) with particles sizes (<40, <100, 80–100, 100–125 μm) was used to determine the effect of the base powder size on shrinkage behavior and sintering characteristics of compacts. TiH_2 powder was blended with corresponding amounts of a master alloy powder, 60% Al-40%V (wt.), with particles size <63 μm to form the required Ti-64 composition. To form the MMC layers the required amounts of reinforcement particles powders were added and mixed before the pressing stage. The size of TiC powder particles was <30 μm. In order to obtain TiB inclusions as a part of composite, we used TiB_2 powder particles <20 μm size expected to chemically transform during the sintering via the following reaction: $TiB_2 + Ti = 2TiB$. For the ML structures forming, the blends for each layer were prepared separately and added to the die before pressing. The die pressing (DP) and sintering technique were used to fabricate the samples. Plates, with the dimension 90 × 90 × 18 mm were pressed at 150 MPa; bars 60 × 10 × 10 mm and cylinders 10 mm diameters, 10–12 mm height were pressed at 150 and 640 MPa. Plates were suitable for ballistic test and bars and cylinders for mechanical tests and structure characterization. Single layer materials were fabricated to test the properties of individual layers. Sintering of all preform samples was conducted in a vacuum furnace (1250 °C, 4 h) followed by slow furnace cooling. Different post-sintering thermomechanical treatments were also conducted.

13.2.3 Structural Characterization of Materials

Structures of materials were characterized using light optical microscopy (LOM), scanning electron microscopy (SEM), electron backscattered diffraction (EBSD), transmission electron microscopy (TEM), x-ray diffractometry (XRD) and x-ray microscopy. Mechanical properties of materials have been evaluated using hardness, microhardness, tensile and 3-point flexural tests. Tensile tests were done on an INSTRON 3376 machine. Following standard protocol, ASTM E8M-04, 3 specimens were tested for individual material, each sample was having the gauge length 40 mm and diameter 4 mm. Three-point flexural tests on the bar samples of sintered bi-layered structures were performed on an INSTRON 8802 following ASTM E290-14 standard. Each value on flexural tests was an average of at least 3 samples tested. The deformation energy was measured based on engineering stress-strain curves obtained on tensile test. Ballistic test was performed in the NATO certified laboratory at Ivan Chernyakhovsky National University of Defense (Ukraine). Seven different bullet types were tested with the bullet weight varied 3.4–10.4 g; speed: 310–930 m/s; kinetic energy 340–3800 J. More details on samples fabrication and characterization protocols used in this project are discussed elsewhere [8, 9].

13.3 Results and Discussion

13.3.1 Structure Characterization of Layered Materials and Some Problems of BEPM Fabrication

Our microscopy results obtained on smaller size layered samples show the grain size of the alloys is generally below 100 μm and for the composite layers is commonly smaller. We observe high structures' compaction and integration between the layers (Fig. 13.2). Details on structures evaluation are reported elsewhere [10, 11]. All layered samples of large dimension (plates) suitable for the ballistic test demonstrate significant linear shrinkage after the sintering. That was result of dehydrogenation of TiH_2 and particles' sintering. Plates with TiC composite, however, were successfully

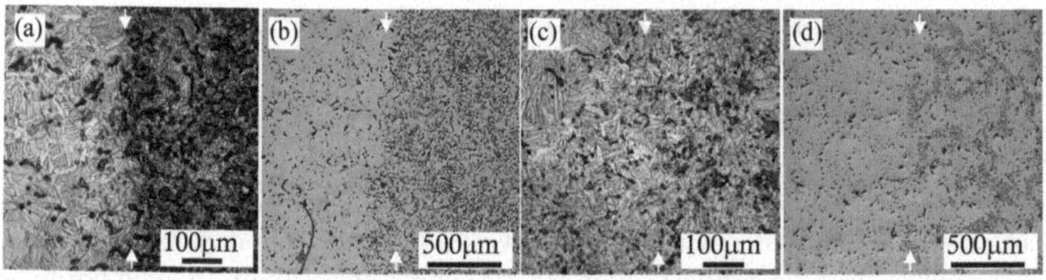

Fig. 13.2 Light optical (**a**, **c**) and SEM (**b**, **d**) micrographs of bi-layered structures fabricated using BEPM with compression at 640 MPa followed by sintering (1250 °C, 4 h). All images show the structure close to the interface (marked with arrows) between the layers of Ti-64 alloy (left) and MMC on its base (right). The MMC in (**a**) and (**b**) 10% Tib; the one in (**c**) and (**d**) has 5% of TiC

made using DP protocol owing to very close values of shrinkage of the alloy and TiC composites layers. Shrinkage values measured on plates were 15.7%, 15.5% and 15.1% for Ti-64, Ti-64 + 5%TiC and Ti-64 + 10%TiC layers correspondingly. Insignificant shrinkage mismatch of individual layers didn't cause cracking or distortion of the plates due to the minor interface stress which most likely was compensated by variation in porosity of adjacent layers [9, 11]. Sintering of ML plates with TiB composite was quite different. Those plates demonstrate very different shrinkage values of individual layers: 13.5%, 11% and 8.5% for Ti-64, Ti-64 + 5%TiB and Ti-64 + 10%TiB correspondingly. Shrinkage mismatch builds up the interfacial stress between the layers during the sintering and causes plates bending, delamination and cracking. The difference in shrinkage behavior of layered plates with TiC and TiB was explained using results on dilatometry study of individual powder compacts during continuous heating [9]. It was noted that TiC has very high melting point (3140 °C) and doesn't undergo changes during the sintering. Whereas TiB is a result of in-situ chemical reaction involving diffusion of reacting elements. Due to the large difference in mutual diffusion mobility of boron and titanium it creates the Kirkendall's porosity effect [10]. The increased porosity of TiB composites causes swelling of corresponding layers creating shrinkage mismatch with the alloy.

To minimize the shrinkage mismatch of individual layers during the sintering of ML structures with TiB several experimental parameters were tested. Our assessment on combined effect of the powder size and the powder hydrogen content shows that acceptable (less than 1%) shrinkage mismatches can be achieved at 150 MPa with proper selection of the TiH_2 powder size used. The ML plates suitable for the ballistic test were fabricated using optimized parameters. All details on structure characterization, observed shrinkage mismatch of layered structures during BEPM fabrication and mismatch minimization are reported elsewhere [9].

13.3.2 *Mechanical Properties of Layered Structures and Anti-ballistic Performance*

Some properties of studied materials are shown in the Table 13.1. Residual porosity of individual layers is strongly affected by the presence and type of reinforcement particles. Ti-64 alloy demonstrates the lowest porosity, which depends on size of TiH_2 powder used for fabrication of ML structures. Porosity of TiC composite is some higher and porosity of TiB composite is the highest. Vickers hardness and tensile properties are pre-determined by the combined effect of residual porosity and the fraction of reinforcement particles. An increasing amount of strengthening particles causes the expected drop in elongation, while the anticipated increase in strength is not observed due to greater porosity and premature fracturing. We observe some increase in Vickers hardness of MMCs. However, the yield stress and ultimate tensile stress values of composites are substantially lower to those values of Ti-64 alloy sintered using optimized parameters (powder size, hydrogen content, compaction pressure) (#7), shown here for comparison.

Our results on 3-point flexure test demonstrated substantial increase in both, the flexure stress and the flexure strain of bi-layered samples compared to a single-layer composite samples as well as the uniform Ti-64 alloy [13]. It was shown that single layer samples of MMC have shown inferior flexure characteristics compared to Ti-64 sample without reinforcement particles. However, when uniform Ti-64 alloy compared to the best bi-layered samples values we see close to 20% improvement of the flexure stress with no loss in the strain value, how it was measured at the same strain rate (10^{-1} s^{-1}) on bi-layered sample with 10% TiC MMC. These results suggest higher deformation energy of bi-layered vs. uniform structures. It is important to note that the order of the layers is essential in the test: the MMC layer must be on the top, where the concentrated load is applied and Ti-64 alloy is at the bottom. When the bar is flipped over the overall flexure performance of the layered structure is far poorer compared to uniform Ti-64 alloy bar. More details are discussed elsewhere [13]. This result implies bi-layered structure of Ti-64//Ti-64 + 10%TiC have potentially better anti-ballistic properties compared to the uniform single-layer Ti-64 and ML structures with TiB prepared using powder approach since it capable absorbing higher energy during deformation. Results on

Table 13.1 Residual porosity, Vickers hardness and tensile properties of individual layers

#	Material	Porosity, %	Hardness, HV	UTS, MPa	Elong., %
1	Ti-64	1.8%	345	1034	7.2
2	Ti-64 + 5TiC	3.8%	333	708	<1
3	Ti-64 + 10TiC	3.6%	373	618	0
4	Ti-64	3.6%	299	969	8.3
5	Ti-64 + 5TiB	4.5%	327	847	0
6	Ti-64 + 10TiB	8%	324	512	0
7 [12]	Ti-64	1.5–1%	320–340	940–1000	10–14

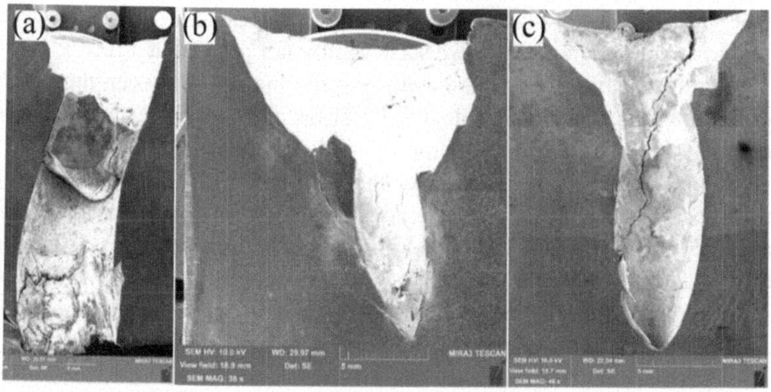

Fig. 13.3 Ballistic test results: (**a**) Cast Ti-64 alloy; (**b**) Bi-layered plate of Ti-64 // Ti-64 + 10% TiB; (**c**) Three-layered plate of Ti-64 // Ti-64 + 10% TiC // Ti-64 + 5%TiC. All plates have about the same thickness 25–27 mm. Test conditions: B32 bullet (Table 13.3), caliber 7.62 × 54 mm, mass 10.4 g, kinetic energy 3500–3800 J. The cast fabricated sample in (**a**) was annealed at 1100 °C for 1 h and cooled with the furnace producing lamellar microstructure [16]

antiballistic test corroborates with the data on 3-point flexure test. We can see substantially better anti-ballistic performance of the layered structures compared to the cast Ti-64 alloy (Fig. 13.3). Bullets of 7.62 caliber pierced the cast plate all the way through and stuck inside of the layered structures without showing damage on the backside of the tested plate.

13.3.3 Anti-ballistic Performance of Bi-layered Structures and Potentials of Its Improvement

Comparing the mechanical characteristics of bi-layered and uniform structures we could expect substantial improvement of anti-ballistic performance of our materials if we have better control over their porosity (Table 13.1). Hardness is one of the most important properties of the front facing layer of the armor as most of the studies suggest and it appears that porosity is undesirable. The data in the Table 13.1 show loss of Vickers hardness with increased porosity, even in the presence of high-moduli particles of TiC and TiB that supposed to contribute to the hardness gain. That is why we attempted hot plastic deformation of sintered ML structures for their more complete densification for the hardness improvement [8]. Ideally such treatment should provide the porosity reduction to about 1–1.5% without imposing uncontrollable grain growth and layers' delamination. Post-sintering hot plastic deformation of different structures was conducted at temperatures not far from the beta-transus of Ti-64 alloy. It was found that the deformation of layered structures using hot rolling was unsuccessful due to significant disparity of the metal flow and degree of accommodated plastic deformation in Ti-64 alloy and adjacent composite layers. It was established that the hot pressing of the plates was more successful. ML samples with TiC particles were successfully hot pressed at 1100 °C to a total deformation degree of 45%, without breaking the integrity between the layers. At such conditions the deformation degree of Ti-64 alloy layer in the ML structure reached 65% that resulted in reduction of total porosity of this layer to less than 1% and decrease in the pore size. After the hot pressing plates were additionally annealed (850 °C, 2 h) for the stress removal. The deformation behavior of the ML plates used MMCs with TiB was fairly different. The hot pressing was performed at 1100 °C and was stopped at 10% of the total deformation due to the cracks formation started within the MMC layers. More details are reported elsewhere [8]. Thermomechanical processing is still ongoing study, but current results show the potentials on mechanical properties improvement of layered structures after the optimized thermomechanical treatment is applied. We expect further improvement of antiballistic performance of ML. We also consider an alternative approach on layered structures fabrication, which separate layers bonding from the sintering. In such approach optimized treatment could be used separately on each layer followed by the layers' bonding using diffusion bonding (DB), hot isostatic pressing (HIP) or friction welding (FW) etc. Potential predicaments of dissimilar structures bonding are that parameters, which control the process, can affect each structure from the couple differently. Some complications can also arise due to presence of reinforcement particles and other defects. To address the feasibility of ML structures fabrication by joining optimized layers we run some studies on DB, HIP and FW of Ti-64 alloy with MMCs on its base. Results of these studies [14, 15] show successful bonding of Ti-64 alloy and TiB/TiC MMCs on its base via different suggested methods. It validates the prospects of joining of Ti-64 alloy and MMC of this alloy with TiC and TiB to form layered structures and opens possibilities of bonding the structures after each layer processed separately to its optimal performance. Such approach can significantly facilitate the hardness increase of MMC layer since it eases the restrains on the amount of the reinforcement phase. Besides, it takes away the concern on shrinkage mismatch between the alloy and composite during BEPM fabrication. Also it potentially allows welding and bonding the armor of any shape and dimension.

13.3.4 Three-Layered Armor Structures and Its Optimization

Three-layered armor structures have advantage over bi-layered since they allow more flexibility in changing the order and the thickness of the layers with different individual properties in order to better resist on specific ballistic impact [7]. Optimization of three-layered structures was done using flexure test performed on 60 × 10 × 10 mm bars. The goal of this assignment was optimization of three-layered structures in order to minimize the number of ballistic trials. Few variables were in question: (1) type of particles (TiC or TiB); (2) the order of the sheets:

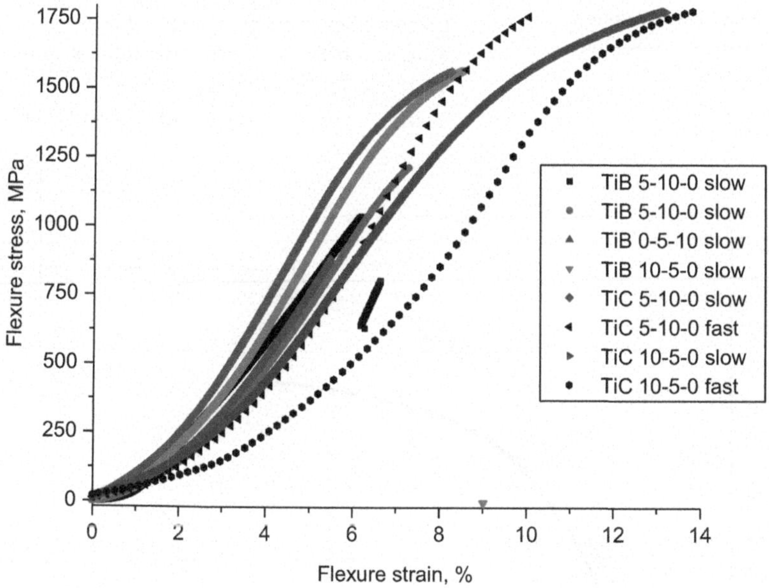

Fig. 13.4 Results on 3-point bend test of three-layered structures. The labels indicate: the type of reinforcement (TiB and TiC); the amount (in vol. %) for the layers at the top-center-bottom of the tested bar (the top layer is where the concentrated load was applied); "fast" and "slow" indicated the strain rate: 10^{-1} s^{-1} and 10^{-3} s^{-1} correspondingly. The thickness of each layer was ~ 3 mm

the hardest, medium hard and ductile with 10%, 5% and 0% of reinforcement correspondingly; (3) thickness of each layer in triplet; and (4) the strain rate.

Results in Fig. 13.4 show that three-layered structures with TiC perform considerably better compared to the structures with TiB. The best performance of three-layered structures is observed in consequence 10-5-0 with the hardest layer on the top of the tested bar. Among all measured structures with TiB performance was the poorest when the alloy Ti-64 layer was used at the top. We also can see consistently higher flexure stress and strain for the faster strain rate used (dotted lines vs. solid green and purple). Last observation points out on potentially good anti-ballistic properties (ultimate stress increase with increasing the strain rate) of considered structures as existing ballistic theories suggest [7].

We also evaluate the optimum thickness ratio of the sheets in three-layered structures with TiC reinforcement. The consequence of the layers 10-5-0, determined earlier was used. Results of this test (not shown) demonstrate that the highest flexure stress/strain is observed when the used ratio was 2:1:3. It suggests that for the best flexure performance the bottom ductile layer should be about half of the thickness of the entire bar and the top hardest layer (10% of TiC) should be twice thicker than the medium hard (5% of TiC).

Table 13.2 Tested samples in the strain rate evaluation test (in grams)

Composition/sample #	1	2	3	4	5	6
Ti-64 + 10%TiC	10	10	8	8	6	6
Ti-64 + 5%TiC	4	4	4	4	4	4
Ti-64	10	10	12	12	14	14

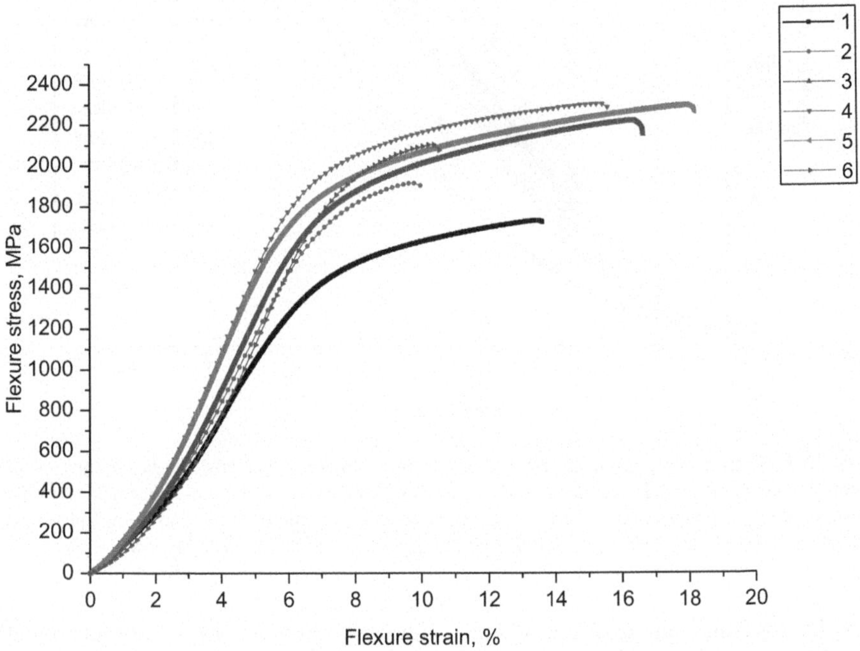

Fig. 13.5 Flexure test results of the three-layered structures in evaluation of the combined effect of the strain rate and the layers' thickness. The number in the label correspond to the sample number in the Table 13.2

Finally, we evaluated the effect of the strain rate on flexure stress-strain behavior of three-layered structures. Even though the optimum ratio 2:1:3 was established in the previous test we wanted to verify it reminds the best when the strain rate is changed. So, basically there were two variables in this test: the strain rate and the layer's thickness ratio. Six tested conditions are listed in the Table 13.2 and results of the experiment are shown in Fig. 13.5. The thickness of each layer in sintered structure was controlled by the mass of the powders used for each layer and it is listed in grams in the label for each curve on the chart. The total mass of entire bar was always the same, 24 g. Samples 1, 3, 5 were tested at 10^{-3} s^{-1} and 2, 4, 6 at 10^{-1} s^{-1} strain rates. Samples #3 and #4 are showing the best combination of strength and ductility and the highest deformation energy at both tested rates. It is also important to note that for those two samples stress is higher at higher strain rate, which points on good anti-ballistic properties. Sample with layer's order 6-4-14 show good result at slow rate (#5), however at fast rate (#6) it performs poorly. Outcome of this test verifies that the ratio 2:1:3 gives the best antiballistic performance and it was used to fabricate three-layered plates for further ballistic examination.

13.3.5 Ballistic Test of Three-Layered Structures

Examination parameters of selected three-layered plates and results of their ballistic examination are listed in Table 13.3 and some details discussed elsewhere [16].

Seven tested bullets' types are grouped into three classes based on their striking factors. The striking factor of the 1st class bullet (PST, PS, LPS) is the kinetic energy of projectile only (steel bullet, not quenched). The 2nd class (PP) is characterized by two striking factors, kinetic energy and hardened core of the bullet (700 HV). And the 3rd class (BZ, B-32) is characterized by three striking factors, kinetic energy, hardened core and incendiary capability. We determined the threshold Specific Kinetic Energy (SKE) per plate thickness required for piercing three-layered plates by different bullets' classes. Those are summarized as followed:

- when SKE is below 8.88 J/mm^3 the 1st class of bullets is not capable of piercing. When the SKE is higher than 10.77 J/mm^3 the plates are getting consistently pierced;
- for the 2nd class bullet the piercing is taking place at SKE exceeding 8.85 J/mm^3.
- for the 3rd class the piercing is not happening at SKE of 3.29 J/mm^3 whereas at 5.19–5.91 J/mm^3 partial piercing (bullets stuck in the plates) can take place.

The SKE calculated in this study was compared to results in two other studies on Ti-64 alloy armor and all three data sets are shown in Fig. 13.6. The armor plates reported in [17] was treated using surface rapid heat treatment (SRHT) that reinforces the surface layer of the armor plate by creating gradient structure. The figure on SKE demonstrates that compared to currently used Ti-64 alloy armor (3) [18] the energy enable the piercing of armor plate is required to be 10-15% higher for Ti-64 alloy armor after its structure is modified by SRHT (2); and it is almost twice (!) higher for the layered structure in this study (1) which incorporate MMC as the front layers.

13.4 Conclusions

Relatively big size plates suitable for ballistic test of ML structures of Ti-64 alloy and MMC on its base with 5–10% (vol.) of TiC and TiB were fabricated using BEPM of hydrogenated Ti. It was found that during the sintering of the layered structures the shrinkage levels of the base alloy layer and MMC with TiC are similar, enabling the successful fabrication of multi-layered structures without requiring optimization of the sintering processing parameters for relatively large plates.

Table 13.3 Ballistic examination of three-layered plates

#	Plate composition/state	Bullet type	Bullet mass, g	Bullet speed, m/s	Bullet energy, J	SKE, J/mm^3	Result
1	8/4/12, as-sintered	LPS	9.6	875	3675	11.01	Pierced
2	8/4/12, as-sintered	PP; PS	3.63.5	919908	15201443	8.858.82	PiercedNot pierced
3	6/4/14, as-sintered	LPS	9.6	880	3717	10.77	Pierced
4	6/4/14, as-sintered	PS	3.5	904	1430	8.88	Not pierced
5	8/6/24, as-sintered	LPS; BZ	9.67.4	867753	36082098	5.693.29	Not piercedNot pierced
6	8/6/24, deformed	BZ	7.4	749	2076	5.19	Partially pierced, cracked
7	6/6/28, as-sintered	BZ; B32	7.410.4	751864	20873882	3.285.91	Not piercedPartially pierced
8	6/6/28, deformed	BZ	7.4	750	2081	5.2	Partially pierced

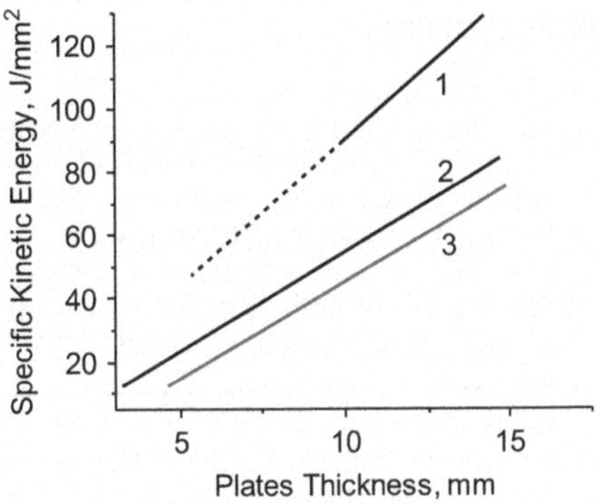

Fig. 13.6 SKE of the 1st class ("soft" core) bullet required to make the armor piercing vs. plate thickness. Chart (1) shows current study results corresponding to the testing condition listed in the Table 13.3, #2 (PS) and #5 (LPS). Chart (2) shows results of previously reported study [17] and (3) of study [18]

The shrinkage level of the Ti-64-based MMC reinforced with TiB is significantly different from that of the alloy layer and requires careful control over sintering parameters. Effective control of shrinkage mismatch between individual layers can be achieved through the careful selection of base powder size and the used amount of hydrogenated Ti powders.

It was found that layered structures fabricated using BEPM protocol with the front facing layer made of MMC and the back facing made of Ti-64 alloy demonstrate superior antiballistic performance compared to the uniform armor plates made using cast and wrought technique of Ti-64 alloy. Superior antiballistic properties of the layered armor structures were explained by their enhanced flexural strength compared to uniform single layer part made of Ti-64 alloy.

The post-sintering thermomechanical treatment, commonly used to increase the densification of the PM products, has major restrictions for the layered structures application. The hot rolling was found to be unsuccessful. The hot pressing at 1100 °C followed by the annealing at 850 °C for 2 h was somewhat promising for porosity reduction of layered structures with TiC MMCs. The effect of hot pressing and annealing on the properties of the layered plates with TiB MMCs was less apparent.

DB and FW can be used for joining sheets of Ti-64 alloy and MMC of this alloy with TiC and TiB to form layered structures. That opens possibilities of bonding the structures after each layer processed separately to its optimal performance. Such approach can significantly improve the properties of individual layer and as a result the entire layered armor. Besides, it takes away the possible concern on shrinkage mismatch between the alloy and composite during the BEPM fabrication of layered components and opens bypass on fabrication of large-sized and complex-shaped structures.

Three-layered armor structures with the layers different by the amount of reinforcement particles were optimized in terms of the layers' thickness and the order. The threshold SKE per plate thickness required for the armor plate to be pierced were determined for three different bullets' classes and found to be superior for ML structures in this study compared to currently used Ti-64 armor.

References

1. Cimpoeru SJ (2016) The mechanical metallurgy of armour steels. Fishermans Bend, Victoria (Australia): Australian Department of Defence; Land Division, Defence Science & Technology Group; Oct. Report No.: DST-Group-TR-3305
2. El-Bitar T, El-Meligy M, El-Shenawy E, Almosilhy A, Dawood N (2017) Thermo-mechanical processing of armor steel plates. Int J Mater Met Eng 11(3):214–220
3. Montgomery JS, Wells MGH, Roopchand B, Ogilvy JW (1997) Low-cost titanium armors for combat vehicles. JOM 49(5):45–47
4. Pettersson A, Magnusson P, Lundberg P, Nygren M (2005) Titanium–titanium diboride composites as part of a gradient armour material. Int J Impact Eng 32:387–399
5. Lutjering G, Williams JC (2007) Titanium. Springer, Berlin/Heidelberg/New York
6. Ivasishin OM, Moxson VS (2014) Low cost titanium hydride powder metallurgy. In: Qian M, Froes FH (eds) Titanium powder metallurgy: past, present and future. Elsevier, Cambridge, MA
7. Crouch IG (2016) The science of armour materials. Elsevier Science & Technology, Cambridge
8. Prikhodko SV, Markovsky PE, Savvakin DG, Stasiuk OO, Ivasishin OM (2018) Thermo-mechanical treatment of titanium based layered structures fabricated by blended elemental powder metallurgy. Mater Sci Forum 941:1384–1390
9. Ivasishin OM, Markovsky PE, Savvakin DG, Stasiuk OO, Norouzi Rad M, Prikhodko SV (2019) Multi-layered structures of Ti-6Al-4V alloy and TiC and TiB composites on its base fabricated using blended elemental powder metallurgy. J Mater Process Technol 269:172–181
10. Prikhodko SV, Norouzi Rad M, Markovsky PE, Savvakin DG, Julian N, Ivasishin OM (2017) 3D imaging of titanium alloys multi-layered structures (MLS) via X-ray microscopy. Microsc Microanal 23(S1):322–323
11. Prikhodko SV, Markovsky PE, Savvakin DG, Stasiuk OO, Norouzi Rad M, Choi C, Ivasishin OM (2018) Characterization of layered structures of Ti-6Al-4V alloy and metal matrix composites on its base. Microsc Microanal 24(S1):2218–2219
12. Ivasishin OM, Savvakin DG, Moxson VS, Duz VA, Lavender C (2007) Ti-2007 science and technology: proceedings, 11th world conference on titanium (Kyoto, Japan) (Niinomi M, Akiyama S et al (eds)), vol 1. Japan Institute of Metals, Sendai, pp 757–760
13. Markovsky PE, Savvakin DG, Ivasishin OM, Bondarchuk VI, Prikhodko SV (2019) Mechanical behavior of titanium-based layered structures fabricated using blended elemental powder metallurgy. J Mater Eng Perform 59:5772–5792
14. Prikhodko SV (2019) Friction welding of Ti-6Al-4V alloy and metal matrix composites on its base. 5th symposium on linear friction welding, March 20–22, Cambridge
15. Prikhodko SV, Savvakin DG, Markovsky PE, Stasiuk OO, Enzinger N, Deley F, Flipo B, Shirzadi AA, Davies HM, Davies PD, Penny J, Bozhilov KN, Ivasishin OM (2019) Layered structures of Ti-6Al-4V alloy and metal matrix composites on its base joint by diffusion bonding and friction welding. Microsc Microanal 25(S2):812–813
16. Ivasishin OM, Markovsky PE, Savvakin DG, Stasiuk OO, Golub VA, Mirnenko VI, Sedov SG, Kurban VA, Antonyuk SL (2019) Microstructure and properties of titanium-based materials promising for antiballistic protection. Prog Phys Met 20(2):288–309
17. Ivasishin OM, Akhonin SV, Savvakin DG, Berezos VA, Bondarchuk VI, Stasyuk OO, Markovsky PE (2018) Effect of microstructure, deformation mode and rate on mechanical behaviour of electron-beam melted Ti-6Al-4V and Ti-1.5Al-6.8Mo-4.5Fe alloys. Prog Phys Met 19(3):309–334
18. Fanning JC (2005) Military applications for β titanium alloys. J Mater Eng Perform 14:686–690

Part III
Sensors and Detectors

Chapter 14
SOLE Project – Demonstration of a Multistatic and Multiband Coherent Radar Network

Giovanni Serafino, Salvatore Maresca, Filippo Scotti, Antonio Malacarne, Antonella Bogoni, Paolo Ghelfi, Leonardo Lembo, Carlo Noviello, Virginia Zamparelli, Gianfranco Fornaro, Eugenio Sansosti, Nicolas Torcheboeuf, and Steve Lecomte

Abstract The aim of the NATO-SPS SOLE project is demonstrating the feasibility and the high performance of a radar network thanks to photonics. Indeed, the coherence offered by photonics makes the proposed distributed radar system capable of an efficient implementation of MIMO processing and ISAR imaging, enhancing the performance in terms of resolution and precision. The advantage of a fully coherent, multistatic radar system here is experimentally proven by a 5-time cross-range resolution enhancement thanks to MIMO processing, and in an efficient focusing in ISAR imaging.

Keywords Photonic radar · MIMO radar · ISAR imaging · Mode-locked lasers

14.1 Introduction

Recently, distributed radar systems have attracted high interest, as a mean to increase the level of system performance, e.g. applying the multiple input-multiple output (MIMO) paradigm [1, 2]. Yet, practical implementations have been rarely

---

G. Serafino · S. Maresca · F. Scotti · A. Malacarne
TeCIP Institute, Scuola Superiore Sant'Anna, Pisa, Italy

A. Bogoni (✉) · P. Ghelfi
PNTLab, Consorzio Nazionale Interuniversitario per le Telecomunicazioni (CNIT), Pisa, Italy
e-mail: antonella.bogoni@santannapisa.it

L. Lembo
Vallauri Institute, Centro di Supporto e Sperimentazione Navale (CSSN), Livorno, Italy

C. Noviello · V. Zamparelli · G. Fornaro · E. Sansosti
Istituto per il Rilevamento Elettromagnetico dell'Ambiente (CNR-IREA), Napoli, Italy

N. Torcheboeuf · S. Lecomte
Centre Suisse d'Electronique et de Microtechnique (CSEM), Neuchâtel, Switzerland

C. Palestini (ed.), *Advanced Technologies for Security Applications*, NATO Science for Peace and Security Series B: Physics and Biophysics,
https://doi.org/10.1007/978-94-024-2021-0_14

proposed, since the attained complexity of such systems was too high, compared to the advantages granted by this kind of solutions. Indeed, to exploit all the advantages they offer, radar networks require a high phase coherence among all the nodes [3], which is very hard to obtain with classical electronic technology. In many cases, signal post-processing techniques can be applied to reconstruct the needed coherence; however, this usually translates in a huge, barely affordable computational complexity, which does not represent only a waste of resources, but can also introduce long time delays, making real-time applications of radar networks unattractive. However, in the last two decades, photonics started penetrating radio frequency (RF) remote sensing systems [4, 5]. Thanks to the guaranteed phase coherence in signal generation [6] and distribution [7], photonics introduces the possibility of going beyond the original concept of stand-alone radar, enabling the feasibility of distributed radars, without requiring an exaggerated increase of the computational load.

The “multiStatic & multiband cOherent radar fLeet for border sEcurity” (SOLE) project aims at the realization of a demonstrator of a 3 transmitters (TXs) by 3 receivers (RXs) full-functional radar network for surveillance purposes, based on the coherent multi-static approach, leveraging on the intrinsic high level of coherence of photonics [8]. The proposed system benefits from the MIMO approach, enhancing the cross-range resolution [2], and enabling the possibility of implementing high-performance imaging algorithms [9]. To this aim, suitable data processing techniques are considered, as well as algorithms for multistatic inverse synthetic aperture radar (ISAR) applications, to obtain a radar with enhanced sensitivity and precision.

Here, we report on the experimental activities for a first validation of the architecture of the proposed system and the related processing. In a campaign of in-field measurements, we tested a preliminary version of the photonic, multistatic coherent radar network, with 2 TXs and 2 RXs, applying MIMO and ISAR processing to the acquired data, showing the increased performance granted by the photonics in a radar network.

14.2 The SOLE Project Radar Network Architecture

The proposed distributed radar system architecture is composed by a central unit (CU) and three sensor peripherals (SPs). The CU and the SPs are connected by means of single mode fiber (SMF) spans. The architecture of the CU is shown in Fig. 14.1a. It can be ideally divided in a transmit (top) and receive (bottom) section. The master optical oscillator is implemented by a mode-locked laser (MLL) with a 600-MHz pulse repetition rate. The transmit section is equipped with three Mach-Zehnder modulators (MZMs) that will operate electro-optical (E/O) conversion of the radar waveforms generated at intermediate frequency (IF) by the digital-to-analog converters (DACs). The employed optical carriers are extracted from the

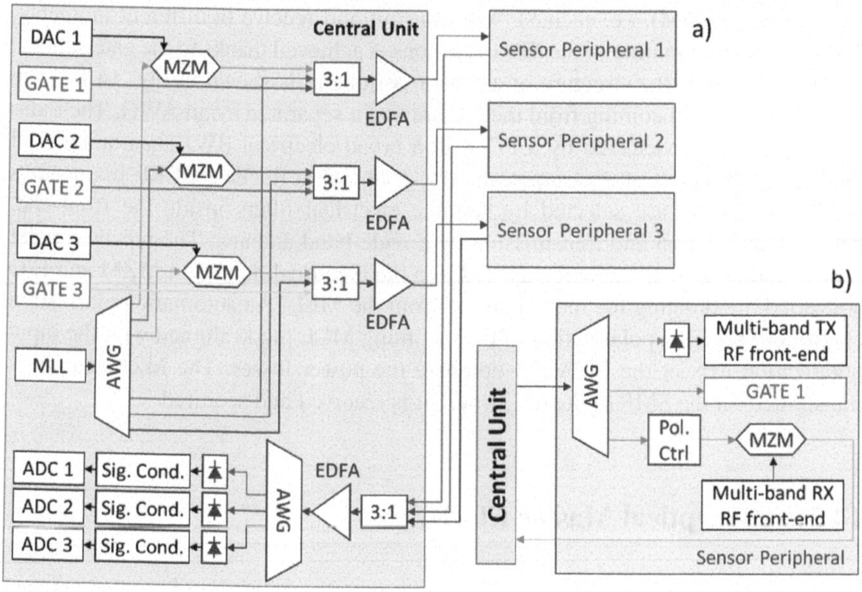

Fig. 14.1 Detailed description (**a**) of the photonics-assisted CU of the SOLE radar network and (**b**) of the structure of the SPs

MLL, which represents the system optical clock, by an arrayed waveguide grating (AWG). The optical carrier is fed into the MZM, where it is electro-optically modulated by the IF signal. The AWG extracts from the MLL spectrum also one line for each SP, that will be employed by the RXs as optical carrier for the E/O conversion of the received radar echoes: for this reason, it is sent unmodulated to the SP. Moreover, a gate to control the transmit and receive operations inside the SPs is generated and converted to the optical domain exploiting the direct modulation of a continuous-wave laser. A 3:1 optical coupler aggregates the radar waveforms converted to the optical domain, the unmodulated optical carrier from the MLL and the E/O converted gate to drive the operation of the SP. These three optical signals are transmitted together through the SMF to one of the SPs, after amplification by an Erbium-doped fiber amplifier (EDFA). This structure is repeated thrice inside the CU, once for each SP.

In the CU section dedicated to the received signals, there is another 3:1 coupler collecting the optical signals coming from the RXs of the three SPs, on three different wavelengths. After an EDFA, another AWG splits the three signals at the input of three photodetectors (PD) for the opto-electrical (O/E) conversion of the radar waveform to base band (BB). The PD outputs, after signal conditioning, are separately digitized by three analog-to-digital converters (ADCs). The SOLE photonic MIMO radar demonstrator will operate on S and X band. Transmission and reception orthogonality between the SPs will be obtained thanks to time division multiplexing (TDM), i.e. each SP will transmit and receive in different moments. The coordination in time of the SPs operations is achieved thanks to the gate signals.

The scheme of the structure of a sensor peripheral is shown in Fig. 14.1b. The three optical signal coming from the CU are again separated by an AWG. The radar waveform is photodetected by a PD with a broad electrical BW. Then, among all the beating products at the output, the one centered at the desired RF frequencies in X and S band are selected by suitable electrical filters inside the front-end, before amplification and transmission by a wide-band antenna. The received radar echoes gathered by the antenna are fed into the electrical input of a MZM and E/O converted, modulating the mode coming from the MLL. An automatic polarization controller keeps the polarization of the incoming MLL mode aligned with the input polarization axis of the MZM, to optimize the power losses. The MZM output is transmitted on the SMF to the CU, where it is received and acquired.

14.3 The Optical Master Clock

In order to optically generate and distribute the signals in the photonics-assisted radar concept, it is advantageous to start with a system clock in the optical domain instead of an electrical clock (quartz, dielectric resonant oscillator, etc.) involving E/O conversion. MLLs are oscillators with unique phase noise purity: the physics of such oscillators, and in particular for those based on a diode-pumped solid-state laser architecture, allows for unmatched phase purity. Since optical fibers and telecom components are used in the radar system, a MLL emitting in a telecom band is necessary. For this purpose, the 600-MHz repetition rate solid-state laser based on Er:Yb:glass gain medium and passively mode-locked with a semiconductor saturable absorber mirror has been realized. A laboratory version of this laser, shown in Fig. 14.2a, emits about 80 mW of output power with soliton pulses of spectral full width half maximum of 12.5 nm (corresponding to a duration of 204 fs) and centered at 1561 nm. The fundamental line of the MLL electrical spectrum, at 600 MHz, is

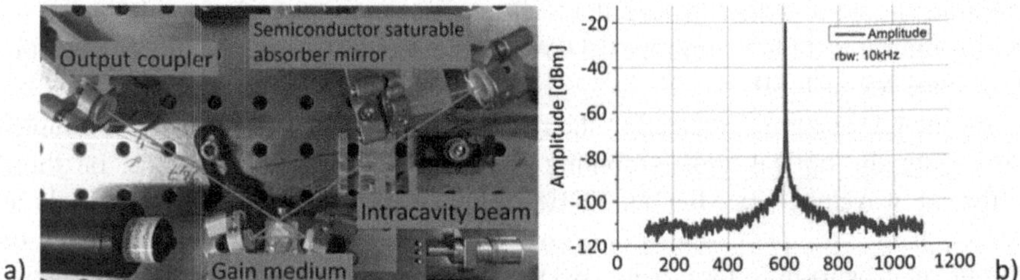

Fig. 14.2 (**a**) Laboratory version of the 600 MHz soliton MLL. The blue lines correspond to the beam path in the standing-wave cavity; (**b**) Lower-frequency component of the electrical spectrum of the MLL, with a spectral purity of around 70 dB

reported in Fig. 14.2b, showing the high spectral purity of such tone, around 70 dB. Now emphasis is made on the packaging of such a laser for reliable and long-term operation.

14.4 Processing and Target Detection

14.4.1 The MIMO Radar Concept and Processing

Unlike generic multistatic radars, MIMO radars are characterized by a high level of coordination between the central and the remote nodes; this unique feature allows to jointly process at the central node the multiple data streams with no need of local pre-processing at the remote nodes [1]. For this reason, they are particularly suitable for surveillance applications. In particular, MIMO radars with widely-separated antennas have the ability to observe the target from different aspect angles [1]. This offers many advantages, like the "super-resolution", i.e. the ability of enhancing the cross-range resolution beyond the limit imposed by the antenna aperture and even higher than the in-range resolution.

Following the general radar signal processing scheme [10], here we will focus our interest mainly on the MIMO fusion and target imaging strategies. In the case of a radar network with M TXs and N RXs, the CU employs N ADCs to digitize the received signals, which are the combinations of the M echo signals from each TXs. With the subscript k, l we indicate the generic radar employing the generic kth TX and lth RX pair, with $k = 1, \ldots, M$ and $l = 1, \ldots, N$. In the signal conditioning stage [10], the N digitized data streams are split into the $M \times N$ individual bistatic channels for data processing. After conditioning, the BB equivalent of the signal received by the lth RX is [1]:

$$r_{k,l}(t) = a_{k,l} \cdot s_k\left(t - \tau_{k,l}\right) e^{j\left[\theta\left(t-\tau_{k,l}\right)-\theta(t)\right]} + n_{k,l}(t), \tag{14.1}$$

where $s_k(t)$ is the signal transmitted by the kth RX, $a_{k,\,l}$ is an amplitude factor, while $\tau_{k,\,l}$ is the time delay proportional to the bistatic distance, being function of the target location (x, y) and the TX/RX positions in the Cartesian plane. The term $n_{k,\,l}(t)$ is modelled as an additive white Gaussian noise stochastic process, while the phase noise $\theta(t)$ models the phase instability caused by the oscillator. When the angular jitter is less than 0.1 rad [3], the coherent MIMO log-likelihood function can be calculated as [1]:

$$\Lambda(t) = a\left|\sum_{k=1}^{M}\sum_{l=1}^{N} e^{-j2\pi f_{IF}\tau_{k,l}} \int r_{k,l}^{*}(t) s_k\left(t - \tau_{k,l}\right) dt\right| + b. \tag{14.2}$$

For each possible target location (x, y) in the monitored space, the statistic $\Lambda(t)$ is computed determining the MN correlations between the received and transmitted signals, while the phase term $e^{-j2\pi f_{IF}\tau_{k,l}}$ compensates the phase shift due to time delay $\tau_{k,l}$. Finally, all the *MN* correlation contributes are summed together coherently. This phase compensation (i.e. coherent MIMO processing) is feasible only when the system architecture can ensure time/phase synchronization. This is the case of the proposed photonic radar network, which employs the same optical oscillator, with very low phase noise [6], for both signal generation and detection. In fact, the temporal jitter of the considered system is in the order of 10^{-12} s, while the angular jitter is in the order of 0.01 rad [8], far better than the operative limit [3].

14.4.2 Multistatic ISAR Processing

ISAR imaging founds on the possibility to gather information of the target from slightly different angular positions by exploiting the relative motion between the sensor and the target. The backscatter echoes from the targets need to be coherently processed to obtain high-resolution images in the cross range direction [9]. Notwithstanding, the target motion which is necessary to produce the angular imaging variation, is also typically responsible of the presence of phase variation that, if not properly compensated, would be source of large defocusing. The target motion shall be, generally, estimated directly from the data and then properly compensated. It can be shown that, for a target characterized by a uniform motion, the bistatic delay can be expressed in the form:

$$\tau_{k,l}(t) = \frac{2R_{k,l}(t)}{c}, \quad R_{k,l}(t) = R_{0kl} + \gamma_{k,l}t + \delta_{k,l}t^2 \tag{14.3}$$

with $\gamma_{k,l}$ and $\delta_{k,l}$ being parameters proportional to the Doppler centroid and Doppler rate associated to each sensor pair (*k*,*l*); in fact, the influence of the Doppler Centroid is mandatory to separate the moving target from the static clutter, and, high value of the Doppler rate allows to achieve sufficient system focusing capabilities, respectively. The block diagram in Fig. 14.3 describes the processing implemented to perform the focusing of ISAR data for either monostatic or bistatic configurations. Core of the procedure is the target motion estimation and compensation, which is based on Doppler analysis. Following this step, an inverse Fast Fourier Transform

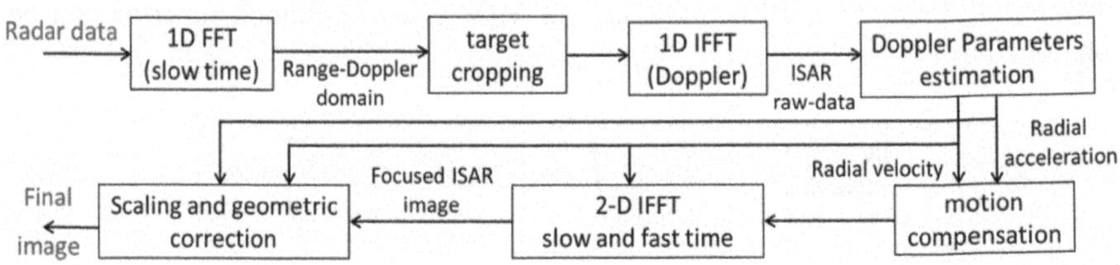

Fig. 14.3 Block diagram of the focusing procedure for monostatic and bistatic ISAR data

(IFFT) is carried out to obtain the focused image: eventually, a scaling is performed in the cross range direction to achieve the final pixel size.

14.5 Experimental Activities and Results

To validate the approach proposed in SOLE project, we present a preliminary test of the multi-static coherent radar network in a real outdoor scenario. Two measurement campaigns have been carried on: the first with extended targets, to validate the imaging capability of the system; the second, with multiple small, closely-spaced targets to demonstrate the attained enhanced resolution. It is important to remark that the system always worked in X band but, between the first and the second field trial, it underwent an upgrade of the employed BW from 20 to 100 MHz.

The scheme of the tested radar network [8] is depicted in Fig. 14.4a. It is composed of 2 TXs $\times$ 2 RXs, connected to the CU through SMF, working as explained in Sect. 14.2. The MLL enables generating RF carriers with extremely low phase noise, upon the O/E conversion of the MLL modes in a PD [6], thus implementing high-quality RF up-conversion and down-conversion. Moreover, exploiting a single MLL for these operations ensures a perfect synchronization of the SPs and maintains the coherence of the signals, which is the most crucial issue in coherent MIMO radars. The radar signal, a linear frequency-modulated upsweep chirp pulsed signal with 100 ns duration, pulse repetition interval (PRI) of 50 μs and BW B, is digitally generated at $f_{IF} = 100$ MHz, whereas the RF output carrier frequency is $f_{RF} = 9.7$ GHz. The large MLL optical spectrum ensures high efficiency, and the wide electrical BW of MZMs and PDs allows for the flexible management of RF signals up to several tens of GHz [6]. The employed antennas at the TXs and RXs are ultra-wideband horn antennas with about 50° half-power beam width (HPBW) aperture and 12 dBi maximum gain. The output power of each antenna was about 100 mW. The optical delay line (ODL) depicted in Fig. 14.4a, consisting of a 1 km-long SMF spool inserted before the TX of SP2, implements the TDM between the two SPs. The detected echoes received by the SPs are sent to the CU, where they are down-converted, low-pass filtered, fed into a two-channel acquisition system and digitized by an ADC at 400 MS/s per channel. An aerial picture of the out-door experimental setup, taken by one of the drones carrying a target, is reported in Fig. 14.4b, showing the geometry of the radar network. The four horn antennas were aligned along a 21-m baseline.

The multistatic ISAR processing, based on Doppler Parameter Estimation Algorithm described in Sect. 14.4.2, has been adopted for testing the imaging capability of the photonic radar network on moving targets. Experiments on extended targets have been conducted employing a single TX and both RXs with $B = 20$ MHz, with a range resolution of 15 m, with an observation interval of 0.1 s which limited the Doppler resolution to 10 Hz. The chosen targets were airplanes taking off in the proximity of the building. In Fig. 14.5 are shown the images of the target before (a, c) and after (b, d) the focusing in the Range-Cross range domain, for the RX1 (a, b)

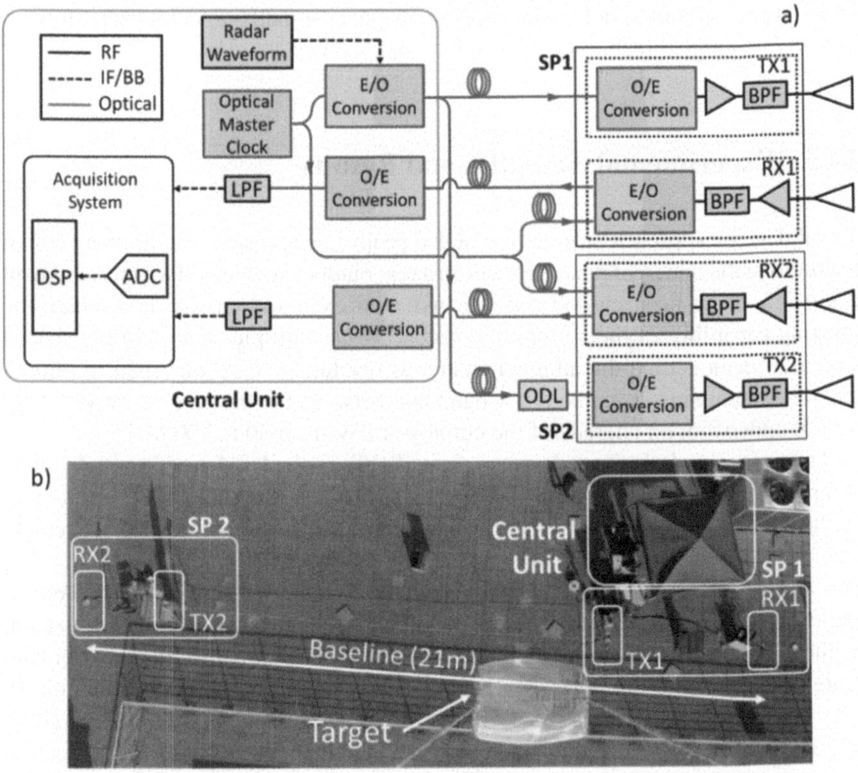

Fig. 14.4 (**a**) Architecture of the proposed Photonic Radar Network. *DSP* Digital Signal Processing, *BPF* Band-Pass Filter, *LPF* Band-Pass Filter. (**b**) Aerial picture of the test field, taken by one of the drones carrying the targets. The target is the metal net cylinder hanging from the drone

and RX2 (c, d). The selected target is moving at radial velocity of about 130 km/h, at a distance from the baseline of around 770 m. Although it is difficult to identify the shape of the airplane, the visual inspection demonstrates the capability of the algorithm to achieve well focused images, despite the short acquisition time, thanks to the stability of the optically generated and distributed signals.

Table 14.1 reports typical quality parameters to evaluate imaging performances for both RXs. As apparent from the results, all the parameters improve after the application of the ISAR imaging algorithm, thus confirming the capability of the system to guarantee good performances. Even though the analysis were carried out on a premature version of the radar system which is not tuned to ISAR imaging, the achieved results are rather promising.

In the second measurements set, carried out employing all TXs and RXs, with a BW $B = 100$ MHz, the two cooperative targets consisted of two cylinders, with 17 cm radius and 50 cm height, made of a tight-mesh metal net and hanging from two mini-drones hovering above the baseline. The two targets were always kept at a

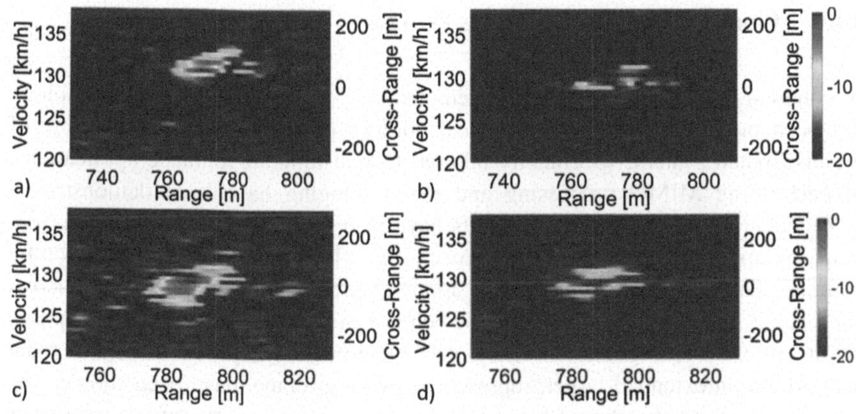

Fig. 14.5 Unfocused (**a**, **c**) vs focused (**b**, **d**) ISAR images of a target took in the first set of measurements, by RX1 (**a**, **b**) and RX2 (**c**, **d**)

Table 14.1 Quality imaging parameters

	RX1			RX2		
Parameters	Unfocused	Focused	Relative	Unfocused	Focused	Relative
Contrast	1.06	1.12	+5.81%	1.02	1.09	+6.54%
Entropy	7.43	7.09	−4.66%	7.61	7.31	−3.99%
Peak	0.30	0.40	+34.96	0.23	0.37	+56.56
Sharpness	0.22	0.37	+69.90%	0.16	0.35	+124.71%

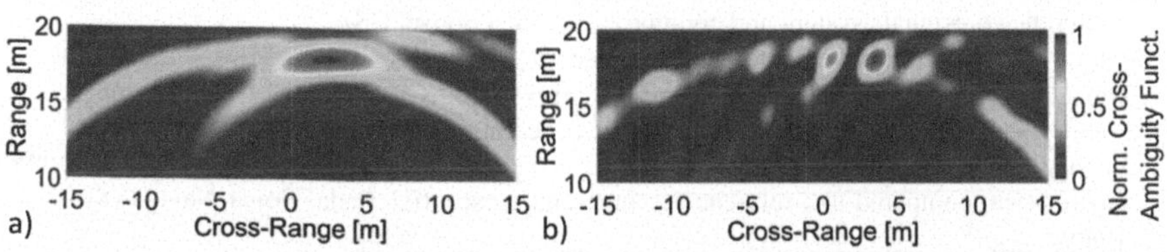

Fig. 14.6 Normalized cross-ambiguity function as processing output of the MIMO detection in the second set of measurements with two closely-spaced targets. (**a**) Non-coherent MIMO with two unresolved targets; (**b**) coherent MIMO where the targets are clearly distinguishable

distance of around 3 m from each other, at 18 m above the baseline. The axes of the cylinders were perpendicular to the baseline, to enhance the radar cross-section seen from the antennas. As we can observe in Fig. 14.6a, the two targets are too close to be correctly separated in the cross-range with a MIMO non-coherent processing. Indeed, being the in-range distance from the baseline of the two targets 18 m, and given the antenna HPBW ~50°, the expected monostatic cross-range resolution is 15.7 m. Figure 14.6b, instead, depicts the output of the coherent MIMO processing. Here, the two targets are correctly separated in cross-range, around 3 m apart from each other, demonstrating a cross-range resolution improvement by a factor of 5.

14.6 Conclusions and Future Works

A coherent MIMO radar network demonstrator based on photonic technology has been presented and tested in a real outdoor environment. The capability of the distributed system, specifically tailored to real-time surveillance applications, of performing MIMO processing and ISAR imaging have been demonstrated. Two cooperative closely-spaced targets have been successfully resolved in cross-range by applying coherent MIMO processing, which enhanced the resolution by 5 times. The superiority of such an approach compared with the non-coherent approach is apparent, demonstrating the benefits of the photonics-assisted proposed architecture. Moreover, the efficiency of ISAR processing by this distributed radar network on an extended target, represented by an airplane, have been shown. The here-presented results should be considered as a starting point: further field trials are planned in the next future with an extended architecture, with one more SP and working at the same time in X and S band, with much more largely distributed antennas, as the final goal of the SOLE project.

References

1. Haimovich AM et al (2008) MIMO radar with widely separated antennas. IEEE Signal Process Mag 25(1):116–129
2. Lehmann NH et al (2006) High resolution capabilities of MIMO radar. 40th IEEE Asilomar conference on signals system and computers, Pacific Groove, USA
3. Pasya I et al (2013) Detection performance of M-sequence-based MIMO radar systems considering phase Jitter effects. IEEE phased array systems & technology symposium
4. Ghelfi P et al (2014) A fully photonics-based coherent radar system. Nature 507:341–345
5. Laghezza F et al (2015) Field evaluation of a photonics-based radar system in a maritime environment compared to a reference commercial sensor. IET Radar Sonar Navig 9(8):1040–1046
6. Serafino G et al (2011) Phase and amplitude stability of EHF-band radar carriers generated from an active mode-locked laser. J Lightwave Technol 29(23):3551–3559
7. Williams P et al (2008) High-stability transfer of an optical frequency over long fiber-optic links. J Opt Soc Am B 25:1284–1293
8. Lembo L et al (2019) In-field demonstration of a photonic coherent MIMO distributed radar network. IEEE radar conference 2019, Boston
9. Noviello C et al (2017) Fast and accurate ISAR focusing based on a Doppler parameter estimation algorithm. IEEE Geosci Remote Sens Lett 14(3):349–353
10. Richards GA (2014) Fundamentals of radar signal processing, 2nd edn. McGraw-Hill, New York

Chapter 15
NORMA: Imaging Noise Radar Network for Covert Air and Maritime Border Security

S. Tomei, D. Staglianò, K. Lukin, V. Palamarchuk, and S. Lukin

Abstract The NORMA project aims at designing and analyzing a imaging noise radar working in network configuration for covert, high spatial resolution air and maritime border surveillance. The main NATO military benefits of the NORMA system consist of the capability of the proposed solution to have an all-weather all-day 24-h covert and LPI surveillance system with high detection, tracking and imaging performance. The system will be also designed to achieve low RCS and high resolution range profile measurements. Particularly, the multistatic and MIMO imaging capabilities of the system allow the target to be observed at different viewing angles providing full shape RCS and imaging and consequently better classification and recognition performance. A technological demonstrator composed of two noise imaging radars in a network will be designed and developed. A system demonstration to assess the performance of the proposed solutions will be performed at the end of the third year of the project. The present paper aims at showing the idea behind the proposed solution and the main advantages that border authorities can benefit of. In particular, the paper will focus on how the noise waveforms will be generated and what radar mode will be used to achieve the desired objectives.

Keywords Noise radar · SAR imaging · Ground noise SAR · Noise waveforms · PRBS · Noise SAR interferometer · Stepped frequency noise radar

---

S. Tomei (✉)
CNIT RaSS: National Interuniversitary Consortium for Telecommunications, Radar and Surveillance System, Pisa, Italy
e-mail: sonia.tomei@cnit.it

D. Staglianò
ECHOES s.r.l., Pisa, Italy

K. Lukin · V. Palamarchuk · S. Lukin
IRE NASU, Kharkiv, Ukraine

C. Palestini (ed.), *Advanced Technologies for Security Applications*, NATO Science for Peace and Security Series B: Physics and Biophysics,
https://doi.org/10.1007/978-94-024-2021-0_15

15.1 Introduction

Given the nature of current evolving threats, border security has become of paramount importance both in civilian and military application. In particular, both air and maritime borders are threatened by unlawful activities which employ new technologies and require the use of more effective surveillance sensors. New technologies such as unmanned platforms employed by drug smugglers require the intervention of border authorities with more effective systems, allowing for an efficient and continuous monitoring of land and sea borders. In particular, such solution should face both civilian and military issues, allowing for all weather/all day operations and covert/LPI (Low probability of Intercept) capabilities.

The solution proposed in this project is a radar network system composed of cooperating wideband noise imaging radars. The capability of the radar to be operative all weather all days 24 h per day is accomplished with the main characteristic of transmitting wideband noise waveforms with imaging capability. It is a powerful solution to have an LPI land and sea border surveillance system also in critical and sensible regions, which are difficult to be monitored by human patrolling missions. It is worth pointing out that LPI is a desired feature especially for military applications that have to deal with antagonistic and hostile forces trying to prevent the surveillance of specific and sensitive areas. The radar is also designed to have an RCS mode for the measurement of target reflectivity and 1D-2D imaging capability, including also detection and tracking techniques for the revelation of slow target motions in sea and strong land clutter, being a very important function for sea border surveillance and mainly for low flying drone detection.

The following sections will focus on the description of the proposed systems based upon Noise Radar Technology [1–4] and software defined noise radar network [5, 6], ground SAR network [7, 8] and ISAR modes [9].

15.2 The NORMA System

The main objective of the NORMA project is to implement an innovative solution that addresses the user needs. According to the end users perspective, the most challenging mission is monitoring protected areas and natural reserves to intercept any illegal activity which might be dangerous for such areas, and the surveillance of harbours or other restricted and sensitive areas in wider regions, which are already monitored by other systems, but need an additional and dedicated monitoring system. Within this mission, the main aim is to detect small vessels, such as motorboats and rubber dinghy and other small targets, such as divers at small ranges from the sensor and small drones used by drug smugglers.

15.2.1 System Architecture

The proposed solution is based on a bistatic/multistatic architecture that offers a number of benefits due to the separation between the transmitter and the receivers. In particular: (1) Enhanced robustness of the overall system introducing a higher level of redundancy; (2). Exploitation of large baselines that increase the signature from low-observable targets; (3) A reduced dynamic range in difficult imaging environments such as urban ones; (4) Robustness against unfavourable geometry; (5) Robustness over shadowing effects; (6) A better detection and tracking capability. When multiple receivers are employed, each one forms a bistatic radar with the transmitter offering the potential for tailored coverage and a richer information source to enable more accurate location, high resolution imaging, target reconstruction and, hence, identification. The price to be paid for these advantages is an increased system and processing complexity. In particular synchronization and beam pointing are more difficult to be implemented.

NORMA consortium has chosen to implement a time-synchronized multistatic system composed of MIMO radars. The reason behinds the choice is twofold:

1. It allows a cheap and easy implementation by using COTS only components (like GPS or GPSDO) so reduces the system cost and complexity and makes its realization feasible in a real application.
2. It allows to operate in different frequency bands thus allowing frequency diversity that along with spatial diversity at a given time may improve the scenario and target understanding.

It is worth pointing out that the chosen architecture is general and scalable in the sense that it can be applied to a general number of radar sensors and regardless of the scenario and application at hand. This points out the potential of the NORMA system that can be further expanded to satisfy the surveillance needs for more complex and challenging missions. In addition, the use of multiple sensors in a network provides a degree of resilience guaranteeing the surveillance even in the case of failure of one of the employed radars, at the cost of reduced performance.

15.3 The Radar Architecture

15.3.1 CNIT RaSS Radar Architecture

The sensor developed by CNIT-RaSS for the NORMA project will consist of a wideband coherent Multiple-Input Multiple-Output (MIMO) radar which exploits noise-like waveforms, as shown in Fig. 15.1. Noise Radar Technology employs noise/random waveforms (NW) as sounding signals and implement coherent reception of radar returns via estimation of correlation between the noise reference (portion of the transmitted signal) and noise radar returns. Due to the wideband

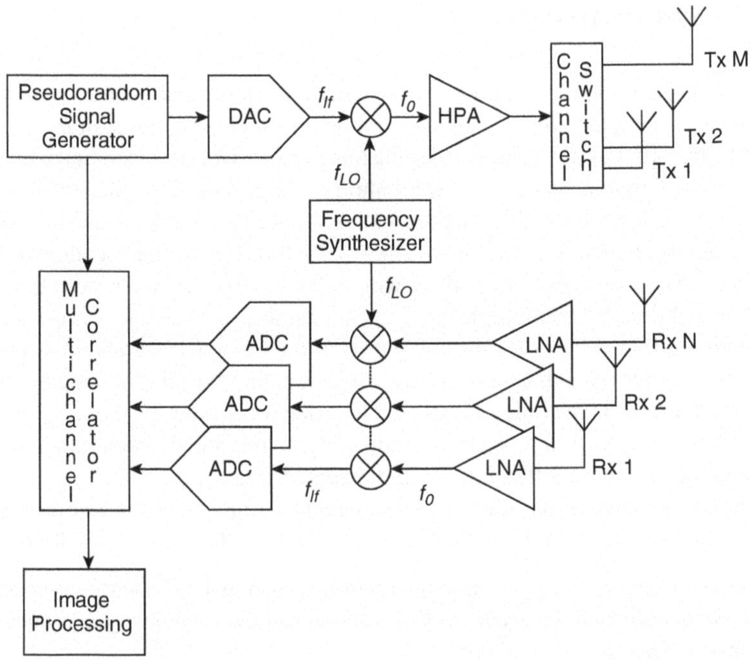

Fig. 15.1 CNIT RaSS MIMO noise radar

capabilities and the coherent architecture, the proposed sensor may be used to form high resolution radar images. If the sensor is moved along a known or at least measured trajectory, it can be used to form Synthetic Aperture Radar (SAR) images of the observed scene. Similarly, by keeping the sensor stationary and exploiting the motion of the target with respect to the radar, it may be used in Inverse Synthetic Aperture Radar (ISAR) mode. In addition, ISAR techniques could be applied to SAR images to focus moving targets captured during the synthetic aperture time.

15.3.2 IRE NASU Radar Architecture

In this section, we briefly describe both the suggested design concept and architecture of the Ground Noise SAR subsystem to be developed by LNDES IRE NASU for the monitoring of maritime and aerial borders aiming at detection and tracking of small RCS targets. This Ground Noise SAR subsystem is part of the NORMA Ground Noise SAR/ISAR system, which is a radar network suitable to be deployed along the borderline, enabling the best coverage of the surveillance area with different viewing angles. A simple demonstrator composed of two Ground Noise SAR systems has been proposed and adjusted to different demonstration scenarios, such as Ukrainian border near Kharkiv and the sea area in front of the

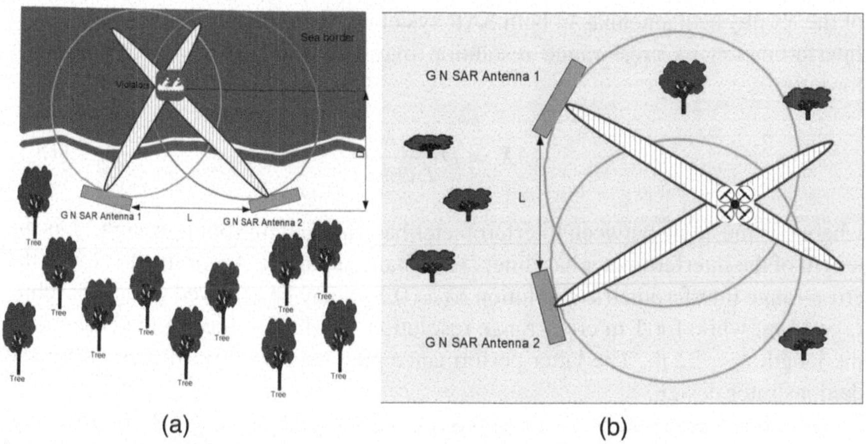

Fig. 15.2 (**a**) Ground Bistatic Noise SAR system deployed on coastal area to detect low cross-section floating targets; (**b**) Ground Bistatic Noise SAR deployed along the aerial border area to detect small RCS aerial targets

Livorno harbor in Italy. These two scenarios are representative of air and sea border surveillance. Transmitted sounding signals in the systems may be either wideband NW or narrowband Stepped Frequency (SF) NW. Typical scenario for detection of low cross-section (RCS) targets near maritime border and aerial border are shown in Fig. 15.2a, b, respectively.

In both cases, range and cross-range resolutions are specified as 0.5 m and 0.5 m at the 1 km distance. The required range resolution $\delta R = \frac{c}{2\Delta f}$ may be readily provided with transmitting a wideband noise signal with at least $\Delta f = 300$ MHz Power Spectral Density (PSD) [1, 2, 10]. The major challenge is providing high azimuth resolution. For the chosen frequency band (13–14 GHz) this will require an antenna aperture of about 44 m. This is impractical for real border scenarios, since it will require design and manufacture of a very long antenna array with about 1 cm spacing between physical antennas. This will require about 4400 *Rx* physical antennas! A novel concept of bistatic ground SAR have been suggested to go around this problem and consists in combining a monostatic SAR operational mode with interferometric SAR operational mode, the details of which will be briefly described below.

Figure 15.2a, shows bistatic version of the suggested Ground Noise SAR deployed on coastal area for monitoring of the sea border. Each Ground SAR performs surveillance of neighboring areas with azimuth (cross-range) resolution provided by an aperture not exceeding 1 m, namely: 22 m at the distance $D = 1$ km. When a target has been detected, the system will perform synthesized beams focusing (may be using antenna arrays mechanical rotation as well) on that target and applying *interferometric technique* to enhance azimuth resolution via coherent processing of the data obtained from different synthesized beams [6–8]. The required interferometric azimuth scan? will be provided by synchronous switching of the *Rx* physical antennas in both SAR systems. According to the theory of Radio Interferometer, its cross-range resolution may be estimated using the following formula:

$$\delta X = D\frac{\lambda}{L\cos(\theta)} \tag{15.1}$$

where θ is the angle between Interferometer baseline and line of target sight; L is the length of the interferometer baseline; D is distance to target. According to (15.1), the cross-range interferometric resolution $\delta X = 0.5$ m may be achieved using a baseline $L = 44$ m, while for 1 m cross-range resolution it will be sufficient the baseline of the length $L = 22$ m. The latter performance may be a good starting point for the demonstrator design.

The above performance estimations of the Ground Noise SAR for maritime targets detection are applicable also to the Ground Noise SAR system for aerial border surveillance shown in Fig. 15.2b. Rx antenna array of both SARs will be composed of several Autonomous Tx/Rx Antenna Array Units mounted on a tripod (Fig. 15.3). Every Tx/Rx Antenna Array Unit will contain a single Tx antenna, and 32 or 36 Rx antenna depending on the available electronic components; I/Q Phase Detectors serve as Receivers; set of slow ADCs for data acquisition from I and Q channels and their sending further to the Central Processing Unit (CPU). We may also use a usual phase detector and apply two times higher sampling frequency to generate I and Q signals digitally preserving the same performance of the reception if this will not slow down the system operation. Every Tx/Rx Unit will be supplied with a standard GPS receiver and Data link & Synchronization Channel if needed.

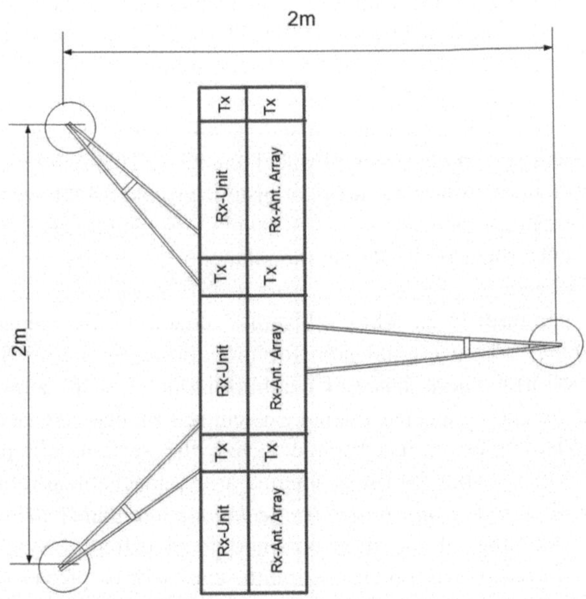

Fig. 15.3 Top view of the Ground SAR composed of 3 Tx/Rx Units and mounted on a tripod

Power supply may be either from a standard power network or, otherwise, from rechargeable battery connected to a solar panel. In general, a MIMO radar exploits an array of M elements on transmit and an array of N elements on receive. In case of perfect orthogonality of the M transmitted waveforms, N receivers may simultaneously acquire the echo from the scene, and a virtual array of MN virtual elements can be extrapolated.

15.3.2.1 MIMO Ground SF Noise SAR

The Tx waveforms orthogonality can be achieved via time division of the transmitted and received signals (transmitting and receiving signals one-by-one) [9–11], or code division provided by orthogonal coded waveform generation. The latter method enables the fastest scan, though it requires application of as many transmitters as number of Tx/Rx antennas to be used. In order to go around that constrain we preferred simplest architecture of MIMO SAR, namely SAR with a single Tx channel and multichannel receiver. Figure 15.4 shows the architecture of such MIMO Ground Noise SAR. Noise waveform with PSD bandwidth, B, is digitally generated in Noise Generator Unit converted up to the central frequency in the Transmitter Unit and is transmitted with the help of Tx antenna onto the scene. Part of this signal is used as the reference inside the Multichannel Receiver.

We suppose to use Vivaldi Antenna as a radiating/receiving physical antenna of 1D Rx antenna array. In addition, dielectric lenses may be used for narrowing the antenna pattern in elevation plane. Figure 15.5 shows schematic construction of the antenna array for a single Autonomous Unit of MIMO Ground Noise SAR.

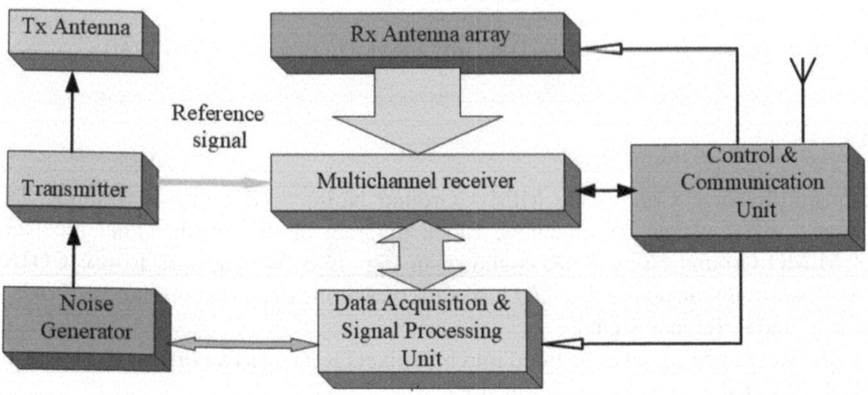

Fig. 15.4 Architecture of MIMO Ground Noise SAR

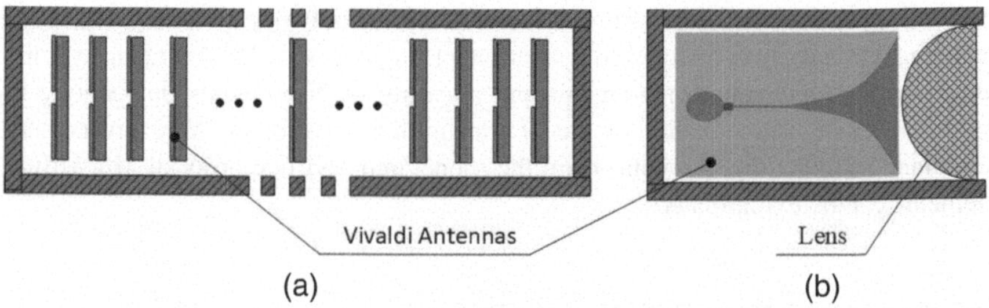

Fig. 15.5 Schematic view of the Tx/Rx Antenna Arrays: (**a**) Front view (no focusing lens); (**b**) Side view: focusing lens may be used for narrowing antenna pattern in elevation

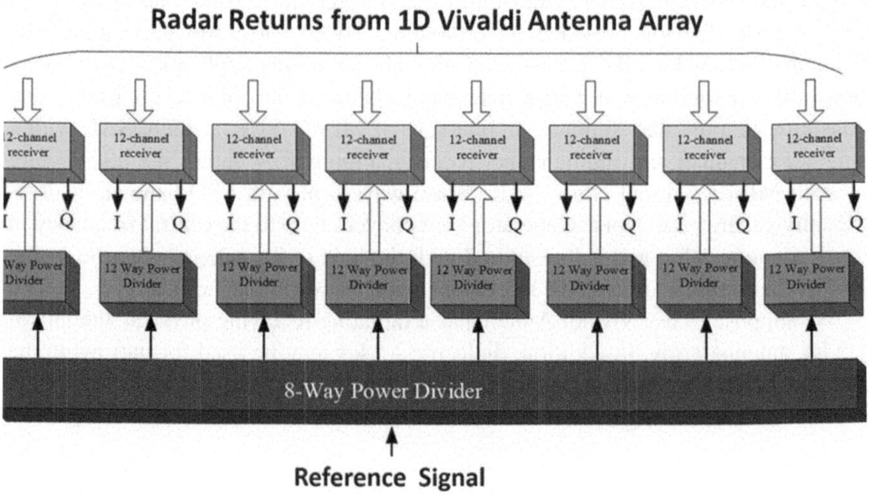

Fig. 15.6 Architecture of multichannel Receiver of the MIMO Ground Noise SF SAR

Tx/Rx Unit

Transmit/Receive Unit of the MIMO Ground Noise SAR contains a single Tx channel and multiple Rx channels. Block diagram of the multichannel receiver of MIMO Ground Noise SAR is shown in Fig. 15.6. We suppose to use COTS electronic components, such as I/Q mixer for measuring of cross-correlation of radar returns and Reference signal.

The Reference signal to be fitted into I/Q mixers is taken as a portion of the Noise SAR Tx signal as it is shown in Fig. 15.7.

The Generator of the Tx noise signal is designed as follows. As the main source of RF power we use VCO with working frequency range from 13 GHz up to 14 GHz. Central frequency of the Tx signal is defined by DC voltage applied. The latter is defined by a digital code from Control Unit which is converted to DC voltage via DAC. Simultaneously, random signal with the given PSD bandwidth

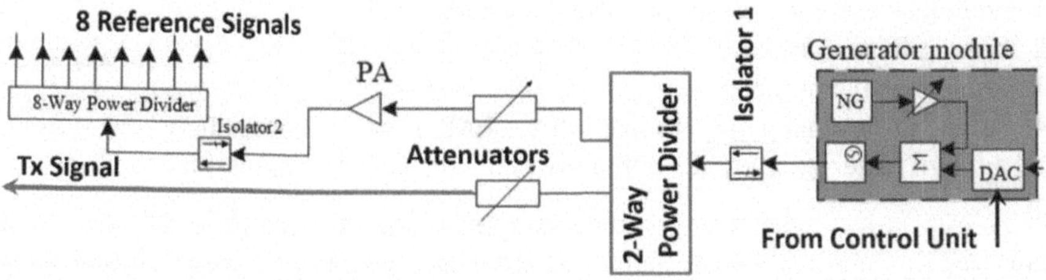

Fig. 15.7 Block-diagram of Noise Signal Generator for MIMO Ground Noise SF SAR with 96 multi-channel Receiver and single channel Transmitter

digitally generated in the Noise Generator (NG) is applied to the VCO as well, which produces noise signal at controllable PSD bandwidth and central frequency. In this way, the Generator module produces Tx noise signal having PSD bandwidth and central frequency both being electronically changed at a high rate. That enables implementation of Stepped Frequency (SF) mode for narrowband Noise signal and thereby design of SF Noise SAR [14, 15].

Suggested architecture of SF Noise SAR provides several advantages when applying SF Noise Radar technique. First of all, it enables analogue (here means in realtime) estimation of cross-correlation between radar returns and the reference signal. Estimation of the phase acquisition by a transmitted signal for a set of frequencies allows reconstruction of range profiles for every Tx and Rx configuration in the SAR. In addition, now we may use a set of slow ADCs in data acquisition unit, which are readily available in multi-channelled and multi-bit implementations. All this makes proposed Ground Noise SF SAR an efficient and flexible instrumentation for real time imaging of the area of interest.

15.4 Noise Waveform Analysis

Within the domain of military radar, in both surveillance and reconnaissance applications, covert surveillance techniques are attracting attention because of very effective countermeasures by hostile forces (for example anti-radiation missiles (ARM) and electronic counter measures (ECM)). To address these trends, past studies have examined signals that display good ambiguity behavior, and concurrent investigations have considered LPI-radar techniques. A promising approach, which combines both, is the use of noise waveforms [1–4]. In this context, significant improvements in military systems can be realized in radio frequency (RF) application areas by taking advantage of emergent Noise Radar Technology (NRT) [1, 2]. for advanced radar design. Noise radar technology uses random noise (or noise-like) waveforms as a sounding signal and coherent processing for radar return reception, enabling essential enhancements of both general and military radar performance, including:

1. no range ambiguity, even for pulse noise radar;
2. better LPI performance,
3. immunity to jamming
4. better electromagnetic compatibility (EMC) performance between sensors through better interference immunity and interference suppression.

A true noise signal can be generated from natural sources such as thermal noise or the aggregation of random electron movements in a mean current flow. A true noise source with infinite bandwidth is unpredictable, which suggests the best LPI performance achievable. In practice, some band limiting will inevitably occur, which has the effect of windowing the signal in the frequency domain. The shape of this window has a significant impact on correlation performance, and a rectangular window for example has high sidelobe levels that can be reduced with alternative windows, at the expense of a slightly wider main lobe. Rather than using a true noise source, a signal that resembles noise in some of its properties may be suitable, while simplifying the implementation. Such signals have been termed pseudo-noise, pseudo-random, noise-like, etc. These signals are deterministic, meaning that their future behavior can be uniquely defined from the knowledge of the current status. The noise-like properties of deterministic signals may be entirely related to the statistical properties being equivalent to a natural, or other desirable, noise source. Predictability may not necessarily be a major issue. Clearly, for cryptographic systems, or LPI sensitive applications predictability is relevant. The predictability of a waveform based on inaccurate knowledge of the system dynamics or its state, is a further discriminator. An important class of noise-like signals comes from chaotic systems, which are deterministic and non-periodic [1, 2, 11]. In these systems, an error in knowledge locally grows exponentially, and so predictability rapidly diminishes. Non-chaotic systems show divergence at a rate less than exponential. The signal properties depend on whether the signal is quantized in time or amplitude. Natural noise sources are continuous in both dimensions, as are continuous time chaotic systems (described by differential equations). Quantization in time or amplitude implies some structural properties that increase the probability of intercept. An implementation using digital circuits is quantized in both dimensions. Digital generation is fundamentally periodic, at best determined by the number of bits used to represent the numbers, but some iterative systems may have initial conditions that lead to short period solutions. Such short solutions must be avoided if the advantages of noise like waveforms are to be achieved. In this paper, two modulated NW are tested: FMCW PRBS (Pseudo-Random Binary sequence) and Stepped frequency PRBS. Pseudo Noise (PN) sequences are deterministic waveforms with a noise-like behaviour, easily generated by using linear feedback shift registers. They exhibit good autocorrelation properties and high spectral efficiency. The best known examples of such waveforms are maximal length pseudorandom binary sequences (*m*-sequences or PRBS). For a maximal length code the train of M pulses $p(t)$ with chip period T_c and amplitudes $a_n = \pm 1$ may be written as follows [12]:

$$p(t) = \sum_{n=1}^{M} a_n c\,(t - nT_c) \tag{15.2}$$

where $c(t)$ is rectangular pulse of duration T_c, its autocorrelation $R_p(t)$ can be expressed by

$$R_p(t) = \begin{cases} (t + T_c)\,\frac{M^2+1}{MT_c}\frac{1}{M} & -T_c \le t < 0 \\ -(t - T_c)\,\frac{M^2+1}{MT_c} + M & 0 \le t \le T_c \\ -\frac{1}{M} \quad otherwise & \end{cases} \tag{15.3}$$

An adequate sequence for target sounding, which will reassure efficient path delay recognition, should have an autocorrelation function with a narrow main lobe and low sidelobes. Traditionally, PRBS are considered as a good option for target sounding. However, if sidelobe level is taken into account, these sequences are not optimal. The sidelobe level for the correlation of PRBS is constant and equals −1/M, or −1 if normalization is not applied. A maximum length PRBS function has length equal to $M = 2^k - 1$, with $k \in \mathbb{N}$. A PRBS is "pseudorandom", because, although it is in fact deterministic, it seems to be random in a sense that the value of an a_n element is independent of the values of any other elements, similar to real random sequences. A PRBS can be stretched to infinity by repeating it after M elements, but it will then be cyclical and thus non-random. In contrast, truly random sequence sources, such as sequences generated by radioactive decay or by white noise, are infinite (no pre-determined end or cycle-period). However, as a result of this predictability, PRBS signals can be used as reproducible patterns (for example, signals used in testing telecommunications signal paths). The PRBS codes can be efficiently used to modulate the instantaneous phase of the FMCW or stepped frequency signal [14, 15] in order to produce a LPI waveform. The concept is very simple, and the basic FMCW or stepped frequency waveform can be used. In such case, here the instantaneous phase of the FMCW waveform is modulated with the addition of the PRBS sequence, which gives an additional term of π radians to the phase only if the PRBS symbol is −1. This randomness gives more variability to the instantaneous bandwidth allocation of the signal, which grows with the noise variance, as can be seen in the spectrograms in Fig. 15.8. In the first case the underlying frequency ramp is still visible, while in the second the noise power masks completely the frequency ramp.

With a 5 MHz noise bandwidth inside a 20 MHz FMCW bandwidth, the first null is still at 0.05 μs, but the sidelobes are lowered to −17.75 dB. With a 30 MHz noise bandwidth, first null is at 0.03 μs with a maximum sidelobe level about of −26 dB, meaning that the increased noise bandwidth has a greater impact on the sidelobe level reduction rather than the range resolution (which is nevertheless improved) (Figs. 15.9 and 15.10).

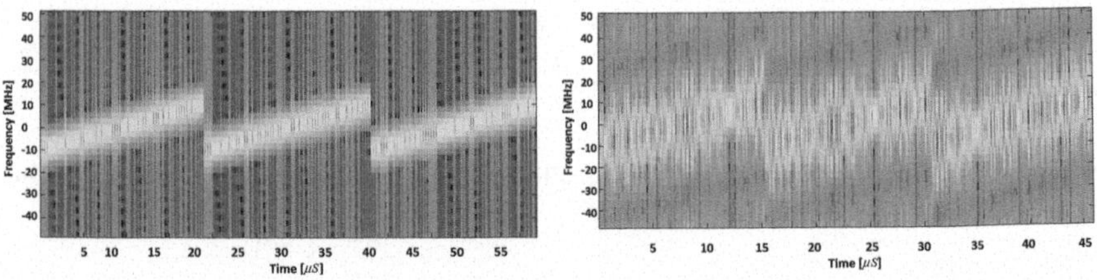

Fig. 15.8 PRBS modulated FMCW waveform, 5 MHz noise bandwidth (left); PRBS modulated FMCW waveform, 30 MHz noise bandwidth (right)

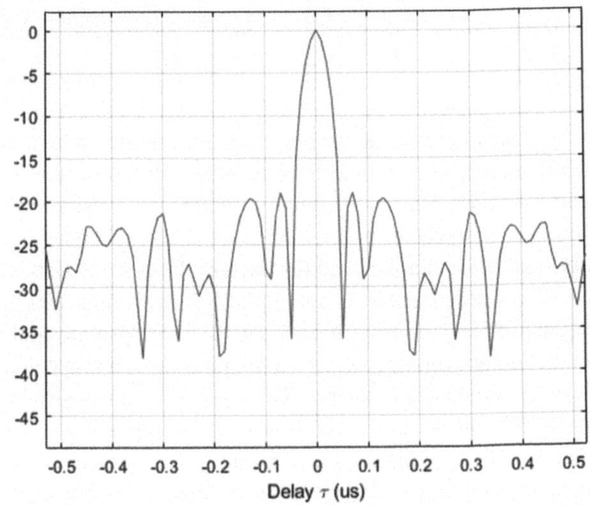

Fig. 15.9 Range ambiguity function of PRBS modulated FMCW sweep, 5 MHz noise bandwidth

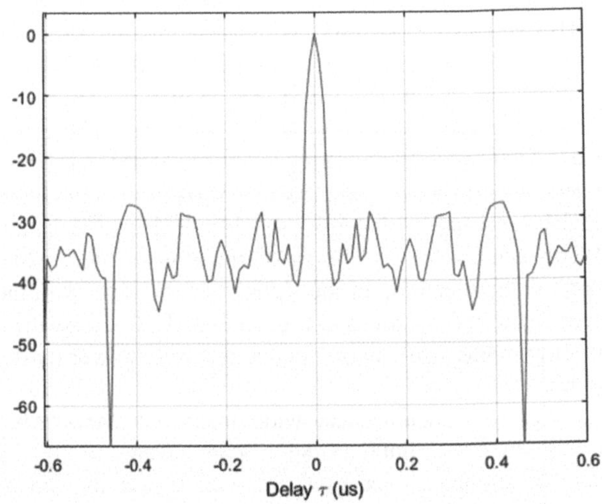

Fig. 15.10 Range ambiguity function of PRBS modulated FMCW sweep, 30 MHz noise bandwidth

15.4.1 *PRBS Modulated Stepped-Frequency Waveform*

As stated above, the instantaneous phase of the Stepped frequency waveform is modulated with the addition of the PRBS sequence, which gives an additional term of π radians to the phase only if the PRBS symbol is -1. The effects on the instantaneous frequency are the same with respect to the case of a FMCW waveform of equal bandwidth (20 MHz), as can be seen in the spectrograms of Fig. 15.11.

The effects of the modulation on amplitude and phase are the rapid switching of the lobes of the sinusoid, in amplitude, while the effect on the cumulated phase are somewhat less severe. Regarding the behaviour in the range ambiguity function (Figs. 15.12 and 15.13), it can be seen that for a 5 MHz noise bandwidth, the first null is located still at 0.05 μs, with a sidelobe level of -16.42 dB, while for a 30 MHz noise bandwidth the first null is still located at 0.05 μs but the first sidelobe amplitude drops to -49 dB.

From this analysis, it is possible to say that PRBS modulated waveforms (both FMCW and stepped-frequency) are well suited to be used in an LPI noise radar. It is to be noted that if the noise bandwidth grows, the performances are better in

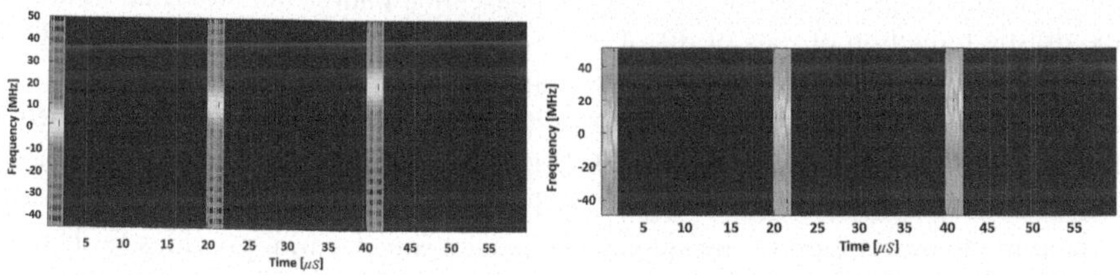

Fig. 15.11 PRBS modulated stepped-frequency waveform, 5 MHz noise bandwidth (left); PRBS modulated stepped-frequency waveform, 30 MHz noise bandwidth (right)

Fig. 15.12 Range ambiguity function of PRBS modulated stepped frequency waveform, 5 MHz noise bandwidth

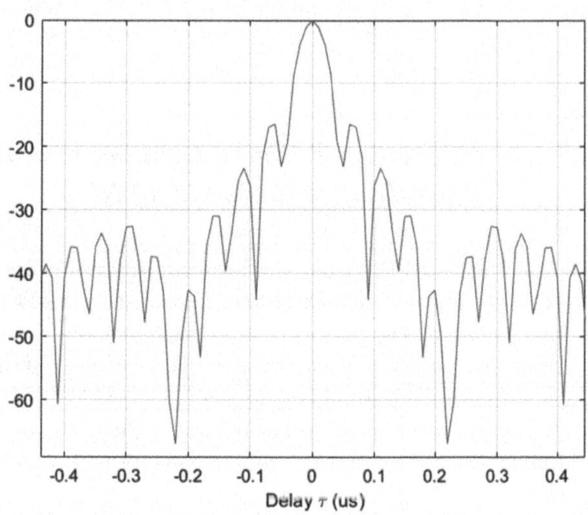

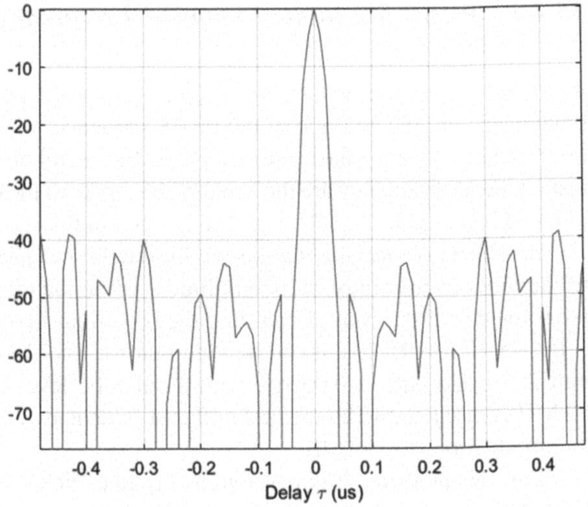

Fig. 15.13 Range ambiguity function of PRBS modulated stepped frequency waveform, 30 MHz noise bandwidth

terms of resolution and sidelobe level. This is due to the additional modulation that the noise-like sequence adds to the original phase modulation, but most importantly, the drastic reduction of tens of dB of the sidelobe level is one of the most notable achievements. At 30 MHz noise bandwidth, it can be seen that the resolution of a PRBS-FMCW waveform has better resolution with respect to a PRBS-stepped-frequency waveform, where latter has better performance in terms of sidelobe reduction. Nevertheless, the PRBS-FMCW is the waveform to be preferred because, as said in the description of the stepped-frequency waveform, it provides far better performance in target detection even if the target is fast-moving, a limitation for the other waveform. In both cases, the LPI is guaranteed, because, while the signal can be received by other receivers, they cannot decode the signal unless they are provided with the "key", which is the PRBS sequence that is used for the phase modulation.

15.4.2 FPGA-Based Digital Generation of Random-Filled Pulse Train with Random PRF

Development of FPGA-based waveform generators for radar applications is relatively easy and contributes largely to the Software-Defined Noise Radar concept implementation. The performance of such generators completely under our control (within the hardware limitations of the platform) and altering or expanding the capabilities of them should not present unsolvable problems. Digital generation of a truly random noise waveform is, of course, the most significant challenge. However, digital generation of pseudo-random waveform of a certain length that is affordable for available FPGA-based prototyping boards [13] should suffice for our short range radar applications.

The developed FPGA-based random signal generator [13] employs the Linear Congruential Algorithm (LCA) to digitally generate pseudorandom filling of pulses with given pulse length (L_p) and repetition period (P_S) values. The LCA is one of the best-known pseudo-random number generation algorithms. LCA is defined by the following recurrence relation:

$$x_{n+1} = (ax_n + c) \bmod m \tag{15.4}$$

As an example, Fig. 15.12 shows the pulse train with pseudo-random filling generated for the following parameters: $a = 173$, $c = 163$, $m = 2^8$.

The Linear Congruential Generator (LCG) with these parameters has a repetition period of 254.2 which is not sufficient for long range applications. However, several simple techniques can be applied to improve this LCG performance. For instance, one may increase parameter m up to 2^{32} and beyond, then use N most significant bits of the result as generator's output, where N is the required signal depth in bits, and this parameter is usually limited by the DAC's depth resolution (Fig. 15.14).

Using aforementioned LCG one may generate random sequence and use it for choosing a new value of the signal period (P_S) for each cycle, achieving in this way pulse train generation with random PRF. Example of such signal is shown in Fig. 15.15.

Finally, aforementioned LCA algorithm may be used both for generation of pulses with random filling and for PRF randomization of pulse train. To make

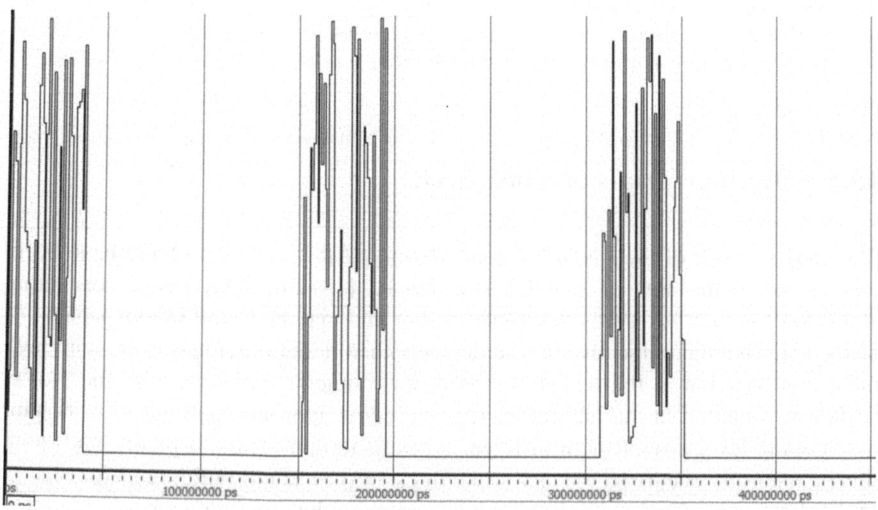

Fig. 15.14 Pulse train with random filling by LCG-generated waveform for $L_p = 30$ μs, $P_S = 150$ μs

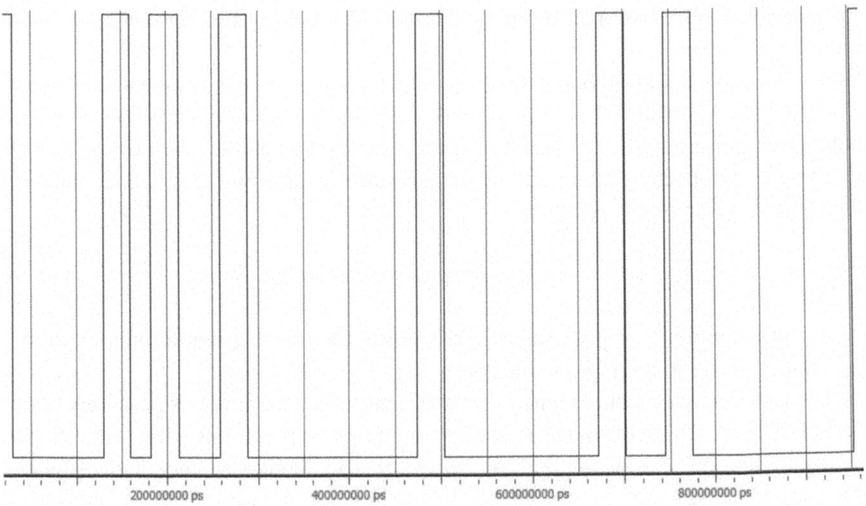

Fig. 15.15 Pulse train with random PRF according to LCG-generated PRF intervals for $L_p = 30\ \mu s$

these two random process really independent one has to use two independent LCGs. Figure 15.16. shows example of the random pulse train generated in this way.

The LCGs developed were successfully tested in the same development board, using VHDL code simulation in ModelSim environment.

The final version of the radar signal generator should be built using evaluation of different types of pulse generators and possibly include several modes chosen during this evaluation. It should be completely controlled by the system synchronizer according to operation algorithms of the MIMO Ground SF Noise SAR [14, 15].

15.5 Outdoor Trials of Noise SAR

The outdoor trials of the LNDES Ground Noise SAR [7, 8] in Ka-band have been carried out in the area of LNDES IRE NASU [18–20]. SAR image formation algorithm for such SAR was developed earlier and may be found in our papers [7, 8, 16, 17]. The main objective was to detect both still and moving objects with low radar cross-section in the realistic ground environment. With this aim, the Noise SAR was mounted in the laboratory room window approximately at 20 m height overlooking level ground with different types of terrain (grass, asphalt, land) and several naturally occurring obstacles like trees, bushes, small buildings, etc. Figure 15.17 shows a photo and simplified top view plan of the imaging area.

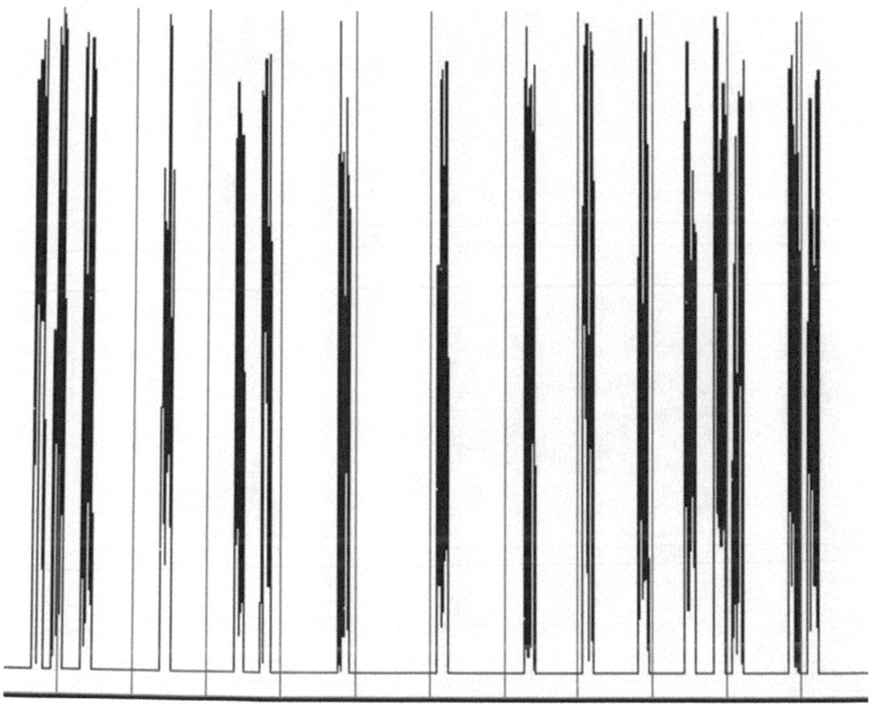

Fig. 15.16 Pulse train with LCG-generated random filling and PRF intervals for $L_p = 30\ \mu s$

We perform continuous imaging of the test field with a moving object: volunteer researcher walking across the field. Figure 15.18 shows the results of Noise SAR imaging. In the figure, the top-left quadrant contains the SAR image computed with the help of SAR imaging algorithm developed; the top-right quadrant contains the SAR image computed with the help of target detection MATLAB script; the bottom-left quadrant contains the photo of the scene and the bottom-right quadrant contains the detection plot produced after application of CFAR algorithm. The detected moving target in each image are highlighted by ellipses.

In this way we have demonstrated that real time SAR imaging (or quasi real time) in combination with efficient detection algorithm enables capability of wideband noise signals and gives a possibility of performing generation of 2D SAR images and producing moving target detection plots needed for the target tracking and identification.

Application of random waveforms gives such benefits as absence of range ambiguity and improving Noise SAR immunity against external electromagnetic interferences and providing high EMC performance [1, 2, 15] .

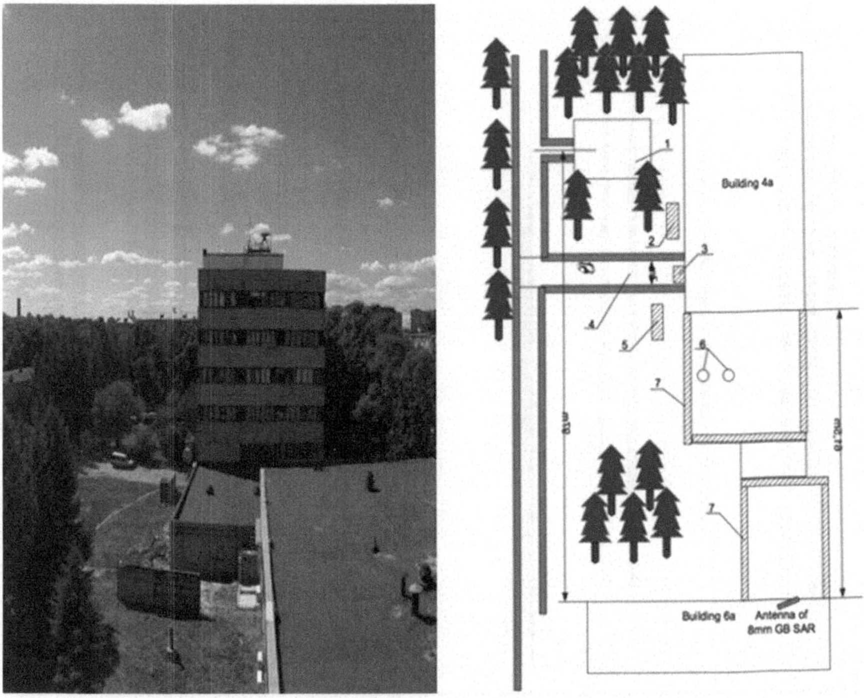

(a) (b)

Fig. 15.17 Test field view as it is seen from Noise SAR location: (**a**) Picture; (**b**) Simplified top view plan

15.6 Indoor Trials of Noise Interferometer SAR Network

We have carried out indoor interferometric experiment to validate suggested approach for enhancing of cross-range resolution of the suggested Ground Noise Radar network deployed in the LNDES room as shown in up-right part of the Fig. 15.19 [19, 20]. Schematic of the experimental setup is shown in its top-left part. Coherent interferometric imaging was performed with the help of bistatic Noise SAR equipment, but using both antennas with pattern synthesis [21, 22] as receive antennas. Step-like motion of the radiating slot and SAR focusing enabled to generate coherent 2D image (Fig. 15.19).

In the experiment, the correlation processing of the received signals has been performed in spectral domain using fast Fourier transform. The sidelobe level can be improved by applying weighting function to frequency spectrums of the both received signals. The image contains amplitude and phase information along coordinates (range and cross range directions). Extracting information on azimuth and range coordinates for every pixel one may readily derive ‘Difference of Arrival’ (DoA) parameters with cross-range resolution defined by Eq. (15.1). The series of

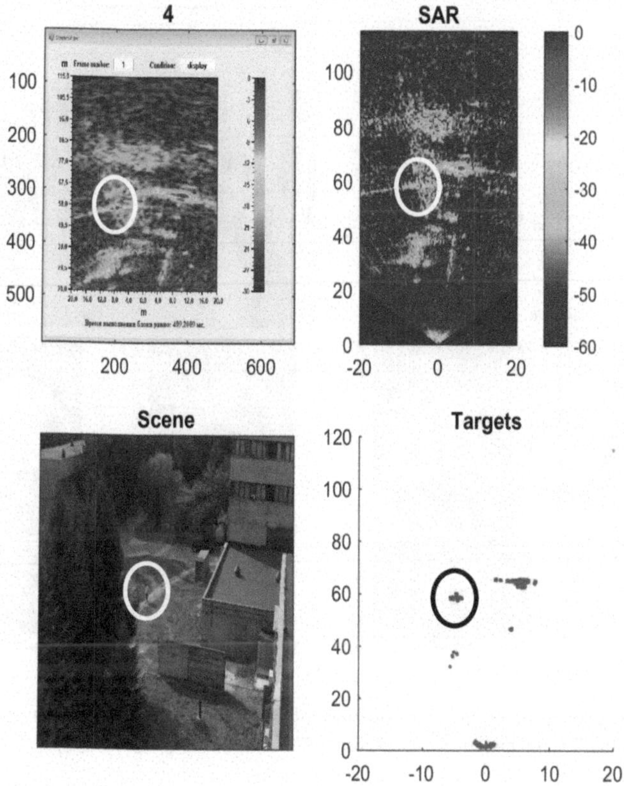

Fig. 15.18 Noise SAR imaging of the test field with a moving target: see details in the paper text above

2D images shown in the bottom of Fig. 15.19 demonstrates good range and cross-range (angular) resolutions. The possibility of coherent interferometric imaging with bistatic Ground Noise SAR network has been demonstrated for a small number of targets in the scene with the known distances being derived from the measurements in regular SAR mode. The cross-range resolution of such operational mode depends on the interferometer baseline and its mutual orientation with the targets.

15.7 Conclusion

The solution proposed within the NORMA project addresses a number of user needs that are not covered by currently employed systems in a scenario of increasing threats affecting border surveillance. The proposed noise radar/SAR network offers enhanced capabilities in border surveillance given the observation of the same area from different viewing angles, allowing the exploitation of multiple baselines and,

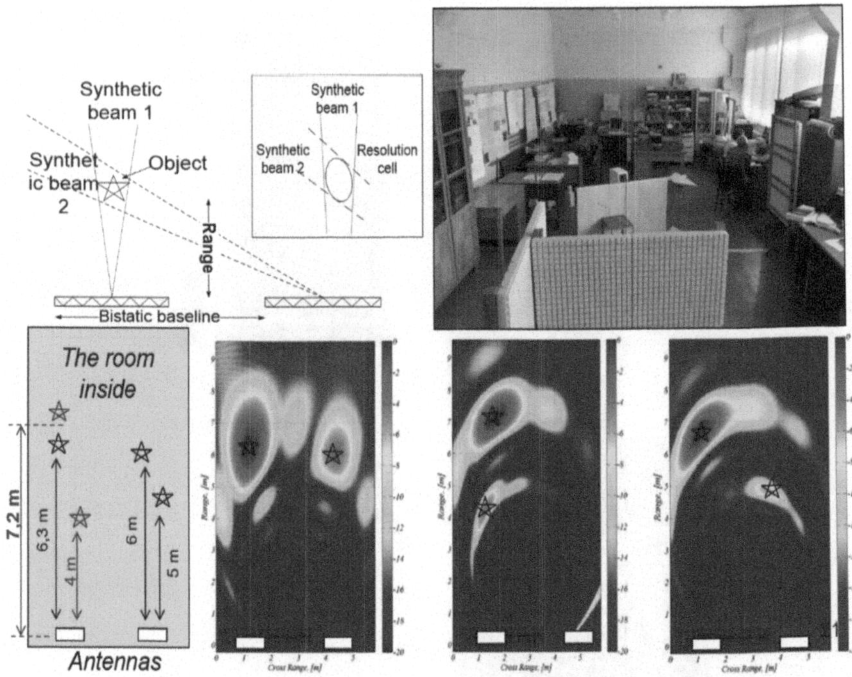

Fig. 15.19 Estimation of angular coordinates of two targets (radiating sources) for various positions with the help of bistatic Ground Noise SAR network

hence, increasing the capabilities of detecting low RCS targets. The redundancy given by the use of multiple radars guarantees a high degree of resilience offering a continuity of surveillance at the cost of slightly reduced performance. Each radar in the network transmits noise-like waveforms that offer LPI capabilities while maintaining detection and imaging capabilities that are essentials for target identification and, hence, proper and tailored response action of border authorities. In addition, along with the spatial diversity offered by the spatial division among the radars in the network, frequency diversity can be applied since only time synchronization is required in the proposed solution. Such degree of freedom can be exploited to enhance the acquired information about the target and, hence the recognition and identification, especially in case of low RCS targets such as small drones and small fast boats, typically used for illegal cross border activities.

Acknowledgment The authors would like to thank all the other team members for their current and future work within the SPS NATO Projects G5465: S. Lischi, D. Petri, R. Massini, E. Giusti, S. Gelli, A. Lupidi, D.Tatyanko, O. Zemlyany, P. Sushchenko, A. Shelekhov, L. Yurchenko, Yu. Shyian, and E. Mishchenko.

References

1. Lukin K (2001) Noise radar technology. Telecommun Radio Eng 55(12):8–16. (Original Lukin KA (1999) Shumovaya radarnaya tekhnologiya. Radiophysika i Electronica 4(3):105–111, in Russian)
2. Lukin K (2005) Noise radar technology: the principles and short overview. Appl Radio Electron 4(1):4–13
3. Kulpa K, Lukin K, Miceli W, Thayaparan T (2008) Editorial. IET Radar Sonar Navig 2(4):229–232
4. Kulpa K (2013) Signal processing in noise waveform radar. Artech House, Norwood, 258 p
5. Lukin K, Moreira J, Vyplavin P, Lukin S, Zemlyaniy O (2013) FPGA based software defined noise radar. Appl Radio Electron 12(1):89–94
6. Lukin KA, Zemlyaniy OV, Tatyanko DN, Lukin S, Pascazio V (2017) Noise radar design based on FPGA technology: on-board digital waveform generation and real-time correlation processing. 2017 18th International Radar Symposium (IRS), Prague, pp 1–7. https://doi.org/10.23919/IRS.2017.8008223
7. Tarchi D, Lukin K, Fortuny-Guasch J et al (2010) SAR imaging with noise radar. IEEE Trans Aerosp Electron Syst 46(3):1214–1225
8. Lukin KA et al (2008) Ka-band bistaic ground-based noise waveform SAR for short-range applications. IET Proc Radar Sonar Navig 2:233–243
9. Berizzi F, Martorella M, Giusti E (2016) Radar imaging for maritime observation. CRC Press, Boca Raton. https://doi.org/10.1201/9781315374253
10. Lukin K (1995) Noise radar with correlation receiver as the basis of car collision avoidance system. 25th European microwave conference, pp 506–507
11. Massaro D, Ardoino R, Grazzini M (2018) An efficient processing architecture for range profiling using noise radar technology. Aerospace 5:4. https://doi.org/10.3390/aerospace5010004
12. Alejos AV, Sánchez MG, Cuiñas I, Dawood M (2010) Wideband noise radar based in phase coded sequences. IntechOpen, Croatia. https://doi.org/10.5772/7173
13. Lukin S, Zemlyaniy O, Lukin K (2015) FPGA based random waveform generators for noise radars. In: Proceedings of the 16th international radar symposium, Dresden, Germany, 24–26 June 2015, vol 2, pp 777–782
14. Lukin K, Vyplavin P, Zemlyaniy O, Palamarchuk V (2011) L-band stepped frequency noise SAR on the basis of arbitrary waveform generator. 3rd international Asia-Pacific Conference on Synthetic Aperture Radar (APSAR), Seoul, South Korea, pp 1–4
15. Lukin K, Tatyanko D, Palamarchuck V, Vyplavin P, Lukin S, Zemlyaniy O, Shiyan YU, Mishchenko O, Yurchenko L, Sushchenko P (2019) Noise stepped-frequency radar for high resolution 2D&3D imaging. In: Capineri L, Turmus E (eds) Explosives detection: sensor, electronic systems and data processing (Publication under the NATO science for peace and security series – 2019). Series B: physics and biophysics. Springer Netherlands, pp XIV, 339. ISBN: 978-94-024-1729-6 (eBook); 978-94-024-1728-9 (Hardcover). https://doi.org/10.1007/978-94-024-1729-6. https://www.springer.com/gp/book/9789402417289
16. Lukin K, Vyplavin P, Kudryashev V, Palamarchuk V (2012) Noise waveform SAR for 2D and 3D imaging. 9th European Conference on Synthetic Aperture Radar (EUSAR-2012), pp 111–114
17. Lukin K, Vyplavin P, Palamarchuk V, Lukin S, Shelekhov A, Zaets N, Vasyuta K SAR tomography for short range applications using MIMO ground based noise waveform SAR. Appl Radio Electron 14(3):257–261
18. Lukin K, Mogyla A, Palamarchuk V, Vyplavin P, Kozhan E, Lukin S (2009) Monitoring of St. Sophia Cathedral Interior using Ka-band ground based noise waveform SAR. European Radar Conference (EURAD), pp 215–217
19. Lukin K, Kudriashov V, Vyplavin P, Palamarchuk V (2014) Coherent imaging in the range-azimuth plane using a bistatic radiometer based on antennas with beam synthesizing. IEEE Aerosp Electron Syst Mag 29(7):16–22

20. Lukin K, Kudriashov V, Vyplavin P, Palamarchuk V, Lukin S Coherent radiometric imaging using antennas with beam synthesizing. Int J Microw Wirel Technol 7(3–4):453–458
21. Lukin K (2003) A novel approach to scanning antennas design. 4th international conference on antenna theory and techniques IEEE, 9 September 2003, vol 1, pp 288–289
22. Lukin K (2010) Scanning synthesized radiation pattern antennas. Radioelectron Commun Syst (Allerton Press, Inc) 53(4):219–224. (Original Lukin KA (2010) Published in Izv Vyssh Uchebn Zaved Radioelektronica 53(4):58–64. In Russian)

Chapter 16
Compact Eye-Safe LIDAR Source for Airborne Laser Scanning – The CALIBER Project

Nadia G. Boetti, Amiel Ishaaya, Mircea Guina, Davide Janner, Daniel Milanese, Diego Pugliese, Antti Penttinen, Antti Härkönen, Omri Moschovitz, Yair Alon, and Federico Leone

Abstract A high power, ultra-compact, lightweight, and low cost eye-safe laser source for a Light Detection and Ranging (LIDAR) system is currently under development in the framework of the NATO Science for Peace and Security project "CALIBER", with the aim to be installed on small Unmanned Aerial Vehicles (UAVs) for surveillance of borders and sensitive areas, with saving in manpower and equipment.

Keywords LIDAR · Phosphate glass · Microchip laser · Glass amplifier · Optical fiber

16.1 Introduction

The security and protection of borders, sensitive infrastructures and strategic sites, such as overseas gas and oil extraction platforms, maritime infrastructures and power plants, is of paramount importance. This capability is directly related

N. G. Boetti (✉) · F. Leone
LINKS Foundation – Leading Innovation and Knowledge for Society, Torino, Italy
e-mail: nadia.boetti@linksfoundation.com

A. Ishaaya · O. Moschovitz · Y. Alon
School of Electrical and Computer Engineering, Ben-Gurion University of Negev, Beer-Sheva, Israel

M. Guina · A. Penttinen · A. Härkönen
Optoelectronics Research Centre, Tampere University, Tampere, Finland

D. Janner · D. Pugliese
Politecnico di Torino – DISAT and RU INSTM, Torino, Italy

D. Milanese
Università di Parma – DIA and RU INSTM, Parma, Italy

C. Palestini (ed.), *Advanced Technologies for Security Applications*, NATO Science for Peace and Security Series B: Physics and Biophysics,
https://doi.org/10.1007/978-94-024-2021-0_16

to current and future security problems of NATO countries, allowing for early hazard detection, protection of strategic infrastructures, and the general securing of Europe's borders (land and sea) (https://www.nato.int/cps/en/natohq/85291.htm).

Current monitoring is mainly done by human patrols securing the perimeter or border, in combination with Charge-Coupled Device (CCD) cameras and various night vision sensors positioned statically in overlooking locations, or on UAVs when a more complete overview of the installments and of the surroundings is needed. The current monitoring methods are limited in terms of continuous coverage and resolution, pose threats to the security forces, and involve high cost (work force and equipment).

Compared to other sensors, LIDAR systems have proved to be more effective and precise for remote sensing applications, ranging from the detection of obstacles for automobiles and aircrafts, to the topographic investigation of landscapes, and to the detection of wind turbulence and hazardous gases or pollutants [4, 12, 13, 16]. Their significant advantage over standard CCD or night vision cameras consists in the ability to measure distances with high accuracy and build a three-dimensional (3D) map of the scanned area.

A LIDAR system includes a laser source or transmitter, optics for precise laser scanning, a sensitive photodetector or receiver, and synchronization and data processing electronics (https://esto.nasa.gov/files/Lidar_TechStrategy_%202016.pdf).

The light emitter is a key element of the system and contributes to the overall system performance. The quality of the sensing strongly depends on the type of source employed, in terms of wavelength, pulse width, average and peak power, which altogether define the precision and reliability of the overall system.

However, high-power eye-safe LIDAR sources that have been used for decades are bulky, heavy, and expensive, which does not make them suitable for installation on small/medium UAVs. For example, the early eye-safe time-of-flight (TOF) LIDAR systems exploited low repetition rate, ns-pulsed, high energy (10's of mJ's) solid state lasers, typically based on Er:glass or Nd:YAG, with an Optical Parametric Oscillator (OPO) wavelength converter operating at the wavelength of 1.5 um [18].

With the recent development of 1.5 μm Er:glass fiber lasers, LIDAR sources became smaller and more lightweight. Here, instead of low repetition rate and high energy pulses, high repetition rates (10–100 kHz) and low energy pulses are used. These lasers are typically built with multiple amplification stages, resulting in high pricing, and can be packed in relatively small size modules.

A flexible and advantageous approach to realize a high-power LIDAR source is the Master Oscillator Power Amplifier (MOPA) configuration. A seed laser with high spectral quality is followed by a single- or multi-stage power amplifier that boosts the output power to attain the needed energy.

The NATO Science for Peace and Security project "CALIBER" (CompAct eye-safe LIdar source for airBorne lasER scanning) aims to develop a compact, lightweight and low-cost version of a LIDAR source, by integrating a SEmiconductor Saturable Absorber Mirror (SESAM) Q-switched microchip seed laser and a new Er:Yb power amplifier in a compact MOPA configuration, which combines high repetition rate and high peak power at the "eye-safe" wavelength of 1535 nm.

In this work we report on the development until now of the key components of the CALIBER laser source, i.e. the seed laser and the power amplifier, as well as the preliminary work towards their integration into a LIDAR source. This source will then be inserted into a LIDAR laboratory system for a range finding proof of concept.

16.2 The Microchip Seed Laser

Pulsed Q-switched microchip lasers [5, 6, 20] are highly compact and rugged laser sources, capable of producing energetic pulses with multi-kW peak power. Q-switched microchip lasers consist of a thin piece of laser gain material (crystal or glass), which is optically contacted to a saturable absorber that is used for initiating the pulsed operation. The saturable absorber can be made of doped glass, crystal, or semiconductor material. The gain element and the saturable absorber will form a monolithic structure, with plane-plane cavity geometry. The necessary laser mirrors are deposited directly on the cavity surfaces, making the laser free of cavity alignment. A laser operation in plane-plane type resonator is typically stabilized by a thermal lens. The laser output parameters, such as pulse width, pulse energy, pulse repetition rate, and spectrum are dependent on the combination of the gain material, saturable absorber, mode diameter, and the laser coatings.

Within the CALIBER project, Erbium-doped phosphate glass gain material ($\lambda = 1535$ nm), 976 nm diode pumping and SESAMs were used. The SESAMs consist of a Distributed Bragg Reflector (DBR) and a Quantum Well (QW)-based absorber section, fabricated on a InP substrate in single step by solid-source Molecular Beam Epitaxy (MBE). In comparison to doped glass/crystal-based absorbers, the SESAMs exhibit certain unique properties: (1) the absorber layers are very thin (only a few micrometers thick), which helps to shorten the laser cavity and consequently the laser pulses; (2) the optical properties of the SESAMs can be tailored by the composition and design of the quantum wells, which makes this approach very flexible and allows customization. The semiconductor fabrication methods also allow high volume and low unit-cost production when necessary.

In CALIBER, 20–100 kHz pulse repetition rate and 1–2 ns pulse duration with 1 kW peak power were targeted. This combination of parameters was chosen to provide sufficient speed of measurements and suitable measurement accuracy and distance. In total, over 100 individual samples with different SESAMs, gain elements and coatings were tested. The laser output parameters obtained for different sample combinations are summarized in Table 16.1. A typical laser spectrum and light output/repetition rate curves are presented in Fig. 16.1a, b, respectively.

The obtained results demonstrate that using SESAM technology it is possible to tailor the output parameters in a wide range, by customizing the SESAM and coating designs. A lifetime test was also carried out at 30 kHz, 2.5 ns, and 500 W peak power, showing no signs of performance drop in 1800 h of operation. Based on these observations we believe that SESAM-based Q-switched microchip lasers

Table 16.1 Laser output parameters tested in CALIBER

	Repetition rate (kHz)	Peak power (kW)	Average power (mW)	Pulse width (ns)
Target parameters	20–100	1.0	10	1.0
Measured results	20	1.6	55	1.6
	50	1.0	63	1.2
	100	0.3	73	2.6

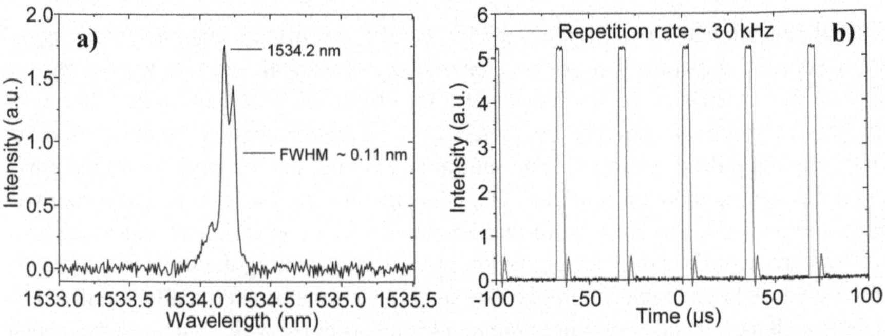

Fig. 16.1 (**a**) Typical laser spectrum and (**b**) light output/repetition rate curves. Repetition rate was measured with a slow photodiode, PDA50B-EC, and therefore not showing the correct pulse width

are well suited for drone/UAV-based LIDAR systems requiring higher performance than common diode-based systems.

16.3 The Waveguide Amplifier

Following the CALIBER project requests of a high degree of compactness while retaining high performance and low cost, the choice for the optical amplifier fell on an Yb/Er co-doped phosphate glass-based waveguide. The use of a glass material shows the advantage, over the other types of materials, of combining: chemical stability, excellent homogeneity, good thermo-mechanical properties and a viscosity–temperature relationship that allows for shaping the waveguide into the form of optical fibers or rods. Moreover, glass displays a wide flexibility of the chemical composition and doping, allowing the preparation of multi-component glasses with properties that can be customized to meet the needs of the different applications [19].

Among the different oxide glass systems, multi-component phosphate glasses are recognized to be an ideal host material for engineering the amplification stage of a pulsed MOPA as they can be doped with a large amount of Rare-Earth (RE) ions (up to 10^{21} ions/cm^3) without clustering thanks to the presence of phosphorus, which introduces non-bridging oxygens in the structure [1, 10]. This enables the realization of an active medium with high optical gain in short length (> 5 dB/cm) and thus mitigating nonlinear optical effects. Moreover, this glass system possesses a large glass formation region, good thermo-mechanical and chemical properties, high emission cross-sections, low nonlinear refractive index, and no evidence of photodarkening even at high population inversion [2, 9, 14].

A highly stable and robust glass host, suitable for the incorporation of a high amount of RE ions and for fiber drawing, was designed and developed for this research. The glass samples were synthesized by conventional melt-quenching method using chemicals with high purity level (99+%).

Three different glasses, named for short YE1 ÷ YE3, were obtained by doping the host material with a fixed level of Er_2O_3 (0.75 mol%) and an increasing level of Yb_2O_3 (ranging from 1.50 to 4.50 mol%) (see Table 16.2).

The fabricated glasses were ground or cut and optically polished to 1 mm-thick samples depending on the characterization technique. First of all, they were thermo-mechanically characterized, then the glasses underwent optical and spectroscopic characterizations such as refractive index measurement, Fourier-Transform Infrared (FTIR) spectroscopy and RE emission spectroscopy resolved in time and frequency. In parallel, the study of suitable cladding glass compositions for the core material was carried out and led to the synthesis of five glass compositions with tunable refractive index.

Single-material rod waveguides were then fabricated starting from the developed phosphate glasses YE1, YE2 and YE3, and using a custom induction heated optical fiber drawing tower. The rods feature a diameter of about 1 mm (see Fig. 16.2a).

A multi-mode optical fiber was also realized by preform drawing, with the fiber preform being obtained by the rod-in-tube technique, using the glass composition YE1 for the core and one of the developed cladding glass compositions for the cladding, with a targeted Numerical Aperture (NA) of 0.11. The optical fiber was fabricated in three different dimensions (core/cladding diameters): 50/125, 78/195, and 100/250 μm.

The quality and morphology of the fabricated optical waveguides were inspected by means of an optical microscope. The cross-section of the multi-mode phosphate glass fiber is shown in Fig. 16.2b.

In order to assess the guiding properties of the fiber, a set of near-field images of the fiber cross-section was measured on a 120 cm-long fiber piece, at the wavelength of 1300 nm, using an end-face coupled fiber pigtailed laser diode source. The light was found well confined in the core.

Table 16.2 Yb/Er co-doped phosphate glass compositions constituting the core of the optical amplifier

Glass name	Er^{3+} [10^{20} ions/cm^3]	Yb^{3+} [10^{20} ions/cm^3]
YE1	1.93	3.86
YE2	1.92	7.69
YE3	1.92	11.50

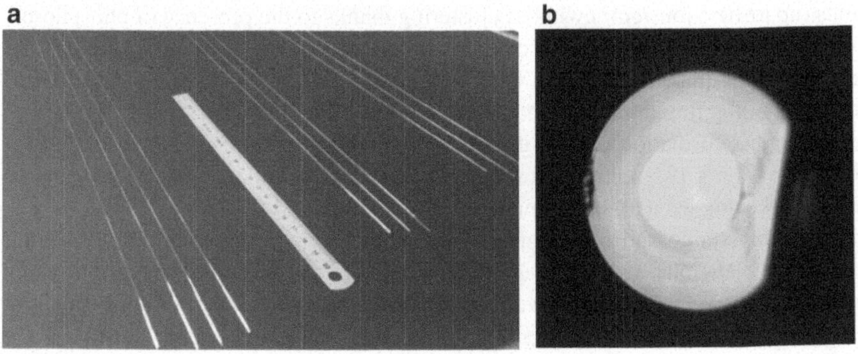

Fig. 16.2 Optical amplifier waveguides: (**a**) single-material rods, (**b**) optical microscopy image of the multi-mode fiber

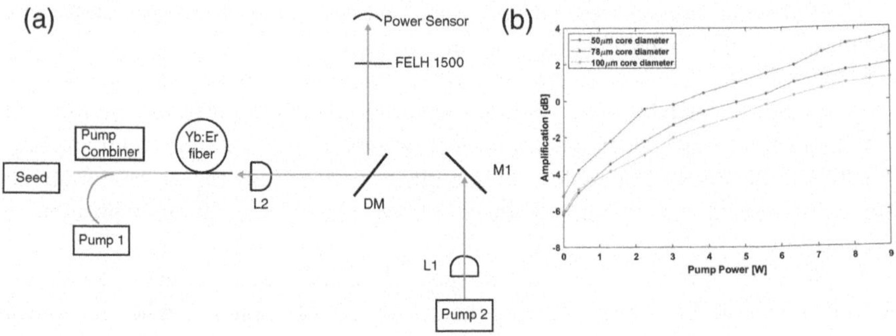

Fig. 16.3 (**a**) Setup used for the Er:Yb phosphate fiber amplification experiments. L1 and L2 are aspheric lens, DM is a dichroic mirror which transmits the 976 nm pump and reflects the 1548 nm of the seed, and FELH 1500 is a ThorLabs longpass filter. (**b**) Seed output power as a function of the total power of pump laser diodes

Fiber losses were measured by the cut-back technique using a fiber length of about 120 cm with a single-mode fiber pigtailed laser diode source at 1300 nm. The attenuation value, as evaluated through a linear least square fitting of the experimental data, was 3.6 dB/m. This value is mainly due to absorption and scattering effects.

The waveguides are currently under testing for amplification using a setup schematically shown in Fig. 16.3a. The initial seed employed was a Continuous Wave (CW) single-mode laser diode (center wavelength of 1548 nm) with an input of 36.5 mW. This seed was connected with a multi-mode pump laser (center wavelength of 912 nm) by means of a pump combiner. The output of the pump combiner (9/125 μm) was butt-coupled to the Er:Yb phosphate glass fiber with a 3-axis precision stage. A counter propagating second pump laser (976 nm with a core of 105 μm) was collimated with a 20 mm aspheric lens (L1), then propagated through a dichroic mirror and focused into the Er:Yb fiber cladding with a second

20 mm aspheric lens (L2). The amplified seed was reflected by the dichroic mirror and through an optical filter into the power meter. The first pump laser (pump 1 in Fig. 16.3a) was limited to 5.5 W, while the second pump laser (pump 2 in Fig. 16.3a) was limited to 3.5 W.

The initial experiments, performed with ~10 cm-long Er:Yb phosphate fibers, showed no amplification. During the experiments, it was observed that the seed amplification increased as the fiber core decreased. Very low amplification was obtained with the 100 μm fiber, while the amplification increased while using the 50 μm fiber. In addition, as the fiber length was shortened down to 3–4 cm, the amplification also enhanced.

The best result achieved up to now is reported in Fig. 16.3b. It was obtained with a 3.5 cm-long section of the 50 μm fiber. This result is equivalent to an amplification of 3.62 dB. It is worthwhile noting that the coupling and propagation losses of the seed in the active fiber were not taken into account, hence the actual amplification is underestimated.

The above reported are very preliminary amplification results with a low power CW seed. The enhancement of the amplification while decreasing the core diameter and the fiber length can be ascribed to the Amplified Spontaneous Emission (ASE) phenomenon. When the seed power is too low, the ASE in the amplifier depletes the gain, and consequently the amplification is low. This is also correlated to the core diameter. With larger diameters, the seed power has to be higher in order to saturate the amplifier and avoid the ASE effects. In addition, it is important to note that a large part of the pump power was not absorbed in the fiber because of its short length (cladding pumping).

Currently, fibers with smaller diameters are being fabricated and will be tested. Once satisfactory amplification will be obtained with these fibers, a pulsed microchip seed source will be employed and the amplification will be measured.

16.4 The LIDAR System

A TOF LIDAR is a system which detects the time a light pulse takes to reach a target and propagate back to the receiver, allowing for the calculation of the distance to the target.

The main goal of the distance measurement is to create a 3D map of the surface. Scanning of an area is achieved by a rotating mirror or a prism, which is synchronized with the laser pulses, allowing for the generation of a 3D map.

Most LIDAR systems exploit a laser source due to its ability to focus high peak powers at relatively long distances. Typically, the laser is pulsed with a repetition rate that varies from few Hz to MHz [3, 11, 15]. While using a high repetition rate in the order of the kHz, the pulse energy is relatively low (~1 nJ), but the peak power is still high [15, 17]. The high peak power enabled operation at long distances.

In order to validate the range finding capability of the developed source, a laboratory setup was built with the aim to be used once the optimal parameters

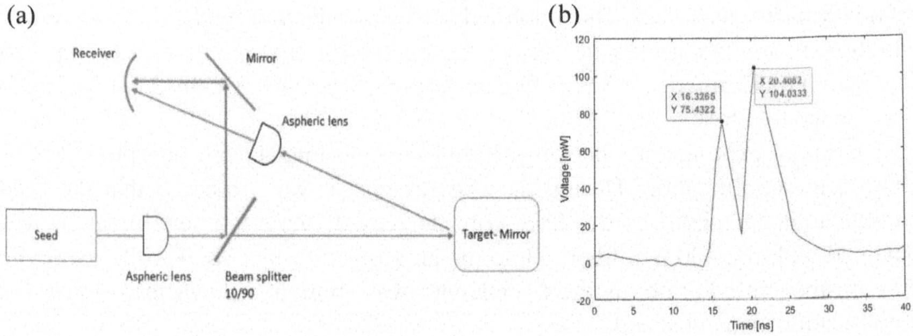

Fig. 16.4 (**a**) Laboratory setup used for distance measurements. (**b**) Distance measurement performed with the home-made laboratory setup

of the amplifier will be found (see Fig. 16.4a). The setup consists of a microchip Q-switched seed prototype, with a pulse width of ~2 ns, which is collimated with an aspheric lens and divided into two beams with a 10–90% beam splitter. The beam with 10% of the power is reflected with a mirror into the receiver, the second beam which is more powerful is sent to the target (a mirror up to now), and the reflected power is collected with a 2-inch diameter aspheric lens into the receiver.

Figure 16.4b represents the two pulses which were detected with the receiver. One can observe that the time difference between the two peaks is 4.08 ns, which is equivalent to 1.23 m, thus indicating that the distance between the beam splitter and the mirror was 61.2 cm. The actual distance was measured to be ~60 cm, corresponding to an error of 1.2 cm. This small error might also be associated to the length measurement of the two optical paths. As clearly visible from Fig. 16.4b, the difference between the two peaks can be determined with sub-ns resolution.

References

1. Boetti NG, Pugliese D, Ceci-Ginistrelli E, Lousteau J, Janner D, Milanese D (2017) Highly doped phosphate glass fibers for compact lasers and amplifiers: a review. Appl Sci 7:1295
2. Boetti NG, Scarpignato GC, Lousteau J, Pugliese D, Bastard L, Broquin J-E, Milanese D (2015) High concentration Yb-Er co-doped phosphate glass for optical fiber amplification. J Opt 17:065705
3. Budei BC, St-Onge B, Hopkinson C, Audet F-A (2018) Identifying the genus or species of individual trees using a three-wavelength airborne LIDAR system. Remote Sens Environ 204:632–647
4. Chen M, Rudd WJ, Hansell J, Pachowicz D, Litvinovitch S, Burns P, Sawruk NW (2019) Er:YAG methane LIDAR laser technology. In: Proceedings of the SPIE 11005, laser radar technology and applications XXIV, 110050Q
5. Gorbachenya KN, Kisel VE, Yasukevich AS, Maltsev VV, Leonyuk NI, Kuleshov NV (2015) CW and Q-switched diode-pumped laser operation of Er,Yb:$GdAl_3(BO_3)_4$ crystal. Advanced Solid State Lasers (ASSL) 2015, Berlin, Germany, paper ATu1A.5

6. Häring R, Paschotta R, Fluck R, Gini E, Melchior H, Keller U (2001) Passively Q-switched microchip laser at 1.5 μm. J Opt Soc Am B: Opt Phys 18:1805–1812
7. https://esto.nasa.gov/files/Lidar_TechStrategy_%202016.pdf
8. https://www.nato.int/cps/en/natohq/85291.htm
9. Jiang S, Mendes SB, Hu Y, Nunzi-Conti G, Chavez-Pirson A, Kaneda Y, Luo T, Chen Q, Hocde S, Nguyen DT, Wright EM, Wang J, Tian W, Nikolajsen T, Peyghambarian N (2003) Compact multimode pumped erbium-doped phosphate fiber amplifiers. Opt Eng 42:2817–2820
10. Lee Y-W, Sinha S, Digonnet MJF, Byer RL, Jiang S (2006) 20 W single-mode Yb^{3+}-doped phosphate fiber laser. Opt Lett 31:3255–3257
11. McCarthy A, Ren X, Della Frera A, Gemmell NR, Krichel NJ, Scarcella C, Ruggeri A, Tosi A, Buller GS (2013) Kilometer-range depth imaging at 1550 nm wavelength using an InGaAs/InP single-photon avalanche diode detector. Opt Express 21:22098–22113
12. Overton G (2017) Lasers for LIDAR: application parameters dictate laser source selection in LIDAR systems, LaserFocusWorld
13. Pulikkaseril C, Lam S (2019) Laser eyes for driverless cars: the road to automotive LIDAR. Optical Fiber Communication Conference (OFC) 2019, San Diego, USA, paper Tu3D.2
14. Qiu T, Li L, Schülzgen A, Temyanko VL, Luo T, Jiang S, Mafi A, Moloney JV, Peyghambarian N (2004) Generation of 9.3-W multimode and 4-W single-mode output from 7-cm short fiber lasers. IEEE Photon Technol Lett 16:2592–2594
15. Ren M, Gu X, Liang Y, Kong W, Wu E, Wu G, Zeng H (2011) Laser ranging at 1550 nm with 1-GHz sine-wave gated InGaAs/InP APD single-photon detector. Opt Express 19:13497–13502
16. Schmitt NP (2017) Research results, lessons learned and future perspective of forward-looking LIDAR for aircraft. 2017 Conference on Lasers and Electro-Optics (CLEO), San Jose, USA, pp 1–2
17. Spinhirne JD (1993) Micro pulse LIDAR. IEEE Trans Geosci Remote Sens 31:48–55
18. Spinhirne JD, Rall JAR, Scott VS (1995) Compact eye safe LIDAR systems. Rev Laser Eng 23:112–118
19. Yamane M, Asahara Y (2004) Glasses for photonics. Cambridge University Press, Cambridge
20. Zayhowski JJ (2013) Microchip lasers. In: Denker B, Shklovsky E (eds) Handbook of solid-state lasers: materials, systems and applications. Woodhead Publishing, Cambridge, pp 359–402

Chapter 17
Microelectronic 3D Imaging and Neuromorphic Recognition for Autonomous UAVs

Franco Zappa, Federica Villa, Rudi Lussana, Dennis Delic, Man Ching Joyce Mau, Jean-Michel Redouté, Simon Kennedy, Daniel Morrison, Mehmet Yuce, Tuncay Alan, Tara Hamilton, and Saeed Afshar

Abstract The article addresses the development of highly sensitive, low-light and efficient, miniature single-photon sensor technology based on Single Photon Avalanche Diode (SPAD) arrays, its integration on a Flash Light Detection and Ranging (LiDAR) system mounted on a custom built multi-rotor Unmanned Aerial System (UAS) platform, for the collection of real time imagery and performance of neuromorphic processing for accurate target detection and classification.

F. Zappa · F. Villa · R. Lussana
Politecnico di Milano, Milan, Italy
e-mail: franco.zappa@polimi.it

D. Delic · M. C. Joyce Mau
Defence Science and Technology Group (DSTG), Department of Defence, Edinburgh, SA, Australia
e-mail: Dennis.Delic@dst.defence.gov.au

J.-M. Redouté (✉)
University of Liège, Liège, Belgium
e-mail: jean-michel.redoute@uliege.be

S. Kennedy · D. Morrison · M. Yuce · T. Alan
Monash University, Melbourne, VIC, Australia

T. Hamilton
Macquarie University, Sydney, NSW, Australia
e-mail: tara.hamilton@mq.edu.au

S. Afshar
University of Western Sydney, Sydney, NSW, Australia

C. Palestini (ed.), *Advanced Technologies for Security Applications*, NATO Science for Peace and Security Series B: Physics and Biophysics,
https://doi.org/10.1007/978-94-024-2021-0_17

17.1 Introduction

An international team of researchers have been developing highly sensitive, low-light and efficient, miniature single-photon sensor technology based on Single Photon Avalanche Diode (SPAD) arrays. A key motivation was to use Silicon CMOS-based processes and advanced 3D-IC manufacturing technologies to miniaturise arrays and digital circuits to realize affordable high definition imaging microchip sensors. Imaging cameras using smart photon sensor SPAD microchips are integral to active electro-optic systems such as 3-D Flash LiDAR (Light Detection and Ranging) for target detection and identification as well as tactical applications requiring imaging in very low light conditions. When coupled with sophisticated machine learning algorithms the work demonstrated accurate detection and classification of land-based targets from a low cost Unmanned Aerial System (UAS).

3-D Flash LiDAR systems, also known as 3-D Time of Flight (ToF) cameras which use 'SPAD array' sensor technology have some advantages over existing LiDAR scanning methods. They have no moving mechanical parts and scanning optics; hence they acquire a 3D depth-resolved image of a scene instantaneously allowing faster image reconstruction, which is especially useful when targets are moving or when large areas need to be surveyed quickly. They also offer an improved SWaP (Size Weight & Power) footprint over scanning systems, which means they can fit easily on power starved and mobile platforms such as UASs.

The project was able to develop state of the art SPAD sensors, successfully integrate and fly a low SWaP Flash LiDAR system on a custom built multi-rotor UAS platform, collect real time imagery and perform neuromorphic processing for accurate target detection and classification. The technology holds the potential to be developed further, both in the ability to image at greater ranges and its application in more challenging environments where targets are camouflaged and/or hidden by obscurants and clutter. The work has effectively showcased how such technology can be carried by low cost UAS's for rapid environmental assessment of threats in low light conditions, thus providing enhanced situational awareness and decision superiority for surveillance and reconnaissance applications.

The four main project Streams are described in the subsequent sections.

17.2 Stream 1: High Density SPAD Array Design and Implementation

The main goal in this Stream was the design of high-end arrays of SPAD detectors, optimized in terms of number of pixels, detection performance, such as noise and sensitivity, and 3D ranging precision. As a first prototyping activity within this

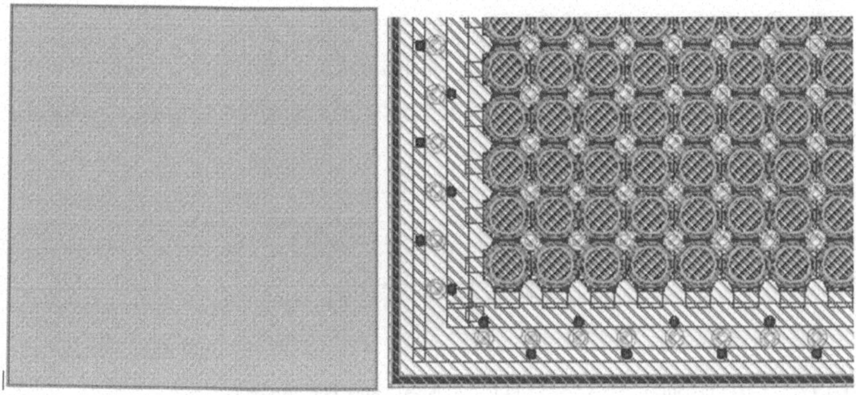

Fig. 17.1 Entire layout (left) of a 256 × 256 SPAD sensor array chip, with overall dimensions of 11 × 11 mm, and (right) zoom in of a corner

project, a set of individual SPAD test structures and miniarrays were designed in the high-voltage CMOS 0.35 μm Fraunhofer IMS technology, to characterise the fabrication process: these designs were based on previously reported prototypes [7]. Next, four different versions of 256 × 256 sensor arrays were fabricated by Fraunhofer IMS. The entire layout is shown in Fig. 17.1: the overall dimensions of the IC are 11 × 11 mm, where each pixel has a size of 40 × 40 μm. All array pixels have 40 μm pitch and comprise a SPAD and space for a TSV (through-silicon-via) connected with the anode, whereas the cathode voltage is common among all pixels. Four 8-inches wafers were fabricated by Fraunhofer IMS. These wafers with the different "SPAD chips" and the wafers of the "SPAD front-end chip" described in next section, were sent to Fraunhofer IZM for 3D-IC assembly using TSVs and bonding.

As a monolithic alternative to the 256 × 256 pixel 3D-IC ensemble imager pioneered in this project, Prof. Zappa's research team also designed a 32 × 32 SPAD ToF imager, which is an improved version of the SPAD array published in [8]. Each of the 1024 pixels contains a SPAD detector, an analog front end (for avalanche sensing, detector quenching and digital pulse shaping), an 8 bit digital counter (for photon-counting), and a 12 bit Time-to-Digital Converter (TDC) covering a full-scale range of 1.2 μs and a least significant bit corresponding to 310 ps. Figure 17.2 shows the chip, with overall dimensions equal to 8 × 8 mm. Figure 17.3 shows the resulting ToF camera, based on the 32 × 32 SPAD imager shown in Fig. 17.2. The weight of the body is about 300 grams, and the dimensions are 6 cm width, 8.5 cm length, and 6 cm height: the camera has been used for LiDAR measurements with a full-scale range up to 200 m and a single shoot resolution of about 5 cm.

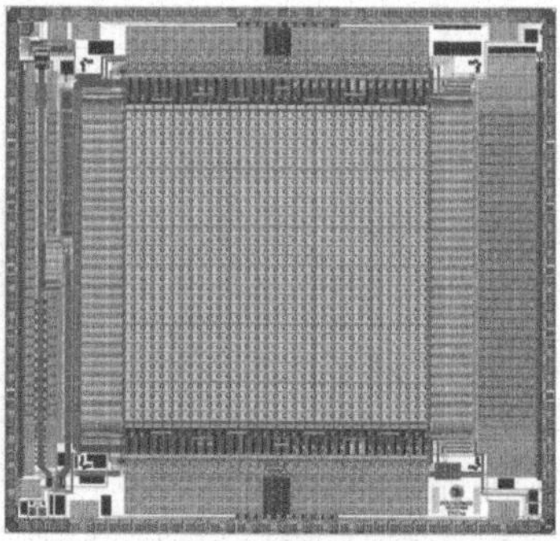

Fig. 17.2 CAD layout of the 32 × 32 POLIMI SPAD and TDC imager

Fig. 17.3 ToF camera designed by POLIMI using the 32 × 32 SPAD and TDC imager

17.3 Stream 2: High Speed Precision SPAD Front-End Design with On-Chip Photon Correlation

Stream 2 dealt with the development of integrated SPAD read-outs and front-ends with a high accuracy and resolution, and reduced chip size. As a first challenge, the "SPAD front-end chip" was designed, forming the read-out connected to the "SPAD chip" in a 3D-IC assembly process (described in the previous section). This chip contains all the necessary electronic circuits for the SPAD quenching, time measurement, photon counting and read-out of the 256 × 256 pixel array: each pixel contains a SPAD frontend, a counter and an interpolator, as well as input signal and clock buffering. Each SPAD front-end pixel fits on a 40 × 40 μm pitch (matched to the "SPAD chip"). In ToF mode and at the beginning of a frame, the chip is triggered so that all pixel counters begin counting from the zero at the global clock rate. Once the SPAD detects a photon event the front-end latches on the first event. The output of the SPAD front-end is synchronized to the global clock which stops the counters, giving a final count value equal to the number of clock cycles between the start of the frame and the first SPAD event. To improve the timing resolution beyond one clock cycle, the interpolator measures the synchronization time from the SPAD event to the counter stopping, giving a negative modifier to the measured time [5]. In photon counting mode, the pixel counter is reset at the start of the frame, but the SPAD front-end is set to self-rearm and outputs a pulse for every detected photon. The dedicated counter, requiring a minimal chip footprint then counts up one value after every SPAD event [4].

A separate 64 × 64 SPAD imager, containing the SPADs and the electronic front-end, was designed using the Silterra 130 nm HV process, measured and implemented into a standalone camera (Fig. 17.4). The camera was measured on the DSTG laser range using an external laser, in collaboration with the DSTG team. The laser range is a 1.5 km field at DSTG Edinburgh (Adelaide) allowing testing of the camera over long distances with a high level of background light noise. The target setup and resulting image from the 250 m target are shown in Figs. 17.5 and 17.6.

Fig. 17.4 (left) PCB interfacing with the Monash02 chips (64 by 64 SPAD array). The packaged 64 × 64 SPAD imager prototype is in the middle of the picture and covered with a glass lid. – (right) Photo of the camera using the 64 × 64 imager chip. A ring of illuminating LEDs placed on the front-panel can be used for short-range imaging

Fig. 17.5 Target setup at the 250 m target

Fig. 17.6 Resulting image of the 250 m target

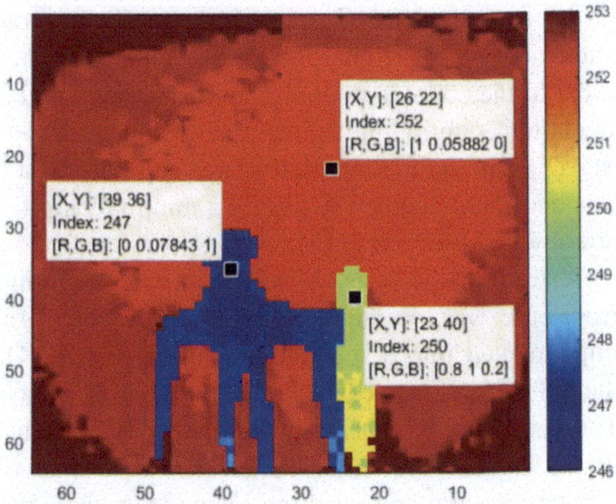

17.4 Stream 3: Neuromorphic Architectures of Event-Based SPAD Arrays and Neuromorphic, Event-Based Processing Algorithms

This Stream focused on developing algorithms and architectures that enable the conversion of SPAD array data into event-based signals which could subsequently be processed by event-based algorithms. These event-based algorithms are inherently neuromorphic, and are thus, low-power, real-time, and, most importantly for this medium, efficient, particularly in data generation, storage, and communications. The Recognition and Tracking System (RTS) algorithms that were developed in this project were based on previous work of the authors [1, 6]. Figure 17.7 illustrates some of the investigated scenarios.

These algorithms were then implemented in the design of a neuromorphic, event-based SPAD – neuroSPAD. In this new implementation, instead of encoding, storing and transferring the ToF data off chip for processing, the calculation of the ToF from the laser pulse is abandoned entirely in favor of a neuromorphic processor that operates directly in the time domain and on the inter-spike intervals within local regions of the SPAD array (Fig. 17.8). The proposed approach motivates the development and hardware implementation of event-based feature extraction algorithms and circuits that generate local sparse event-based representations from the non-sparse event based data and in this way drastically reduces the I/O requirements of the overall system [2].

The output of the neuroSPAD is, as stated above, significantly different to the interfaces used on other SPAD designs. An event-based processor has been developed to take the sparse, event-based output of the neuroSPAD and perform hardware efficient and low-power recognition on the resulting 3D image.

Figure 17.9 shows the functionality of the event-based processor. In Fig. 17.9a we see the raw SPAD output: using the neuroSPAD architecture a small (local) section of the SPAD image is represented with local binary events (Fig. 17.9b).

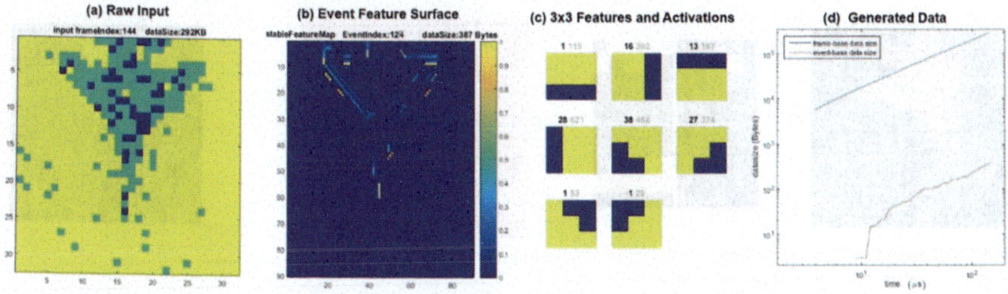

Fig. 17.7 Example of data generated from planes in motion in front of the SPAD camera: (**a**) raw input – (**b**) "surface" of the events (returning photons) – (**c**) subsequent 3×3 features extracted by the RTS algorithm – (**d**) amount of data generated by the developed event-based methodology (red curve) vs. traditional frame-based approaches (blue curve). It can be seen that the developed event-based method generates several orders of magnitude less data than traditional methods making it much more attractive for full integration as well as implementation in an autonomous system

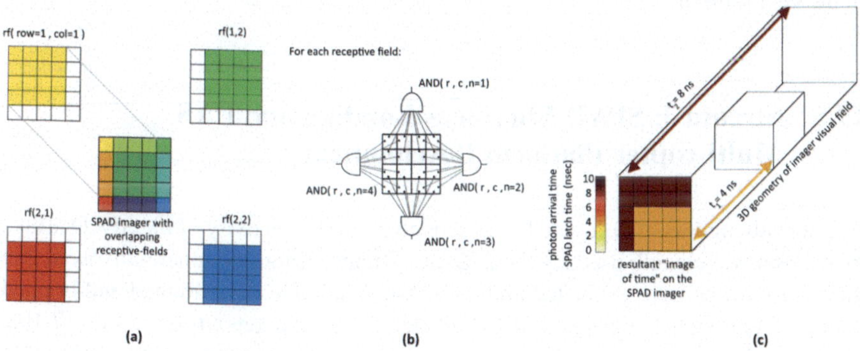

Fig. 17.8 Illustrative example of a 5×5 imager with four 4×4 receptive fields, (**a**). Each receptive field is connected to four AND gates – (**b**). An example 3D visual scene and the resultant SPAD timing pattern is shown in (**c**). For illustrative purposes we show a small imager with $5 \times 5 = 25$ pixels (instead of the chip's 128×128). This imager uses four 4×4 overlapping receptive fields (same as chip). Each receptive field has four ANDs (same as chip). One of the four ANDs take as input the left/west 8 pixels of the 16 pixels of the 4×4 patch. Another AND takes as input the right/east 8 pixels, another the lower/south and another the top ones/north. If we consider the case shown in (**c**) with the 5×5 imager viewing a scene where the background is far away and there is a box in the foreground that is seen by pixels (r = 3:5 and y = 2:5), then the lower/south AND of rf(1, 2) (green) and rf(2, 2) (blue) will latch which can be expressed as AND (1, 2, 3) and AND (2, 2, 3) latching at t = 4 ns

The event output is temporally decayed, either linearly, exponentially, or in discrete time and then the dot-product is taken with a feature map – "layer 1" (Fig. 17.9c). The feature map is learned over time based on the 3D images that are presented to the neuroSPAD. The "learning" of the feature map can be disabled so that the feature map is set and no longer changing and thus representative of the images/objects that the system needs to classify. The resulting layer 1 surface (Fig. 17.9d) which comes from the dot product of the decayed input and the feature map can then be classified.

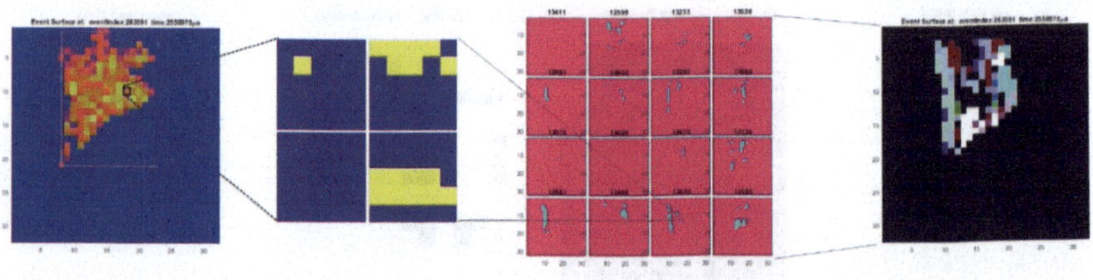

Fig. 17.9 Functionality of the neuromorphic, event-based processor. (**a**) Raw SPAD output. (**b**) Local SPAD event output. (**c**) Layer 1 activation. (**d**) Layer 1 surface

Measurements of the combined NeuroSPAD (or a conventional SPAD can be used in its place), RTS algorithm and neuromorphic, event-based processor, are presented in the next section.

17.5 Stream 4: SPAD Microlens Bonding and UAS Multi-copter Platform Development

A first challenge was to provide better sensitivity and selectivity of the SPAD arrays to the laser wavelength used for the LiDAR system. Various commercially available microlens arrays were evaluated and those found suitable were aligned and bonded on top of a previously designed SPAD imager (32 × 32 pixels), refer to Fig. 17.10a. This was done to improve the fill factor and hence sensitivity of the sensor. In addition, custom manufactured micro-lens array technology and techniques were developed which could be directly patterned to the SPAD array imaging sensors, see Fig. 17.10b, c.

A second challenge tackled in this Stream was to interface SPAD-based sensor technology (operating in ToF mode LiDAR mode) with neuromorphic-like image processing algorithmic techniques, to perform classification of imaged targets in real time, with improved target detection and recognition performance [9]. Indoor laboratory experiments of a low powered LiDAR system (a 100 mW 660 nm Coherent CUBE diode laser in conjunction with a 32 × 32 SPAD camera) and NVIDIA TX2 embedded board was used for data capture and image processing to verify operation of the real time classifier. Classification of 4 different model airplanes imaged by 32 × 32 SPAD camera with accuracy up to 98.7% was achieved. The power consumption of the embedded board was measured to be 5.1 W, which is ideal for low SWaP UAS applications [3].

Finally, DSTG was able to successfully integrate and fly a low SWaP, high powered Flash LiDAR system on a custom-built multi-rotor UAS platform (Fig. 17.11). In collaboration with partners a custom octo-copter multirotor aircraft was designed and developed. It could achieve a maximum take-off weight of 40 kg with full propulsion redundancy. All the components of a high-powered LiDAR system were

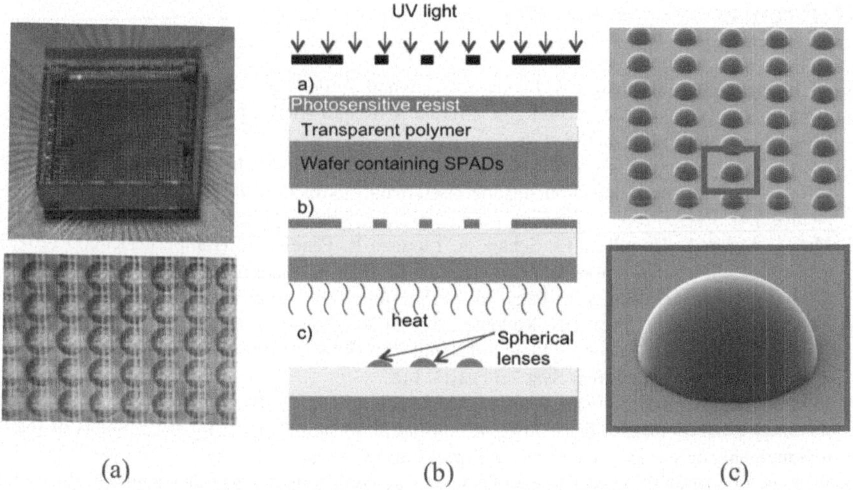

Fig. 17.10 (**a**) (Top) Commercial microlens array aligned and bonded to a 32 × 32 SPAD array sensor. (Bottom) Magnified top view showing each lens aligned directly above each SPAD detector. – (**b**) Customised polymeric microlenses were developed using lithographical techniques and applied directly to SPAD array sensor. Alignment +/− 2 μm possible. – (**c**) SEM image of an individual microlens directly patterned on the SPAD array sensor

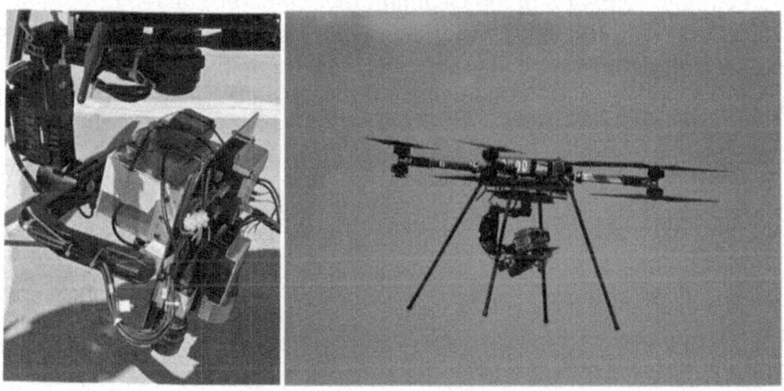

Fig. 17.11 (Left) – Photograph of the LiDAR system integrated and balanced onto a 3 axis Ronin Gimbal. (Right) – Complete system successfully operating in-flight

then integrated so it would fit on a 3-axis Ronin gimbal and be balanced (i.e. not prone to oscillations or drift). This included: a 40 mJ 532 nm Arete laser, optics, 32 × 32 SPAD camera, optical camera and power electronics. The total payload weight (including gimbal) was measured to be 7.5 kg. Various successful tests were conducted to ensure the gimbal was controllable, balanced and immune to vibration. The power conversion, wireless communication and kill switch (for eye safety precautions and if the UAS was no longer obeying commands) hardware and corresponding software interfaces to enable integration of the LiDAR fitted gimbal to the UAS platform were developed at DSTG.

References

1. Afshar S, George L, Thakur CS, Tapson J, van Schaik A, de Chazal P, Hamilton TJ (2015) Turn down that noise: synaptic encoding of afferent SNR in a single spiking neuron. IEEE Trans Biomed Circuits Syst 9(2):188–196
2. Afshar S, Hamilton TJ, Tapson J, van Schaik A, Cohen G (2019) Investigation of event-based surfaces for high-speed detection, unsupervised feature extraction, and object recognition. Front Neurosci 12:1047
3. Mau J, Afshar S, Hamilton TJ, Schaik A, Lussana R, Panella A, Trumpf J, Delic D (2019) Embedded implementation of a random feature detecting network for real time classification of time-of-flight SPAD array recordings. In: SPIE defence + commercial sensing conference
4. Morrison D, Delic D, Yuce MR, Redouté J-M (2019) Multistage linear feedback shift register counters with reduced decoding logic in 130 nm CMOS for large scale array applications. IEEE Trans Very Large Scale Integr Syst 27(1):103–115
5. Morrison D, Kennedy S, Delic D, Yuce MR, Redouté J-M (2018) A triple integration timing scheme for SPAD time of flight imaging sensors in 130 nm CMOS. In: Proceedings of IEEE international conference on electronics, circuits and systems, pp 13–16
6. Sofatzis RJ, Afshar S, Hamilton TJ (2014) The synaptic kernel adaptation network. In: IEEE international symposium on circuits and systems (ISCAS), pp 2077–2080
7. Villa F, Bronzi D, Zou Y, Scarcella C, Boso G, Tisa S, Tosi A, Zappa F, Durini D, Weyers S, Paschen U, Brockherde W (2014) CMOS SPADs with up to 500 μm diameter and 55% detection efficiency at 420 nm. J Mod Opt 61(2):102–115
8. Villa F, Lussana R, Bronzi D, Tisa S, Tosi A, Zappa F, Mora AD, Contini D, Durini D, Weyers S, Brockherde W (2014) CMOS imager with 1024 SPADs and TDCs for single-photon timing and 3-D time-of-flight. IEEE J Sel Top Quantum Electron 20(6):364–373
9. Woods WF, Delic DV, Smith BW, Swierkowski L, Day GS, Devrelis V, Joyce RA (2019) Object detection and recognition using laser radar incorporating novel SPAD technology. In: Proceedings of SPIE 11005, laser radar technology and applications XXIV, vol 1100504

Chapter 18
WITNESS: Wide InTegration of Sensor Networks to Enable Smart Surveillance

Alessandro Mattiacci, Marco Cosentino, Walter Matta, Carlo Maria Medaglia, Andrei Braicov, Ivan Budanaev, Mircea Petic, Vasileios Argyriou, and Mahdi Maktab Dar Oghaz

Abstract In the recent years the need of advanced security systems has grown in an unpredictable manner. We saw how modern cities are very susceptible to terrorist attacks and the population asks to feel safer. In this perspective WITNESS project was designed, to help security forces to preview terrorist attacks and to facilitate disaster recovery scenarios. In this paper we break down some aspects of modern surveillance systems within the context of the Wide InTegration of sensor Networks to Enable Smart Surveillance (WITNESS) NATO project. A brief description of the architecture of the system and the methodology are presented, as well as several use case scenarios which WITNESS will cater for.

Keywords Surveillance · Security · UAV · Drone · Sensors · Computer vision · Data fusion

18.1 Introduction

Information technology develops at a fast pace, and its progress influences the evolution of many other fields of science. It is difficult to say nowadays if there are any aspects of our modern world which have not yet been affected in some way

---

This research is sponsored by the NATO SPS Programme under the funding NATO SPS G5437 – WITNESS – Wide InTegration of sensor Networks to Enable Smart Surveillance.

A. Mattiacci (✉) · M. Cosentino · W. Matta · C. M. Medaglia
Link Campus University, Rome, Italy
e-mail: a.mattiacci@unilink.it

A. Braicov · I. Budanaev · M. Petic
Tiraspol State University, Chisinau, Republic of Moldova

V. Argyriou · M. M. D. Oghaz
Kingston University, London, UK

C. Palestini (ed.), *Advanced Technologies for Security Applications*, NATO Science for Peace and Security Series B: Physics and Biophysics,
https://doi.org/10.1007/978-94-024-2021-0_18

by the IT revolution. One such area of today's society is the urban surveillance and security. Although security is paramount in its core concept, the importance of this area has substantially increased since the launch of the Global War on Terrorism military campaign in 2001. This can easily be observed with the shift of military technologies into domestic applications and the confluence of internal and external security [21].

Modern surveillance systems consist of many individual components, and represent a very complex architecture. Most common problems that arise when building such products are related with data transfer speed and reliability, data analysis and automation of processes. The widely-spread high bandwidth mobile networks and attack resistant cryptosystems provide solutions to the first two of the above requirements. Computer vision algorithms permit to analyze streams of data and detect predefined patterns to spot hostile or abnormal behaviour and automatically send signals to supervisors at the monitoring stations.

Present paper addresses this specific area of modern security – surveillance systems. This article is written in the framework of the Wide InTegration of sensor Networks to Enable Smart Surveillance (WITNESS) project which has the aim to design a system for urban surveillance and security to help detect, prevent and give an efficient response to terrorist threats and attacks.

The aim of this paper is to describe the methodology of the research in the project, starting with the architecture of the surveillance system.

The article is structured as follows. First, we start with the state of the art of similar research topics. Then we will describe our main aspects of the system architecture. More details are given in the following sections, concerning sensor models.

18.2 Related Works

The topic of smart surveillance is very popular nowadays. Its idea came from the practical issues where video surveillance was used to monitor municipal buildings, banks, train stations, etc. Surveillance becomes more and more important because of increasing number of exceptional situations that require a high attention to the people and society [19]. Smart surveillance is the use of automatic video analysis technologies in video surveillance applications [12]. Automatic video analysis technologies usually take into account both GPS and telemetry data. As video data also contains noise, video sequences are usually divided to be processed into smaller pieces. A separate analysis of smaller video sequences allows one to create an overview of the whole video.

The current available smart surveillance infrastructure allows to configure and implement the algorithms for recording and filtering video data streams from interconnected objects on the Internet. One of the examples of state-of-the-art infrastructure are RFID frameworks. They are elaborated in such a way that they provide the functionality to monitor space (2D or 3D), identify suspicious events, and react by generating appropriate responses to the situation. RFID frameworks have been developed as a result of researchers' efforts [17] and commercial demands. Examples of open-source RFID frameworks include Mobitec [20], AspireRfid [5, 6], and the Fosstrac project [9, 10] that provides free infrastructure deployments.

Another approach would be to use Wireless Sensor Networks (WSNs). These were initially used as surveillance in military conflict zones. The first implementations of WSNs used distributed sensor networks (DSNs) technology. Just as the first sensors were quite large, their applicability was reduced as well as due to their limited wireless connectivity. Currently the sensors are significantly smaller and cheaper. This led to the implementation of sensor networks for monitoring apartments, the environment, and the use of body sensors. WSN is considered one of the most prospect technologies of the present century [5, 8].

The variety of WSNs platforms is great. There is a platform that only addresses the system as a network of sensors. Other platforms work with devices and other sensor networks connected to the WSNs. There are WSN development and monitoring systems that have limited extensibility, for example Moteview [14] and [11]. The following tools provide development and/or programming environments for WSNs systems: Hourglass, SenseWeb, jWebDust [4] and GSN [1]. A more detailed description of the architectural particularities of the WSNs systems can be found in [4].

The effectiveness of WSNs in the surveillance process is acknowledged. However, there are approaches that seek to improve the use of WSNs by combining them with Unmanned Aerial Vehicles (UAVs) in surveillance. UAV is a solution in situations where it is necessary to fly over dangerous areas without endangering people's lives.

Paper [2] presents a project example describing the interaction between WSNs technology and UAV tools used for border surveillance. The UAV in the given case is considered a quadcopter.

Research focused on the use of quadcopter in terrestrial surveillance is focused on identifying cost-effective solutions and preserving the same functionality. Usually a quadcopter is driven by the means of proprietary framework APIs from a laptop or a PC. A quadcopter is useful in reading the altitude to the ground, as well as in measuring air temperature, humidity and gas composition [3].

18.3 Use Case Scenarios

WITNESS proposes an innovative technological solution to incidents and accidents that may occur in an unpredictable urban scenario characterized by crowded scenes with potentially complex structured man made surroundings. The typical scenarios WITNESS will cater for are those where an incident or accident has caused disruption in the normal 24/7 operation of a public space (for instance a metro, a railway or a bus station). In such environments a nominal flow of people can be expected and, therefore, normal behaviour can be predicted. We also envisage that the public spaces of interest will be monitored by fixed cameras and by police forces. Some examples of possible case studies follow:

- Natural disasters: a disaster is a serious disruption of the functioning of a community or a society involving widespread human, material, economic or environmental loss and impacts. Disasters caused by natural or technological reasons are identified as NaTech.
- General disruption of individuals: public areas witness the presence of intoxicated individuals, usually disruptive in small groups. Such events may have an effect on the normal flow of people, for instance in a metro or railway. Such individuals may start pushing one another and other people in the surroundings.
- Public events related to holidays, manifestations or protests where there is an abnormal accumulation of people with high density. As noticed from latest terrorist attacks, these types of scenarios are a very attractive target for terrorists [15, 16].
- Accident: this describes a category of events that may be caused by the failure of electrical power – for instance delaying metro or trains, or by physical accidents happened to individuals, including suicide attempts. In such case an entire station may be closed and events may more or less slowly affect the entire area.
- Incident: this could be caused by a terrorist threat, even including hostage situations.

Specific use case scenarios of WITNESS system applications would include monitoring of the sport events, large-scale peaceful demonstrations and violent protests. During these events, the abnormal behavior under study is sought to be well-organized groups of people who aim to destabilize public order by taking hostage among civilians, using guns, fumigants or explosives. Depending on the event type, the size of monitored area can span between 10,000 to 300,000 meters square. The average occupancy of the monitored area is projected to be up to 20 thousand people.

18.4 WITNESS Architecture

WITNESS implements a distributed multi-layered architecture to satisfy the operational requirements of a situational awareness and decision making system. This approach facilitates building a flexible and pervasive enough product, ready to be automatically reconfigured and quickly redeployed when needed. The cornerstone of the system is the data, which is collected by a predefined set of sensors – wearable by police forces deployed on the grounds or sensors installed on UAVs and police vehicles. In this way, the layer responsible for collecting data in WITNESS system represents a “sensor fusion grid” that leverages data from multiple heterogeneous sensor nodes, including cameras, microphones and drones. A schematic diagram of the system’s architecture is depicted in the figure below (Fig. 18.1).

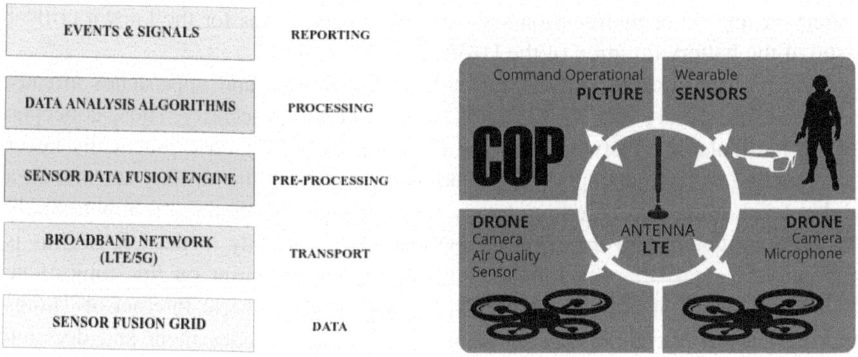

Fig. 18.1 WITNESS architecture

Data originating from different sources dispersed in multi-dimensional space (police forces on the field, quadcopters in the air, lags and losses during data transfer synchronization) need to be gathered and pre-processed before being submitted for analysis. This is a difficult task to address, since information received scales exponentially in terms of location-space, time and means (multi-source information). Recent advances in ICT technologies can boost the efficient acquisition, fusion and integration of information from the above sources. Sensor data fusion component will be implemented to address the above problem. Data fusion technologies involve the fusion of multi-sensory data to estimate the position, speed, attributes and identity of the detected and flagged targets, e.g. a person, a vehicle or an object in the operation area.

It is widely accepted that the data transfer between system components of the above architecture requires a dedicated broadband communication infrastructure (e.g. LTE, 5G technology) to be deployed in a very short time, so as to be promptly used to support the communications among security forces deployed on the field and used to gather and to process awareness data coming from monitoring devices (e.g. wearable sensors or mobile nodes) and to perform the command and coordination of the forces. In this regard, the underlying infrastructure plays a critical role in the security, processing, flow, supporting information requirements throughout the operational forces.

Within the proposed methodology, WITNESS ensures that data and information are delivered to the right place on time and optimally encoded for use by their intended recipients to take the appropriate actions at the right time. The security and safety of this information will be granted by the dedicated LTE cell. This architecture is a key enabler of Net-centric Enable Capability (NcEC) and is essential for "information superiority" and "decision superiority".

WITNESS will adopt a breadboard architecture enabling plug&play integration of the various components. The architecture also enables the integration of third-party components. Furthermore, this architecture will give the possibility to assign a task to a UAV and to reallocate it to another one when the first drone will have low battery or any other malfunction scenario. This need arises for the known critical issue of the battery duration of the UAVs.

This project aims to explore technological concepts and approaches already proposed for military operational fields for use in civilian security. Those concepts, and in particular the Network Enabled Capabilities (NEC), have practically never been applied in civilian applications, and are new even to the military sectors. At a global level, however, there are already focus groups questioning on how to apply the NEC philosophy to the security applications (especially in the USA), thus is imperative to start the development of a European blueprint on this topic. One of the technological impacts of WITNESS will be the generic Internet-of-Things approach towards creating a civil security C2 situation assessment and decision aiding framework.

18.5 Methodology

Taking into account the current approaches in this field and the use cases for the system to be built the WITNESS research project will follow the scheme presented in Fig. 18.2.

In order to obtain a more accurate result we need a complete supply of important parameters for the proper determination of the instant situation. The parameters are

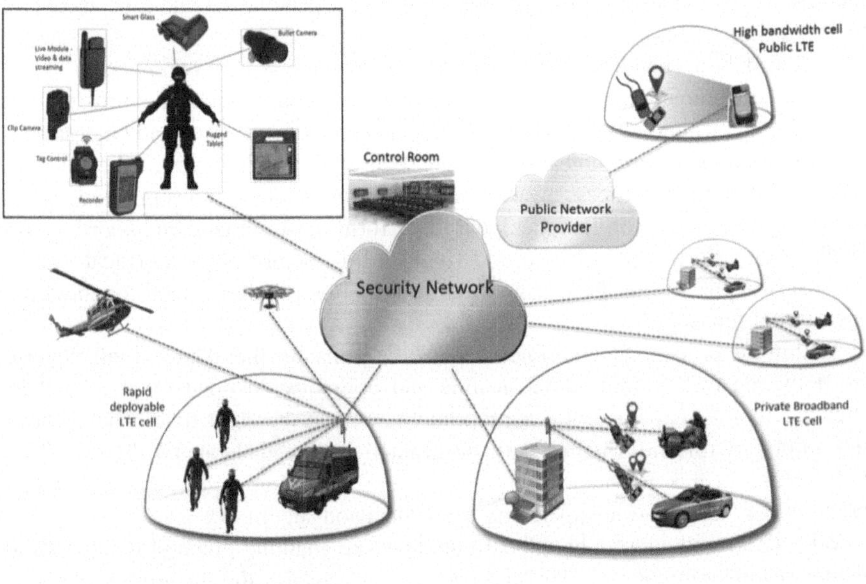

Fig. 18.2 WITNESS concept

monitored through a RGB/Thermal camera, a directional microphone and an Air Quality Sensor (CO_2).

The UAV solution for the architecture described will be a quadcopter, DJI Mavic 2 Enterprise Dual [7], which allows porting of several types of sensors. Flight time – up to 31 min. A dedicated server will be used to build the Artificial Intelligence application that will process the data provided by the sensors.

The integrated thermal camera is useful for detecting people in low visibility conditions (total darkness, fog or smoke). It can see through smoke or light mist, it does not require additional lighting, the image being obtained due to temperature differences between the target object and the environment.

Air pollution sensor is useful for gas detection [18].

Wearable sensors are to be used to monitor vital parameters of police agents. Furthermore, system can use the data provided by a rugged smartphone. This way, the monitoring system relies on a large set of different types of sensors. The volume and complexity of data supplied by these sensors dictates the necessity to develop a data fusion algorithm, which will collect, serialize and normalize the data, and pass it forward into the system pipeline.

Finally, a demonstration will be done using a simulator to let the security forces be ready for some events that may be planned in advance.

18.6 CV: Crowd Management and Monitoring

Large gatherings such as sporting events, music concerts, cultural festivals, and amusement parks are always characterized by crowded environments. Crowd dynamics, based on the behavior of individuals in a crowd vary. In fact, people move with some purpose and intentions might be completely different between or among individuals, making identification of behavior patterns hard to comprehend. However, the behavior of a crowd is widely understood to have collective characteristics which can be described in general terms. This because in a very crowded and cluttered space even individual behavior tends to be influenced by the behavior of people nearby. Crowd behavior may vary from a peaceful to hostile. Nonetheless, the safety and security of an individual in a crowded environment are of utmost importance for the person themselves and security bodies in charge of keeping orderly behavior and prevent accidents, incidents and disruptions such as police, security operators and guards.

One of the objectives of NATO project is to promote crowd safety and security using aerial imagery. For many years surveillance cameras and nvr systems were the only commonly practiced solution for crowd monitoring and management, with televisions observed by a member of the security staff, looking for problems or conditions such as various forms of congestion, disruptions in crowd movement and flow, stampede or problems arising from individual incidents not related to a crowd, such as fight, a theft, the outbreak of a fire, or potential act of terrorism. In these approaches, human observers are normally employed to monitor multiple camera streams during or after the event. Although these solutions can be practical for monitoring small scale crowds with a limited number of cameras, scalability, reliability, and cost-effectiveness of these solutions are questionable in large scale events.

Recent advancements in parallel computing and GPU technology diminished this computational barrier and allowed complex models such as deep neural networks to flourish. The rise of data-driven approaches such as deep learning, simplified modeling tasks by eliminating the need for domain expertise and hard-core feature extraction. The emergence of deep learning in the last decade has boosted the performance of image classification techniques and has started having a positive impact on crowd behavior analysis. CNN(Convolutional Neural Networks) have gained ground in crowd monitoring and behavior analysis. It has been shown that such models are not only able to achieve state-of-the-art performance for the visual recognition tasks, in which they were trained, but also the learned representation can be readily applied to other relevant tasks.

Deep learning methods are very data-hungry: they require a large amount of annotated training data, often very limited and some times unavailable. Furthermore, deep learning models rely on costly parallel computing facilities and GPU technology.

In this project we used deep learning to model complex crowd behaviors and characteristics. Several deep learning based crowd analysis algorithms including crowd counting, localization, and density estimation, crowd flow analysis, crowd anomaly detection, fight detection, and object detection have been developed to be used for crowd management using aerial images.

18.6.1 Applications and Use-Cases

Crowd counting and density estimation are of great importance in computer vision due to its essential role in a wide range of surveillance applications including crowd management and public security [23]. However, the presence of drastic scale variations, the clutter background, and severe occlusions make it challenging to generate high-quality crowd density maps. In this project we use powerful CNN to handle the challenging situations in crowds mainly by fusing multi-scale or multi-context information to improve the feature representations. CNN (shown in Fig. 18.3) which specifically inspired by the biological visual cortex is a type of deep neural networks, most commonly applied to analyzing visual imagery. Unlike traditional fully connected deep neural networks, a CNN is able to successfully capture the Spatial and Temporal dependencies in a given image through the application of relevant spatial filters.

In dense crowds, individuals heads are the only reliably visible body part in an image as the other body parts are usually occluded by the crowd. Hence, majority of the CNN based approaches rely on heads as the only discriminant feature of the

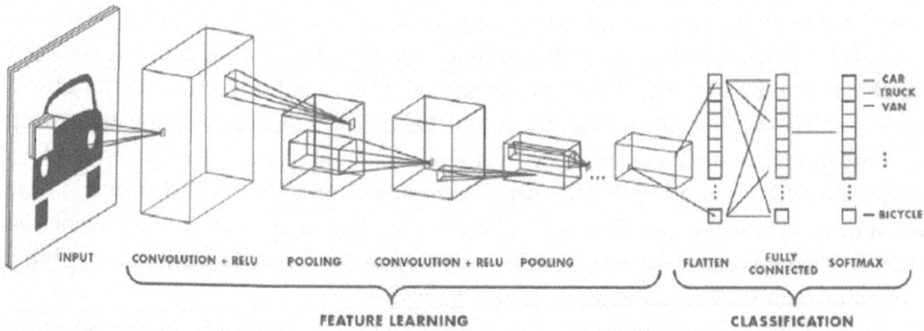

Fig. 18.3 A typical CNN architecture

Fig. 18.4 A Sample crowd scene along with its corresponding crowd density map

image. As shown in Fig. 18.4, CNNs can generate crowd density maps which not only localize the crowd but also estimates its size and density.

Crowd flow is another informative metric in crowd management and behavior analysis. There is a critical capacity where flow begins to decrease as the crowd's density increases. A dense crowd with high flow magnitude poses a serious safety threat and might lead to a human stampede. In visual surveillance, optical flow algorithms have become an important component of crowded scene analysis. The application of optical flow allows crowd motion dynamics of hundreds of individuals to be measured without the need to detect and track them explicitly, which is an unsolved problem for dense crowds. We employed CNN to measure the optical flow in the crowd. The algorithm takes two frames at given time intervals to measure the crowd flow magnitude and direction and generates the flow map which localizes the flow intensity. Crowd flow can provide valuable information about crowd behavior in locations such as stadiums and parades where smooth flow or crowd is essential.

Analysis of crowd behavior in public places is a critical objective for video surveillance [22]. We are often reminded how time consuming it is for police officers or forensic investigators to search through hours of video looking for an incident. It is clear that there would be a substantial advantage if some form of image processing system could be applied to the video signal, to automatically spot the anomalous crowd behavior as they arise, and to trigger a suitable response in a reasonable time. This project offers state of the art deep learning techniques to create a model capable to handle overwhelming complexity of crowds and detect anomalous crowd behavior. The proposed anomalous crowd behavior detection solution uses deep learning models which are also deep over temporal dimensions. These deep models, which are addressed as RNN (Recurrent neural network), are rolled over the time domain. A visual feature extractor (CNN) runs in tandem with a model that can learn to recognize and synthesize temporal dynamics (LSTM) for tasks involving sequential data such as anomalous crowd behavior.

Object detection is breaking into a wide range of industries, with use cases ranging from personal security to productivity in the workplace. One of the most important applications of object detection is crowd safety and security management. Object detection algorithms can be utilized for identifying missing objects or individuals in the crowd, vehicular traffic or even early warning of a potential terrorist attack by identifying unattended suspicious objects in public. We use state-of-the-art CNN models capable to identify a wide range of objects such as persons, vehicles, suitcase, trees, animals, etc.

18.7 Data Analysis and Data Fusion

In a system with a multi-sensor grid, consisting of inertial, audio, visual, location, chemical, and biosensors, an abundant volume of data is generated on the fly. Analysis of this data represents a crucial component of WITNESS.

The stated above goals of the WITNESS system, i.e. detection and prevention of terrorist attacks and other disasters, when approached from the data science point of view, fall into two main categories: classification and anomaly detection. The classification part is responsible for spotting predefined scenarios like gunshots, armed person or armored vehicle detection. Anomalies represent those type of scenarios which are detected as a statistical outlier and/or have not been categorized prior to deployment on the ground.

Denoting the representation of a single sensor observation by the tuple $O := \langle p, t, x \rangle$, where x is the measurement recorded by the sensor in the position p and time t, we collect the following data vector produced by the sensor grid: $S_t := \{O_{1 \leq j \leq n}\}$, with n as the total number of sensors in the grid. Consequently, the data produced by the system in time interval $(t_i, t_{i+1}]$ is given by the time series $S_{t_i, t_{i+1}} = \{S_{t_i < t \leq t_{i+1}}\}$. In modern data analysis, one must make sure to avoid the most common pitfalls. These are related to choosing the right interval to collect data and addressing the problem of noise induced by sensor uncertainty (poor sensor data and quality). The solution we chose is similar the way the Adam algorithm [13] addresses the learning rate optimization problem in machine learning – we substitute the collected observations in $S_{t_i, t_{i+1}}$ by its moving average giving more weight to most recent data:

$$\overline{S_{t_i,t_{i+1}}} = WMA_{S_{t_i,t_{i+1}}} = \frac{\sum_{t=t_i,k=1}^{t_{i+1},\Delta} k \cdot S_t}{\Delta(\Delta+1)\,/\,2}, \Delta = t_{i+1} - t_i.$$

Adding weighted moving averages smooths out induced noise and allows detecting patterns even if some of the data supplied by some sensors came delayed.

Due to the high dimensionality of the data provided by mutually complementary sensors, we apply data analysis on multiple groups of features concurrently. Obtained insights are combined together with processed visual and audio data to refine estimates and assessments regarding the current state of the environment. This approach is called Data Fusion and proves to have better performance and more accurate results. Among other components described here, it also consist of the following aspects: sensor modeling, management, control and optimization, statistical and probabilistic methods, neural networks, situation and impact (threat) refinement modules.

We use distribution-based approach to detect and flag outliers, a distance-based approach to detect how far a tuple of a pre-defined group of observation is far from the set's centroid, and, where applicable, we apply naive Bayes to combine the outputs. One example of a data fusion algorithm output is the detection of a bomb explosion based on the group of audio, chemical (smoke), and visual sensors. This is a good example, as the data produced by the smoke detection sensor will arrive delayed in comparison to other sensors. Another example would refer to the safety of the officer on the premises. Obtaining information about his state is based on position, velocity and heart rate parameters.

18.8 Conclusions

WITNESS provides an opportunity for companies to increase the competitiveness of the WSN (wireless sensor network) industry by developing novel sensors so far never exploited in this field.

One of the project objectives is the definition of tools, technologies and methods that will facilitate the countering to attacks and critical situations. This study will lead to some conclusions about the state of the art resources and, hopefully, to the definition of new methods that will improve the ability to develop a monitoring solution in short time. The results obtained will be the base for future development that will improve a easily deployable and secure system.

References

1. Aberer K, Hauswirth M, Salehi A (2006) The global sensor networks middleware for efficient and flexible deployment and interconnection of sensor networks. LSIR Report 2006-006
2. Berrahal S, Kim J-H, Rekhis S, Boudriga N, Wilkins D, Acevedo J (2016) Border surveillance monitoring using Quadcopter UAV-Aided wireless sensor networks. J Commun Softw Syst 12:67–82

3. Borah DR, Debnath L, Gogoi M (2016) A review on quadcopter surveillance and control. J Eng Technol 4(1):116–119
4. Chatzigiannakis I, Mylonas G, Nikoletseas S (2005) jwebdust: a java-based generic application environment for wireless sensor networks. In: In Proceedings of the IEEE international conference on distributed computing in sensor networks (DCOSS). Lecture notes in computer science (LNCS), pp376–386, Springer
5. Consortium O (2019) AspireRFID OW2 Project, https://projects.ow2.org/view/aspire-rfid/. Accessed on 05 Sept 2019
6. Dimitropoulos P, Soldatos J (2010) RFID enabled fully automated warehouse management: adding the business context. IJMTM 21:269–288
7. DJI (2019) DJI Mavic 2 Enterprise Dual, https://www.dji.com/it/mavic-2-enterprise. Accessed on 05 Sept 2019
8. Yinbiao S, et. al. (2014) Internet of things: wireless sensor networks, technical report p 78, Wireless sensor networks project team. International Electrotechnical Commission, Geneva
9. Floerkemeier C, Lampe M, Roduner C (2007) Facilitating RFID development with the accada prototyping platform. In: IEEE international conference on pervasive computing and communications
10. Fosstrak (2019) Fosstrak: open source RFID software platform, http://www.fosstrak.org. Accessed on 05 Sept 2019
11. Ritter H (2019) Short presentation of ScatterWeb Software and Hardware for next generation wireless networks, http://files.messe.de/cmsdb/001/14916.pdf. Accessed on 05 Sept 2019
12. Hampapur A, Brown L, Connell J, Pankanti S, Senior A, Tian Y (2004) Smart surveillance: applications, technologies and implications, vol 2, pp 1133–1138
13. Kingma DP, Ba J (2014) Adam: a method for stochastic optimization. arXiv preprint arXiv:1412.6980
14. MoteWorks (2012) MoteWorks getting started guide. Revision G, Technical report, MEMSIC, Inc, McCarthy Blvd, Milpitas
15. of State, U. D. (2016) Country reports on terrorism, Technical report, U.S. Dept. of State
16. of State, U. D. (2016) Global terrorism index, technical report, Institute for Economics and Peace
17. Prabhu BS, Su X, Ramamurthy H, Chu C-CP, Gadh R (2005) WinRFID – a middleware for the enablement of radio frequency identification (RFID) based applications. In: Mobile, wireless and sensor networks: technology, applications and future. Wiley, Hoboken
18. Robot D (2019) Gravity: UART Infrared CO2 Sensor, https://www.dfrobot.com/product-1565.html. Accessed on 05 Sept 2019
19. Saha S, Neogy S (2014) A case study on smart surveillance application system using WSN and IP webcam. In: 2014 applications and innovations in mobile computing (AIMoC), pp 36–41
20. technology Center M (2019) MobiTeC – open source RFID middleware 1.0, http://mobitec.ie.cuhk.edu.hk/rfid/middleware/. Accessed on 05 Sept 2019
21. Wilson D (2012) Military surveillance. Routledge, New York, pp 269–276
22. Xu D, Yan Y, Ricci E, Sebe N (2017) Detecting anomalous events in videos by learning deep representations of appearance and motion. Comput Vis Image Underst 156:117–127
23. Zhang Y, Zhou D, Chen S, Gao S, Ma Y (2016) Single-image crowd counting via multi-column convolutional neural network. In: Proceedings of the IEEE conference on computer vision and pattern recognition. IEEE, pp 589–597

Chapter 19
Laser Ablated Graphene/Polymer Based Sensors: Relating Composite Morphology and Sensor Properties

Radmila Tomovska, Jadranka Blazevska Gilev, Yvonne Joseph, and Radek Fajgar

Abstract Offering new materials for gas sensing performance with low cost, simple preparation, high sensitivity, selectivity and reproducibility at ambient working conditions are the main objectives of this project. Using relatively low cost, water-borne polymer system with excellent physical and chemical properties reinforced with different carbon based nanomaterials was considered as the most appropriate. The polymer matrix consisted of copolymer made of methyl methacrylate and butyl acrylate, synthesized by polymerization in dispersed media techniques, such as emulsion or miniemulsion polymerization. The polymer particles obtained are usually functionalized using small amount of functional monomers, such as glycidyl methacrylate, acrylamide or acrylic acid, which improve compatibility with the carbon nanomaterial. Additionally, to improve their distribution within the polymer matrix and sensing properties of the composite materials, post-treatment by laser ablation is performed.

Keywords Polymer nanocomposites · Graphene based polymer hybrids · Graphene nanoribbons · Silver doped reduced graphene oxide · Gold doped reduced graphene oxide · Gas sensors

---

R. Tomovska (✉)
POLYMAT, University of the Basque Country, Donostia-San Sebastián, Spain

Ikerbasque, Basque Foundation for Science, Bilbao, Spain
e-mail: radmila.tomovska@ehu.es

J. B. Gilev
Faculty of Technology and Metallurgy, University Ss. Cyril and Methodius, Skopje, North Macedonia

Y. Joseph
Institute of Electronic and Sensor Materials, TU Bergakademie Freiberg, Freiberg, Germany

R. Fajgar
Institute of Chemical Process Fundamentals, Czech Academy of Sciences, Prague, Czech Republic

C. Palestini (ed.), *Advanced Technologies for Security Applications*, NATO Science for Peace and Security Series B: Physics and Biophysics,
https://doi.org/10.1007/978-94-024-2021-0_19

19.1 Introduction

The atmospheric pollution nowadays creates conditions of continuous exposure of humans to various air toxins, resulting in serious health problems like respiratory disease and poisoning. Toxic gases like CO, NH_3 and NO_2 are one of the most common air pollutants that can irritate skin, corroded lung tissue and respiratory tract and in the worst case, cause mortal danger [1, 2]. Consequently, new pathways toward novel gas sensor materials with improved performance are required. Therefore, in the last decade with the rapid progress of detection techniques, the investigations are oriented to obtain sensors with higher sensitivity to determine low concentration, selectivity along with faster response time and reproducibility [3]. However, other properties as thermal and mechanical resistance, operating temperature, durability and service life of the sensor are also of large importance.

This project (G5244) is focused on development of polymer/graphene composites gas sensors, based on two techniques: on one hand production of water borne composites by emulsion polymerization techniques, and, on the other, ablation of these composites by IR or UV lasers and preparation of electrically conductive thin films. In the following, the most relevant results of the first investigation line obtained throughout the 2 years of investigation are shown.

19.2 Experimental

Multiwalled carbon nanotubes (CNT, length = 5–15 μm; diameter = 10–30 nm) and graphene nanoribbons (GNR, with length 2–15 μm and width 40–250 nm) were acquired from Sigma Aldrich, whereas, graphene nanopowder (G, 8 nm flakes) or graphene oxide (GO) (4 mg/mL aqueous dispersion with 80% monolayer GO) from Graphene Supermarket (USA). GO was reduced by using Ascorbic Acid, procedure explain elsewhere [4]. Hybrid G/CNTs filler was prepared by mixing G and CNT in different weight ratios (G: CNT of 10:1; 5:1; 1:1; 1:5 and 1:10) by sonication in water in presence of SDS using Branson 550 instrument, Danbury, CT at 70% of power output and 50% duty cycle for 90 min efficient time, under mixing with magnetic stirrer of 200 rpm.

For waterborne polymer and *in situ* composite preparation miniemuslion technique was used, for which materials and methods are explained elsewhere [4, 5]. Methyl methacrylate (MMA) and butyl acrylate (BA) in 1:1ratio were copolymerizaed. 1 wt% of functional monomer glicidyl methacrylate (GMA), acryl amide (AA) or hydroxyethyl methacrylate (HEMA) was added in order to functionalize polymer chains and to compatibilize them with the carbon nanomaterials. Additionally, composites were prepared by emulsion mixing technique (blending of aqueous dispersion of both phases), where polymer and filler aqueous dispersions were mixed in appropriate ratios.

Wide range of the polymer composites are synthesized, all of them based on MMA/BA in 1:1 ratio, in order to obtain polymer matrix with glass transition temperature bellow room temperature. This will allow formation of a continuous film from polymer dispersion at ambient conditions upon water evaporation. In the presented results G/CNT; rGO/CNTs or GNR were used, in concentration range of 0.2–1 wt%. The sensing properties of these composites were related to their morphology. Apart of these results, we present synthesis of rGO nanoplatelets decorated with Ag and Au nanoparticles and their use as substrate for Surface Enhanced Raman Scattering (SERS) measurements.

The films from the neat polymer and the composites were casted in a silicon molds at 25 °C and 55% relative humidity for various characterizations. For sensing measurements by QCM, special substrates were used, onto which the hybrid latexes were deposited by means of drop-casting with 10 μl pipette and dried at room temperature for several days. Material dispersed Spin-coating technique was also used, with the primary rotation rate of 600 rpm for 20 s and the second rotation rate of 3000 rpm was applied for 6 s in order to spread the material and make a homogenous thin film, respectively. The film thickness measured with confocal microscope. The results show that the film thickness of all samples was equal to or less than 10 μm with the accuracy of ±0.3 μm.

19.3 Result and Discussions

19.3.1 Morphology of Composite Materials

Morphologies of the composite films were investigated by SEM imaging of the cross-section area of the films. For the preparation of the composite blends rGO/CNT mixtures were used, and for the *in situ* processes G/CNT, because in case of G/CNT phase separation occurred during the blending process.

In Fig. 19.1, the SEM cross-section area of the composite blends containing 1 wt% of rGO/CNT with ratios of 10:1, 1:1 and 1:10. Figure 19.1 presents the cross-section morphology of the blend composites containing 1 wt% rGO/CNT filler. One may observe dark areas corresponding to polymer surrounded by the rGO/CNT white structures. The presence of the filler around neat polymer areas, often corresponding to the size of polymer particles (300 nm) is known as morphology of segregated network, typical for waterborne composites [6], which allows obtaining mechanical or conductive percolation at low filler contents. The segregated network is more pronounced for higher content of CNT in rGO/CNT, mixture (1:10), probably due to better flexibility of the CNT than rGO.

In Fig. 19.2 representative images of the *in situ* composites containing 1 wt% hybrid G/CNT filler in two ratios (1:1 and 10:1) are shown. The polymer matrix is black and the grey structures represent the hybrid G/CNT. The zoomed areas presents the TEM image of the neat G/CNT fillers. The composite with 1:1 G/CNT

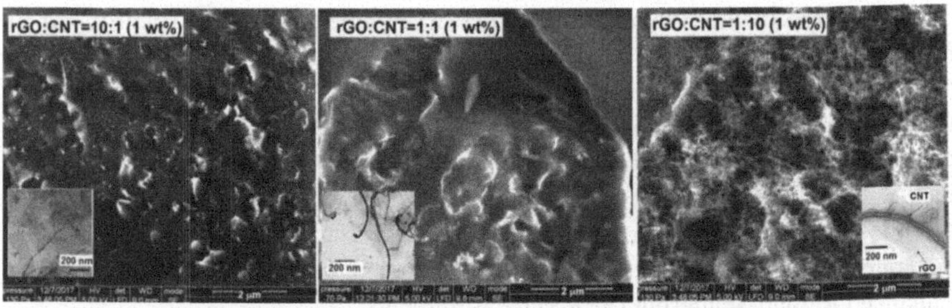

Fig. 19.1 Morphology of blend film containing 1 wt% hybrid filler of different ratios. Inset: TEM images of the neat rGO/CNTs (scale bar 200 nm)

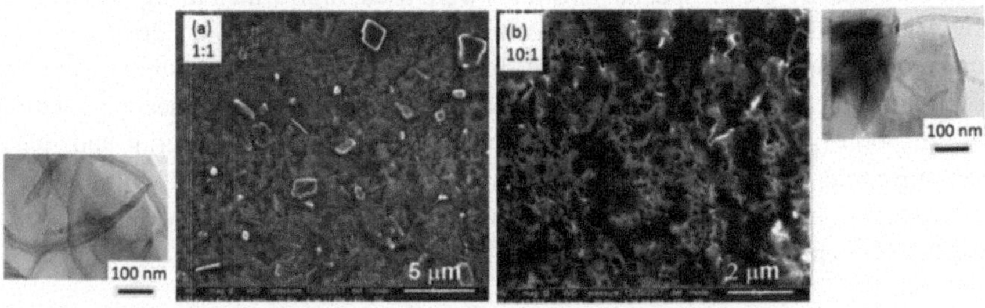

Fig. 19.2 SEM images of composite containing 1 wt% G/CNT hybrid filler with ratio of: (**a**) 1:1; and (**b**) 10:1. Zoomed areas: TEM images of hybrid filler with ratio 1:1 (left) and 10:1 (right)

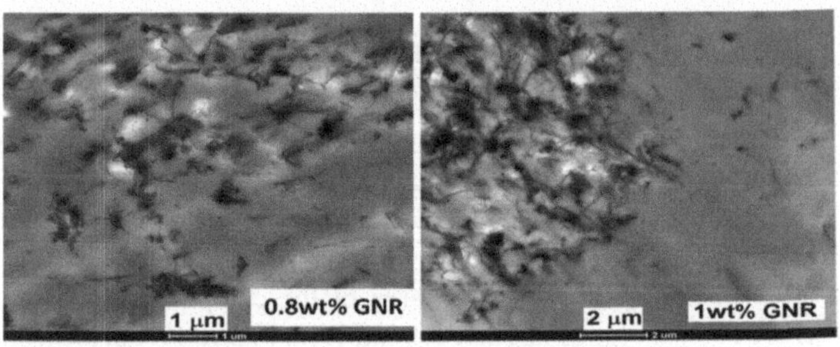

Fig. 19.3 TEM images of blend composites

(Fig. 19.2a) presents larger aggregates, probably due to higher aspect ratio of CNT (in 1:1 their content is higher) that is more difficult to stabilize colloidally. Figure 19.2b presents composite containing 10:1 hybrid with nice distribution of the hybrid filler, without any aggregation, similar to segregated network morphology.

Figure 19.3 represents the TEM structure of the composite blends obtained by mixing of aqueous dispersion of MMA/BA/AAm with that of GNR. Areas rich of GNR may be observed in both images, coexisting with polymer film in which few GNR are well distributed.

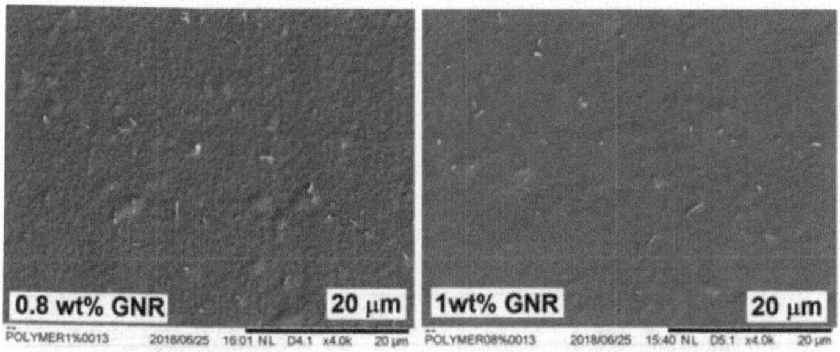

Fig. 19.4 SEM image of the top surface of *in situ* composite film with different content of GNR

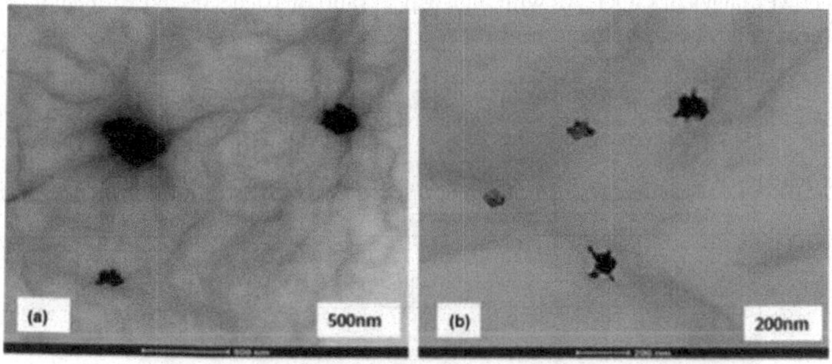

Fig. 19.5 TEM images of GOS_Au with low aspect ratio of 1, decorated with Au nanoparticles under lower (**a**) and higher (**b**) magnification

In situ composites with GNR have improved distributions of GNR within the polymer matrix, without any visible aggregation observed (Fig. 19.4). GNR are visible as a 3–4 μm long needle like white structures embedded within the grey matrix. It is notable that the GNR are much more rigid than the rGO and CNT, therefore no indication of segregated network of GNR within the matrix is presented. The surface is quite rough, coming from the individual polymer particles (in average 300 nm each).

The electrical conductivity of the composites was of order 10^{-5} S/m. Various strategies have been used to improve it, between them the graphene was surface modified by nanoparticles of Ag, Au, Pt, etc. In the following, TEM images of two types of rGO platelets decorated with gold nanoparticles are presented. The first platelets type GOS has low aspect ratio of 1 and is characterized by high purity. The second platelets type, GOG has high aspect ratio of 10 and contains about 4% impurities (heteroatoms N, O, S). Their TEM images are shown in Figs. 19.5 and 19.6. The GOS_Au has anchored few aggregated particles of Au, whereas GOG_Au has anchored large number of particles with wide distribution of particle sizes. The presence of heteroatoms (4%) onto GOG, during the synthesis likely provided nucleation points for Au nanoparticles.

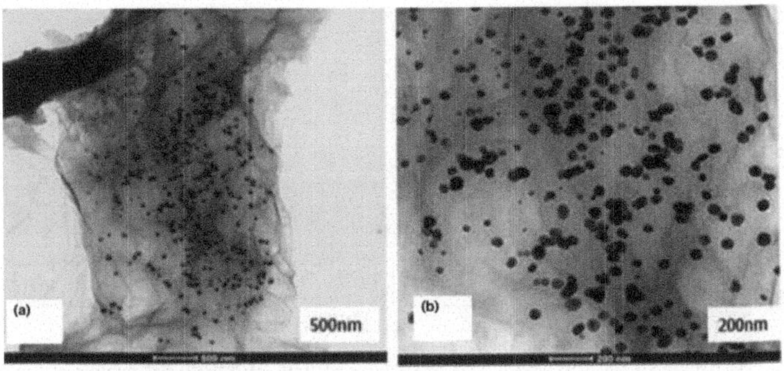

Fig. 19.6 TEM images of GOG_Au with high aspect ratio of 10, decorated with Au nanoparticles under lower (**a**) and higher (**b**) magnification

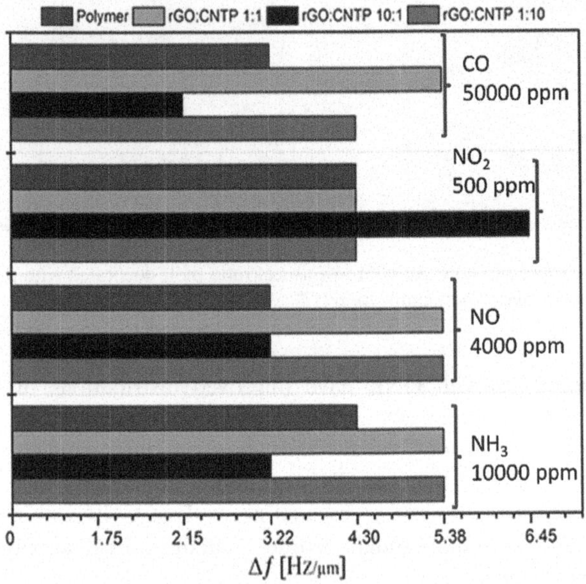

Fig. 19.7 Sensing response of composite blends of 1% rGO/CNT towards various gases with different concentrations normalized to the film thickness

19.3.2 Sensor Characteristics of the Presented Materials

Figure 19.7 illustrates the response of blend rGO/CNT composites towards NH_3, NO, NO_2, and CO gases diluted with nitrogen.

The neat polymer shows certain response to the analyzed gases, which was improved by adding of the rGO/CNT hybrid likely due to the introduction of number of oxygen functional groups that improve the interaction with the analyzed gases. Except for NO_2, for all other gasses 1:1 and 1:10 rGO/CNT composites provided

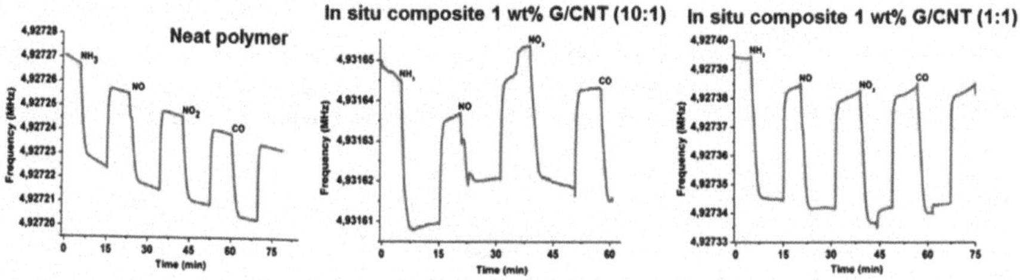

Fig. 19.8 Sensing characteristics of the neat polymer and *in situ* composites containing 1 wt% G/CNT on different gases. Concentration in (ppm): NH_3 (10000); NO (4000); NO_2 (500); CO (50000)

the highest frequency changes. Higher content of CNT seems to be preferable for sensing properties. Besides that, the morphology of these composites (Fig. 19.1) indicates that, for more developed segregated network the performance is improved.

The representative results for sensing characteristic of neat polymer and the *in situ* composite with 1 wt% hybrid G/CNT (10:1 and 1:1) for series of gases are shown in Fig. 19.8. Surprisingly, the neat polymer presented an excellent response to all of the analyzed gases. Addition of the 10:1 G/CNT did not introduced significant improvement, despite nice distribution of the hybrid within polymer matrix (Fig. 19.2b). However, when 1:1 G/CNT were added, significantly improved response was obtained for all of the gases. XPS analysis (results not shown here) demonstrated that CNT are more oxidized than G. Taking into account that 1:1 nanocomposite contains higher quantity of more oxidized CNT than 10:1 composite, this may be the probable reason for the improvement. The presence of small aggregates of G/CNT additionally may contribute to the improved performance.

The morphology of 10:1 nanocomposite presents larger aggregates of G/CNT, which may create surface defects, known to increase the gas adsorption capability [7] due to addition of adsorption sites to the nanocomposites, increasing the sensitivity towards the analyzed gases. The adsorption of CO, NO and NO_2 on the material surface is physical, therefore, the sensitivity is attributed to the existence of oxygen functional groups on the surface of the material, which should be augmented by addition of the G/CNT hybrids. In case of NH_3, additionally H-bonding gas-nanocomposite may be established increasing the adsorption; therefore the best results were obtain for this gas.

The reproducibility (Fig. 19.9) was checked for NH_3 gas, since all materials showed the best sensitivity towards this gas. Surprisingly, an increasing frequency with respect to the initial one was noticed after gas desorption, which indicates material mass loss. With exception of this drawback, relatively good sensing reproducibility was obtained. The 1:1 G/CNT nanocomposite with the best sensor activity had standard deviation of the frequency change of 9 Hz. However, the 10:1 G/CNT nanocompoite has the lowest standard deviation error of 1 Hz. In all cases nice responses were obtained.

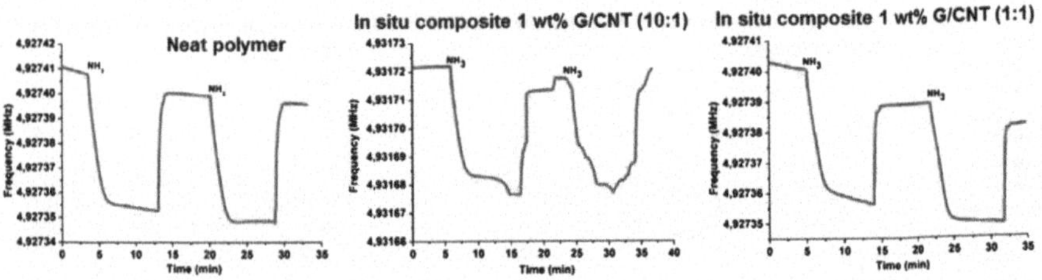

Fig. 19.9 Reproducibility of sensing measurements for neat polymer and *in situ* G/CNT composites in case of NH_3 with 10.000 ppm concentration

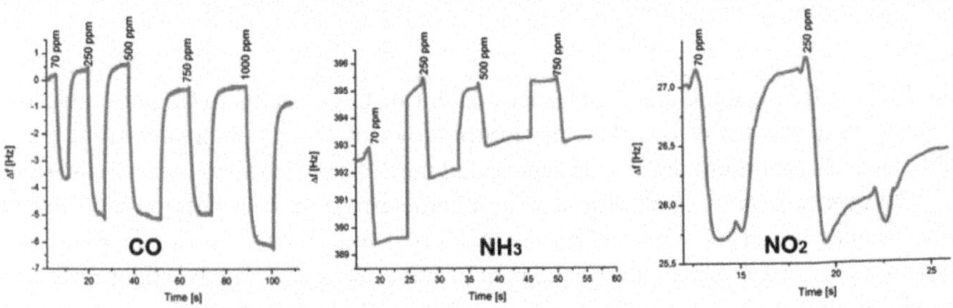

Fig. 19.10 Sensing results of blend composites with 1 wt% GNR

The sensing characteristics towards various gases of 1 wt% GNR blend nanocomposites are presented in Fig. 19.10.

In general, GNR nanocomposites have shown higher sensitivity than all other investigated composites, therefore, gas concentration in much lower range were used (70–1000 ppm). This can be due the GNRs narrow channel width and highly dense edge defect sites. The desorption was complete and the frequency returns to its original value, ascribed to weak Van der Waals bonds established between the analyzed gas molecules and the films. Moreover, higher and faster drop in the frequency was detected for NH_3 gas, which means that these systems are keen to absorb more NH_3 molecules compared to CO and NO_2.

The results obtained for the reproducibility of the nanocomposite exposed to 250 ppm for CO and 70 ppm for NH_3 are presented on Fig. 19.11, showing good performance. TEM images (Fig. 19.3) show that the composite films contain areas reach in GNR and the other areas which contain few GNR. The irregularities presented onto the composite surface, forming two different phases, seem to promote interactions with the gases (Fig. 19.11).

The *in situ* composites containing 1 wt% GNR, presented as well very good sensing properties for the three investigated gases in concentration range of 70–1000 ppm (Fig. 19.12). For CO gas, a dependence of frequency drop on concentration was observed, which may be used in creating of quantitative sensor. For NH_3 and NO_2, the signals were equal for all the concentrations, creating a concentration independent sensor. Such characteristic could be useful in preparation of sensors for

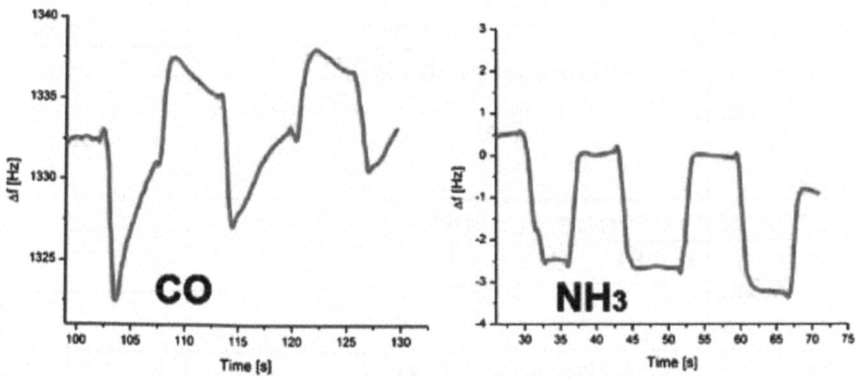

Fig. 19.11 Reproducibility of sensing for the 1 wt% GNR blend composites

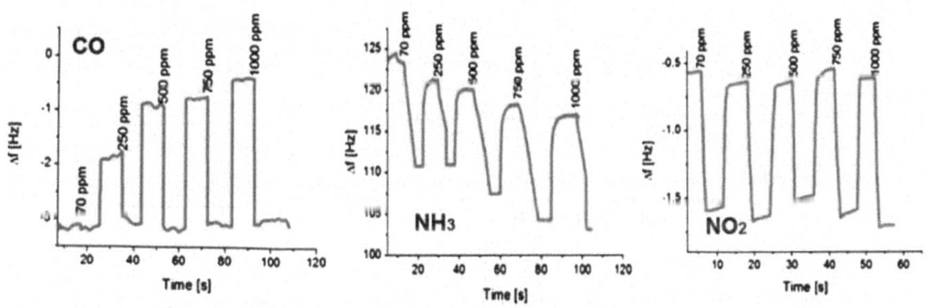

Fig. 19.12 Sensing results of *in situ* composites with 1 wt% GNR

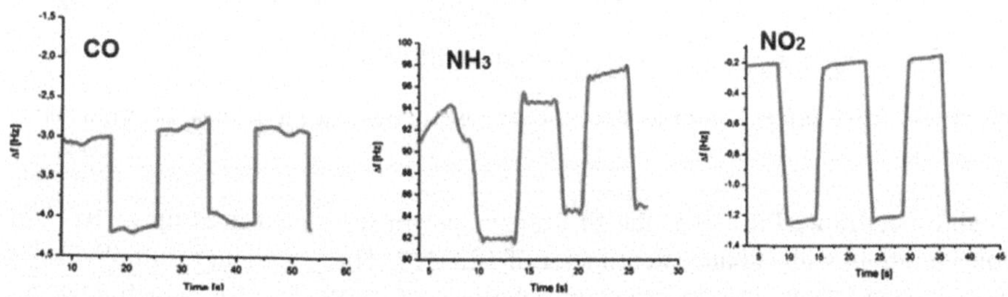

Fig. 19.13 Reproducibility of sensing measurements of *in situ* composites (1%GNR)

selective detection of gas mixture. Reproducibility of these measurements was as well excellent for all gases, as shown in Fig. 19.13.

Apart of these results, the decorated graphenes with noble metal nanoparticles were used as substrate for SERS. Fluorescein (Fl) was used as a probe molecule. The results obtain for Au nanoparticles are shown in Fig. 19.14.

The Fl concentration was changed in range 10^{-1} to 10^{-10} M. Onto GOG_Au (high aspect ratio, Fig. 19.6), Raman spectra of Fl were measured up to concentration of 10^{-4} M, giving rise to enhancement factor of 900. In case of GOS_Au

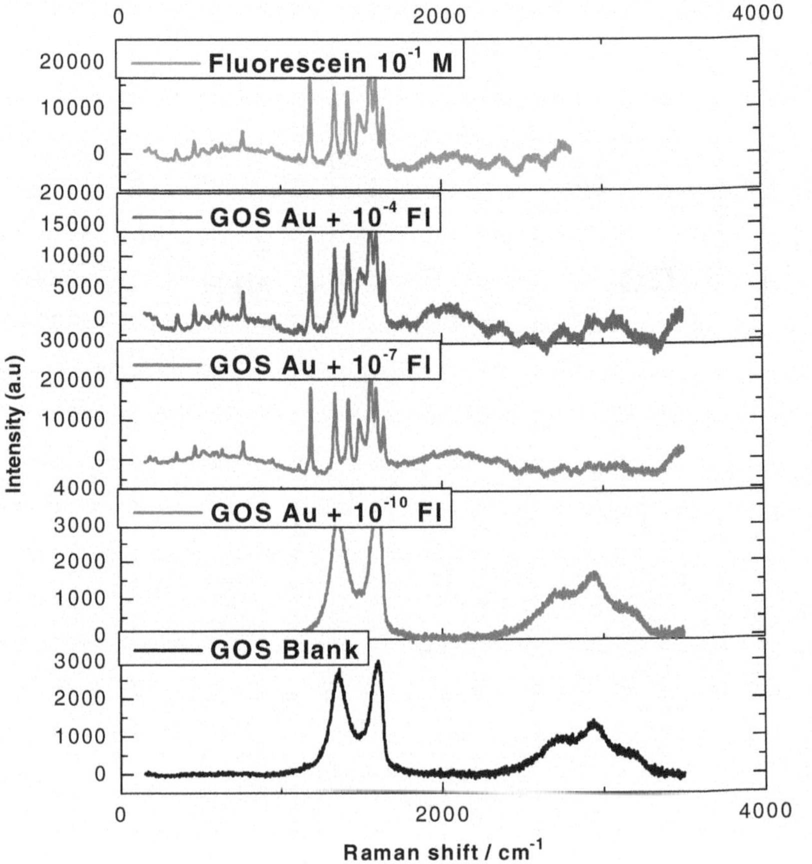

Fig. 19.14 SERS results. Raman spectra of Fl on GOS_Au substrate for different Fl concentration

(high aspect ratio, Fig. 19.5) the Fl Raman spectra were measured up to 10^{-7} M concentration, with enhancement factor of $122 \cdot 10^4$. This excellent result obviously was obtained from the observed morphology of GOS_Au, for which few Au nanoparticles were anchored onto the rGO surface but they were presented in aggregates.

19.4 Conclusions

In conclusion, wide variety of novel composite materials were produced based on different type of carbon nanofillers: graphene or reduced graphene oxide mixed with CNTs, or GNR. The materials presented interesting morphologies and promising sensing properties towards various toxic gases. With respect to sensing properties, the analysis to now using QCM method show that the nanocomposites based on GNR presented the most sensible properties towards toxic gases CO, NH_3 and NO_2. It was found that the addition of nanofillers positively influenced the performance of sensing toxic gases. Analyzing the composite morphology of these composites, it was found that presence of non-idealities (aggregates, different phases) promote further the sensitivity. Apart of that, reduced graphene oxide nanoplatelets were surface modified with noble metal nanoparticles. These hybrids have shown promising results as substrates for SERS measurements, identifying Fl molecules in concentration as low as 10^{-7} M, with excellent enhancement factor of order of 10^4.

References

1. Wang T, Huang D, Yang Z, Xu Z, He G, Li X, Hu N, Yin G, He D, Zhang L (2016) A review on graphene-based gas/vapor sensors with unique properties and potential applications. Nano-Micro Lett 8:95–119
2. Qi P, Wang Z, Wang R, Xu Y, Zhang T (2016) Studies on QCM-type NO_2 gas sensor based on graphene composites at room temperature. Chem Res Chin Univ 32(6):924–928
3. Procek M, Stolarczyk A, Pustelny T, Maciak E (2015) A study of a QCM sensor based on TiO_2 nanostructures for the detection of NO_2 and explosives vapours in air. Sensors 15:9563–9581
4. Prosheva M, Aboudzadeh MA, Leal GP, Blazhevska Gilev J, Tomovska R (2019) High-performance UV protective waterborne polymer coatings based on hybrid graphene/carbon nanotube radicals scavenging filler. Part Part Syst Charact 1800555:1–11
5. Pérez-Martínez BT, Farías-Cepeda L, Ovando-Medina VM, Asua JM, Rosales-Marines L, Tomovska R (2017) Miniemulsion copolymerization of (meth)acrylates in the presence of functionalized multiwalled carbon nanotubes for reinforced coating applications. Beilstein J Nanotechnol 8:1328–1337
6. Arzac A, Leal GP, de la Cal JC, Tomovska R (2017) Comparison of water-borne polymer/graphene nanocomposites. Macromol Mater Eng 302:1600315
7. Lin T, Lv X, Hu Z, Xu A, Feng C (2019) Semiconductor metal oxides as chemoresistive sensors for detecting volatile organic compounds. Sensors 19(2):233

Chapter 20
Based on Nanocomposite Resonant Photonic Crystal Structures for Sensing Applications

Tatiana Smirnova, Pavel Yezhov, Volodymyr Hryn, Oksana Sakhno, Volodymyr Fito, and Andriy Bendziak

Abstract The approximate theory of resonance phenomena in waveguides with volume gratings that allows obtaining analytical dependencies of the spectral position of guided-mode resonances on parameters of a waveguide structure was developed. Verification of the obtained results was carried out by the rigorous coupled-wave method (RCWA). Sensitivity of sensors based on waveguide gratings was estimated. Nanocomposite sensors were fabricated by holographic method. The characteristics of the sensor were measured. A modified method for refractive indices measurement was proposed. The sensitivity of the sensors for standard and modified measurement methods was compared. It was established that the proposed method allows increasing the sensitivity of the sensor by an order of magnitude.

Keywords Polymer nancomposite · Photonic crystals · Resonant Properties · Polymer sensors

20.1 Introduction

Detection of small amounts of toxic species and contaminants is critical for environmental monitoring, human health improvement as well as prevention of biological and chemical warfare threats. Therefore development of new highly sensitive and selective sensors has a high priority. We study photonic crystal (PC)

T. Smirnova (✉) · P. Yezhov · V. Hryn
Coherent and Quantum Optics Department, Institute of Physics of NASU, Kyiv, Ukraine
e-mail: smirnova@iop.kiev.ua

O. Sakhno
Fraunhofer Institute for Applied Polymer Research, Potsdam-Golm, Germany

V. Fito · A. Bendziak
Lviv Polytechnic National University, Lviv, Ukraine

C. Palestini (ed.), *Advanced Technologies for Security Applications*, NATO Science for Peace and Security Series B: Physics and Biophysics,
https://doi.org/10.1007/978-94-024-2021-0_20

sensors based on volume gratings recorded in a nanocomposite by a holographic method.

Periodic waveguide structures formed by periodic refractive index modulation possess the resonant phenomena termed guided-mode resonance, at which phase matching of the wave diffracted by the periodic structure with the eigenmode of the waveguide occurs. Under the resonant conditions, series of strong peaks appear in the reflection spectrum of structure. The central wavelength of these peaks can be changed by varying PC structure parameters and environment permittivity. Therefore, waveguide PC structures can be used as sensors of analytes deposited on PC surface by monitoring the resonance wavelength shift. Resonance conditions also results in the strong enhancement of the field in the waveguide and near its surface (local field). The excitation of local field can promote high enhancement of fluorescence and Raman scattering of analytes (enhancement effect). In this work, we investigate the label-free sensor using wavelength shift of resonant peaks.

The main types of polymer PC structures that are investigated now are structures with corrugated surface (e.g. [1, 6, 9, 10]). Their fabrication procedures (UV or electron beam lithography, nanoimprinting, hot or UV embossing, reactive ion etching, etc.) are complicated and rather expensive. In addition, it is difficult to achieve the homogeneity of the grating over the entire surface with the mentioned methods. Relief inhomogeneity causes spectral widening of resonance peaks, decreases of reflection coefficient and, consequently, reduces the sensitivity of the sensor.

We proposed, as an alternative, waveguide PC structures based on nanocomposites with volume periodic modulation of permittivity. This approach has a number of advantages. Volume 1D, 2D structures can be fabricated by a one-step holographic lithography method that enables the production of large-size PC structures with excellent homogeneity, easily varied symmetry and period, and flat surface. For structure fabrication, we used original photosensitive polymer-based nanocomposites containing inorganic nanoparticles (NPs) developed in the Institute of Physics (Ukraine) and Fraunhofer Institute for Applied Polymer Research (Germany) [7, 8].

The goals of our work are (i) theoretical analysis of resonant properties of volume PC structures and sensitivity of sensors on their base; (ii) fabrication of the waveguide PC structures in nanocomposite by holographic method; (iii) characterization of the label-free sensor.

20.2 Theoretical Analysis of the Resonant Properties of PC Structures: Development of the Approximate Theoretical Model

In case when a plane wave with a wavelength λ falls normally on a grating of period Λ then under waveguide resonance, in which the reflection coefficient of the grating is equal to one, the condition must be satisfied as follows:

$$\beta(\lambda_0) \cong 2\pi/\Lambda, \tag{20.1}$$

where $\beta(\lambda)$ is a constant propagation of waveguide mode for given wavelength.

Equation (20.1) is valid, because the amplitude of the grating refractive index modulation is much smaller than the average grating refractive index. However, in the case when the plane wave falls on grating at the angle θ, the condition of the waveguide resonance is written as follows:

$$\frac{2\pi}{\lambda} sin\theta \pm \frac{2\pi}{\Lambda} \cong \pm\beta(\lambda, n_a), \tag{20.2}$$

where θ is the propagation angle of the plane wave in the air, n_a is refractive index of the surrounding research medium. It should be noted that the incident wave can pass through the air, research medium, grating, substrate and again the air.

The constant propagation is about $15\mu m^{-1}$ for our waveguide parameters, consisting of the grating, substrate and researched medium. The difference between the right and left parts of Eqs. (20.1) and (20.2) is less than 0.001.

Let's expand the right part of Eq. (20.2) near the point λ_0 into the Taylor series; as a result we obtain the next equation:

$$\beta(\lambda) = \beta(\lambda_0) + \frac{d\beta}{d\lambda}\delta\lambda + \frac{1}{2}\frac{d^2\beta}{d\lambda^2}(\delta\lambda)^2 + \ldots \tag{20.3}$$

We abridged ourselves to the three terms of the constant expansion of the expansion and substitute the Eq. (20.3) into (20.2) as follows:

$$\frac{2\pi}{(\lambda_0 + \delta\lambda)} sin\theta \pm \frac{2\pi}{\Lambda} = \pm\left[\beta(\lambda_0) + \frac{d\beta}{d\lambda}\delta\lambda + \frac{1}{2}\frac{d^2\beta}{d\lambda^2}(\delta\lambda)^2\right] \tag{20.4}$$

Taking into account Eq. (20.1), finally it was obtained:

$$\frac{2\pi}{(\lambda_0 + \delta\lambda)} sin\theta = \pm\left[\frac{d\beta}{d\lambda}\delta\lambda + \frac{1}{2}\frac{d^2\beta}{d\lambda^2}(\delta\lambda)^2\right]. \tag{20.5}$$

The analysis of the Eq. (20.5) allows us to conclude that, in the case of the deviation of the incident plane wave from the normal to the angle θ, there are two resonant peaks. The one is at the wavelength shorter than λ_0 and the second one is at the wavelength longer than λ_0. In fact, Eq. (20.5) splits into two equations as follows:

$$F_1(\Delta\lambda_1) = \frac{2\pi}{(\lambda_0 + \delta\lambda_1)} sin\theta - \left[\frac{d\beta}{d\lambda}\Delta\lambda_1 + \frac{1}{2}\frac{d^2\beta}{d\lambda^2}(\Delta\lambda_2)^2\right] = 0 \tag{20.6a}$$

$$F_2(\Delta\lambda_2) = \frac{2\pi}{(\lambda_0 + \delta\lambda_2)} sin\theta + \left[\frac{d\beta}{d\lambda}\Delta\lambda_2 + \frac{1}{2}\frac{d^2\beta}{d\lambda^2}(\Delta\lambda_2)^2\right] = 0 \tag{20.6b}$$

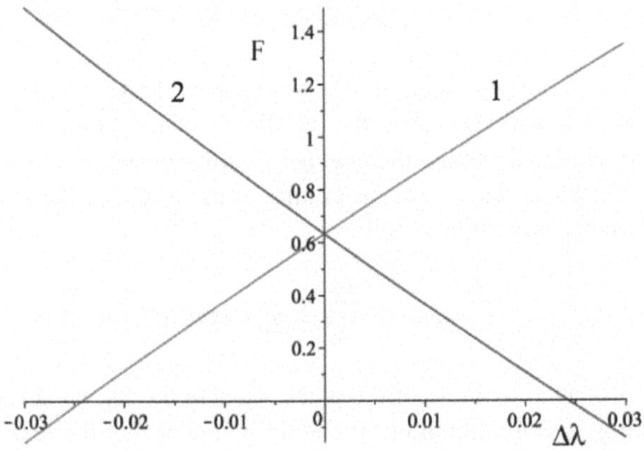

Fig. 20.1 Dependencies $F_1(\Delta\lambda_1)$ (1) and $F_2(\Delta\lambda_2)$ (2) at angle of incidence on grating is equal 3.5 angular degrees

Equations (20.6a) and (20.6b) are convenient to solve graphically. The first and second derivatives will be founded by numerically from the solution of the waveguide equation using the method described in Ref. [2]. The periodic structure was created with the following parameters: the average grating refractive index of 1.525, the refractive index of the glass substrate of 1.515, the modulation of the refractive index of the grating medium of 0.017, the grating thickness of 1.3 μm, the grating period of 0.395 μm. The resonant wavelength is 0.6053418 μm for these parameters at the normal incidence. Thus, the first $d\beta/d\lambda$ and the second derivatives are equals $-26.158647765\,\mu\text{m}^{-2}$ and $86.637\,\mu\text{m}^{-3}$ for this wavelength, respectively. The dependences $F_1(\Delta\lambda_1)$ and $F_2(\Delta\lambda_2)$ are shown in Fig. 20.1, where the roots of Eqs. (20.6a) and (20.6b) are determined by the intersection of the curves with the abscissa.

Thus, we obtained $\Delta\lambda_1 = -0.0242603\,\mu\text{m}$ and $\Delta\lambda_2 = 0.0242637\,\mu\text{m}$. Therefore, the predicted resonance wavelengths are equal $\lambda_1 = \lambda_0 + \Delta\lambda_1 = 0.5810815\,\mu\text{m}$, $\lambda_2 = \lambda_0 + \Delta\lambda_2 = 0.6296055\,\mu\text{m}$, respectively. The wavelengths calculated by RCWA for an infinite grating [5], at which the reflection coefficients are close to unity take values of 0.581138 and 0.629689 μm respectively. It can be seen that predicted wavelengths are very close to the wavelengths determined from the spectral dependences of the reflection coefficient calculated by RCWA when the incident beam is deviated from the normal.

The spectral dependence of reflectance (R) on wavelength at normal incidence and angle of incidence of 3.5° when n_a=1 is shown in Fig. 20.2.

The spectral half-width (full width at half-maximum) of the reflection peaks depends on the angle of incidence. The half-width of the spectrum is 0.0018 nm at zero beam incidence on the grating (the refractive index of the substrate is of 1.45) and increases to 0.0025 nm the angle of 30° in the air.

Theoretical investigations were shown that sensor sensitivity increases when the angle of incidence on the grating increases. It also increases when the refractive

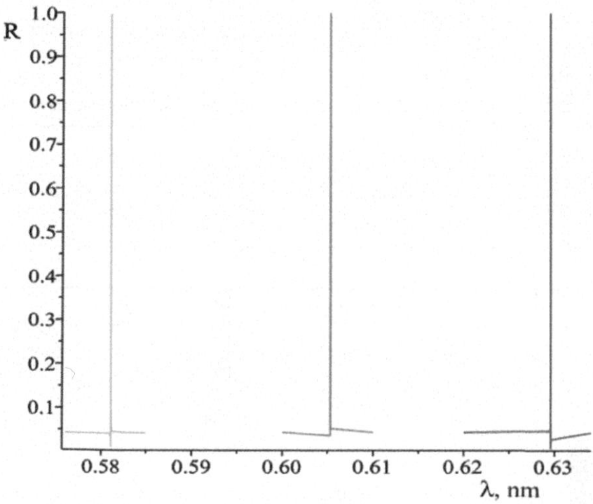

Fig. 20.2 Spectral dependencies of the reflectance (R) at the incident angles of a plane wave of 0° (central peak) and 3.5° (right and left peaks)

index of the researched medium approaches to the average grating refractive index. It was also found that PC structures with a thickness (d) of 1–2 μm, period (Λ) $\leq$ 400 nm and refractive index amplitude modulation (n_1) ≥ 0.01 are the most suitable for effective operation of the sensors.

20.3 Fabrication of PC Structures by Holographic Method

The organic inorganic UV photosensitive nanocomposite for fabrication of volume PC structures was prepared using our earlier developed technology [8]. The holographic nanocomposites are generally consisting of organic monomers of different functionality (as an organic low refractive index (n) matrix) and inorganic nanoparticles, NPs (as a high n additive). A mixture of mono-functional, isobutyl acrylate (IBA, n = 1.476) monomer and multi-functional acrylate monomer (Sartomer SR444, n = 1.481) was employed for preparation of a monomer bland using 75% wt.% IBA and 25% SR444, respectively. The photoinitiator Irgacure 1700 (Ciba) 1.5 wt.% was added to the monomer blend to provide the material sensitivity to UV (355 nm) light. The inorganic NPs series X green ($LaPO_4$ doped with rear-earth elements) from Fraunhofer Zentrum für Angewandte Nanotechnologie CAN (Hamburg, Germany) were used. The average NPs size was found of about 3–4 nm. According to the manufacturer's information the current NPs possess specific gravity of 5 g/cm^3 and the refractive index of about 1.7. The weight fraction of the inorganic core was found to be of about 85 wt.%. The concentration of the NPs

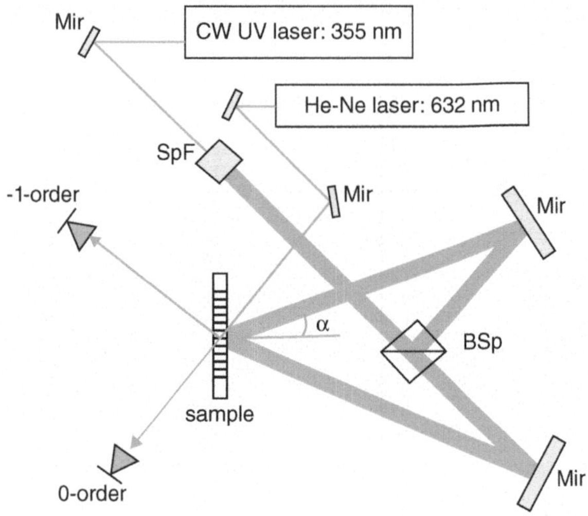

Fig. 20.3 Schematic presentation of the holographic setup: SpF-spatial filter and telescope unit, Mir – mirrors, BSp – a beam-splitter, dividing the input laser beam on two mutual recording beams

was about 25 wt.%. The fabrication method of recording layer with the thickness 1–2 μm and high film quality was developed and described in [3].

A conventional two-beam interference scheme based on an DPSS Genesis-SLM laser (Coherent) emitting at $\lambda_{rec} = 355$ nm (s-polarization) was used for holographic fabrication of PC volume structure (Fig. 20.3).

Two laser beams of about 15 mm diameter and equal intensities produce the interference pattern in the intersection plane (position "sample" in Fig. 20.3). The spatial period of the grating (Λ) was 395 nm. Because of the symmetric beam configuration, the fringes of the gratings are perpendicular to the substrates. The grating is formed due to diffusion NP redistribution in polymer matrix during exposure. The grating formation was monitored in real time by diffraction of a He–Ne laser beam (s-polarization, $\lambda_t = 633$ nm), placed at the Bragg angle corresponding to the period of the grating. The samples were exposed until the steady-state value of the diffraction efficiency was achieved (typically for 80–150 s). The recording intensity (I) of the exposing beams was varied from 10 to 50 mW/cm^2. The diffraction efficiency (η) of the grating was determined as $\eta = I_{dif}/(I_{tr} + I_{dif})$, where I_{tr} and I_{dif} are the intensity of the zero diffraction order (transmitted beam) and the first diffraction order (diffracted beam). Such estimation allows excluding the Fresnel reflection in the substrates as well as scattered light and linear absorption of the layer.

The Bragg diffraction conditions are satisfied for the gratings with the specified parameters. Only the 0th and -1st diffraction orders were observed in the diffraction pattern. Therefore, we used Kogelnik's formula [4] for calculation of the refractive index modulation amplitude, $n_1 = \lambda cos\theta_B asin(\eta)^{1/2}/\pi d$, where θ_B is the Bragg angle within the material, λ is the wavelength of the reading beam.

We fabricated a series of volume waveguide PC-structures, which were used for further researches. Their characteristics are follows: $d = 1.25 \div 1.70\,\mu$m, $n_1 = 0.012 \div 0.017$.

20.4 Investigation of the Label-Free Sensor Based on the Nanocomposite PC Waveguide Structures

In order to study the properties of the sensor with volume nanocomposite PC structure a special fluid cell was fabricated. The scheme and the photo of the cell are shown in Fig. 20.4a,b. As shown in Fig. 20.4 the substrate with grating was mounted into a cell placing the grating in contact with analyte medium. The liquid was injected into the cavity of the cell with a syringe, located in the lower part of the cell. The second syringe in the upper part of the cell provided air displacement at the cuvette filling.

To change the angle of incidence the cell was installed into the goniometric stage. The halogen lamp with maximum intensity in the red part of the spectrum (of about 670 nm) was used as a light source. The spectral position of the long-wave peak in the specular reflection spectrum (the zero order diffracted beam) depending on the refractive index of the studied liquid, n_{an}, was measured.

The range of n_{an} variation for which the measurements were performed was chosen taking into account the waveguide-gratings characteristics. At $n_{an} > 1.5$, the analyte refractive index approaches the average refractive index of the waveguide-grating, n_{gr}=1.525. The system becomes unstable; the conditions for propagation of guided-modes and, correspondently, the conditions for the excitation of resonances are violated. The interaction of liquids with PC gratings was previously investigated. Only such liquids were selected, which in direct contact with the grating for 4 hours did not lead to the destruction of the grating or to the change of its parameters (diffraction efficiency, thickness).

It was found from theoretical estimation that the sensitivity of the sensor, defined as $S = \Delta\lambda/\Delta n$, increases with increasing angle of incidence. In addition, the long-wave resonance is more sensitive to the changes in the refractive index than the short-wave one, *i.e.* the spectral shift of the long-wave resonance peak, when

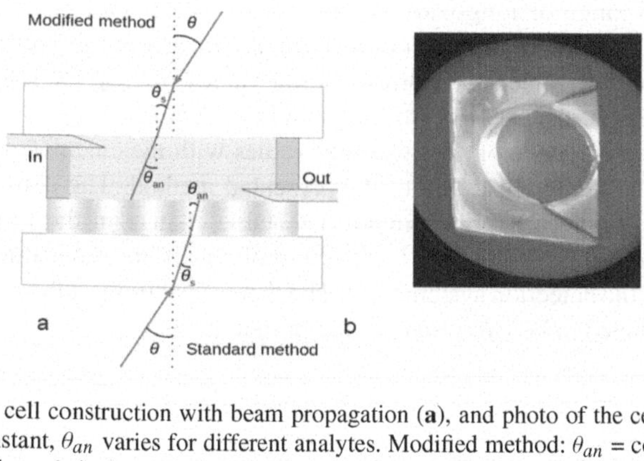

Fig. 20.4 Fluid cell construction with beam propagation (**a**), and photo of the cell (**b**). Standard method: θ = constant, θ_{an} varies for different analytes. Modified method: θ_{an} = constant, θ varies for different analytes θ_s is the angle in substrate

the refractive index changes, exceeds the shift of the short-wave peak. Therefore, the following measurements conditions were chosen for experimental studies. For different analytes the wavelength of the long-wave peak in the reflection spectrum was determined. The only difference of the observed specular reflection spectra for different analytes was the position of the resonance peaks. The half-width of the resonance peaks in all cases did not exceed 0.012 nm. The reflection coefficients were varied in the range of 10–15%. It should be noted that the value of the reflection coefficient is not fundamental by measuring the refractive indices of analytes.

In the first case light irradiation with TE polarization entered the cell from the side of a substrate, on which the PC grating is located (standard measurements). The angle of incidence in air, θ, was 20°. The dependences of λ_{res} on n_{an} were measured and theoretically calculated. There is a good agreement of the measurement results with theoretical ones. The $\lambda_{res}(n_{an})$ dependence is nonlinear, so the sensitivity of the sensor, $\Delta\lambda_{res}/\Delta n_{an}$, depends on the refractive index of the analyte and increases when its value approaches the average refractive index of the grating. The defined sensor sensitivity in this case varies within 0.4÷12 nm/RIU. One pixel shift in the peak position corresponds to a shift of 0.012 nm, which can be measured with our measurement system. As a result, a minimum detectable change in the refractive index, Δn_{min}, also varies in a range 0.0279 ÷ 0.001 RIU.

The sensitivity of sensor can be increased by modifying the measurement procedure. If the radiation enters the cell from the side of test analyte, by changing the angle of incidence θ, it is possible to measure the resonance wavelength at a constant angle θ_{an} of the radiation propagation in analytes with different n_{an}. The theoretically calculated dependences of the resonance wavelength on the refractive index of the medium for different θ_{an} shown that in this case the sensitivity of the measurements also arises with increasing θ_{an}. Herewith a long-wave reflection peak is more sensitive to n_{an} changes.

In the pointed range of n_{an}, the dependence $\lambda_{res}(n_{an})$ can be approximated by a straight line Fig. 20.5. The slope of this line determines the sensitivity of sensor. When $\theta_{an} = 15°$ and $1 \leq n_{an} \leq 1.5$, the range of θ is 1° ÷ 22.844°. It corresponds to the spectral range of long-wave resonances of 708.344 nm ÷ 761.730 nm. By calculating the value of θ for each of analyte and measuring the reflection spectra for each θ, we determined the corresponding values of λ_{res} for long-wavelength peaks. The measurement results are shown in Fig. 20.5.

There is a coincidence of the measured values with the calculated theoretically. It confirms clear the applicability of the presented method. The defined value of S is equal to 122 nm/RIU. The minimum detectable change in the refractive index, Δn_{min}, was found as 1×10^{-4} RIU. It is important to note that this limitation is determined by the detection system used. The theoretical limit of the refractive index change is predicted to be two orders of magnitude lower.

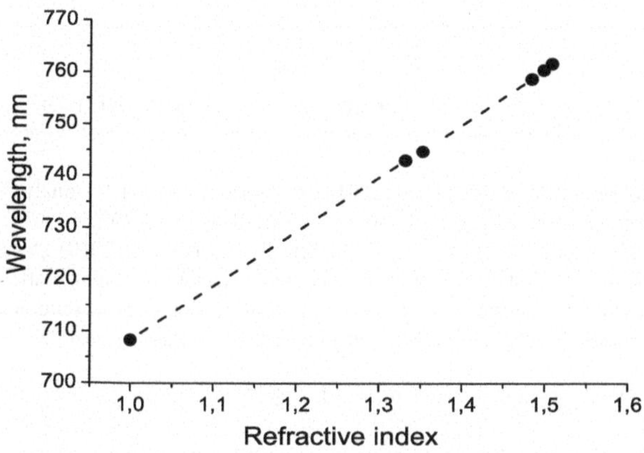

Fig. 20.5 Experimentally measured spectral positions of resonance peaks (circle dots) obtained for the angle in liquid 15° and dependence of resonance wavelength on refractive index calculated by RCWA method (dash line)

20.5 Conclusions

Approximate methods for determining the parameters of the created grating as sensitive sensor element for measuring the refractive index of the research medium has been developed. It is possible to calculate the resonance wavelengths for different angles of incidence using obtained approximate equations and knowing λ_0.

The exact values of the resonant wavelengths at different angles of the incidence on the grating were found by the RCWA method. They are very close to the obtained values of wavelengths using developed approximate method.

Suitable organic-inorganic nanocomposites were prepared. Optimization of the exposure conditions allows obtaining the PC structures with thickness $<2\,\mu m$ and the refractive index amplitude modulation $0.01 \div 0.017$.

The characteristics of the sensor based on nanocomposite PC waveguide structure were examined. A modified measurement method was proposed. The modified method provides the sensor sensitivity 122 nm/RIU and minimal detectable change in the refractive index Δn_{min} of about 1×10^{-4} RIU.

Acknowledgments This work was supported by the grant in the framework of the a Multi-years Project from the SPS Programme – SPS G5351 "Science for Peace and Security" "Nanocomposite Based Photonic Crystal Sensors of Biological and Chemical Agents" (Project 12/18-52).

References

1. Chaudhery V, George S, Lu M, Pokhriyal A, Cunningham BT (2013) Nanostructured surfaces and detection instrumentation for photonic crystal enhanced fluorescence. Sensors 13:5561–5584
2. Fitio VM, Romakh VV, Bobitski YV (2014) Numerical method for analysis of waveguide modes in planar gradient waveguides. Mater Sci (Medžiagotyra) 20:256–261
3. Hryn V, Sakhno O, Bendziak A, Fito V, Yezhov P, Smirnova T (2019) Development of the waveguide photonic crystal structures formed by distribution of nanoparticles in polymer matrix. In: Fesenko O, Yatsenko L (eds) Nanophotonics, Nanooptics, Nanobiotechnology, and Their Applications. NANO 2018. Springer Proceedings in Physics, vol 222. Springer, Cham, pp 73–85
4. Kogelnik H (1969) Coupled wave theory for thick hologram gratings. Bell Syst Tech J 48:2909–2947
5. Moharam MG, Gaylord TK (1983) Rigorous coupled-wave analysis of grating diffraction – e-mode polarization and losses. J Opt Soc Am 73:451–455
6. Nair RV, Vijaya R (2010) Photonic crystal sensors: an overview. Prog Quantum Electron 34:89–134
7. Sakhno OV, Goldenberg LM, Stumpe J, Smirnova TN (2009) Effective volume holographic structures based on organic-inorganic photopolymer nanocomposites. J Opt A Pure Appl Opt 11:024013
8. Sakhno OV, Smirnova TN, Goldenberg LM, Stumpe J (2008) Holographic patterning of luminescent photopolymer nanocomposites. Mater Sci Eng C 28:28–35
9. Yaremchuk I, Tamulevičius T, Fitio V, Gražulevičiūte I, Bobitski Y, Tamulevičius S (2013) Guide-mode resonance characteristics of periodic structure on base of diamond-like carbon film. Opt Commun 301–302:1–6
10. Zhuo Y, Cunningham BT (2015) Label-free biosensor imaging on photonic crystal surfaces. Sensors 15:21613–21635

Chapter 21
Hand-Held Gamma Detector Based on High-Pressure Xenon Gas: Determination of Charge Carrier Lifetime in Xe

V. Kutny, A. Rybka, A. Pudov, A. Abyzov, S. Sokolov, L. Davydov, S. Melnikov, G. Kholomyeyev, K. Kutny, I. Kolodiy, R. Austin, and U. Rohatgi

Abstract The primary goal of the work is the design and construction of the hand-held gamma detector based on high-pressure xenon (HPXe). The primary use of the detector with spectrometric capabilities is the detection and identification of radioactive materials in the normal stream of commerce at rail terminals, seaports, border crossings. The advantages of the detectors are high resolution, light weight, robustness, radiation and temperature stability. A major part in reaching the project's final goal is creation of a Xe purification facility. Xenon gas compressed up to pressure ~50 bar and sometimes doped with hydrogen, methane or others gases, which is the working medium for the high-resolution HPXe detectors, needs to be very pure. The purity level can be measured directly or, alternatively, the gas usability in gamma-spectrometers can be evaluated indirectly through the charge carrier (electron) lifetime measurements. Typically, lifetime determination methods involve the measurement and analysis of individual ionization pulses from particles registered in a Xe-filled ionization chamber. In the present paper, we studied the HPXe electron lifetime by using measurements in a cylindrical ionization chamber and developed the respective analytical charge transport model. Our results showed that it is possible to determine electron lifetime in the cylindrical configuration. The parallel plate electrode configuration was studied for comparison. The measurements were conducted in a two-electrode configuration for a range of pressure values (5–50 bar) for the $Xe+0.25\%H_2$ gas mixture and pure Xe.

V. Kutny (✉) · A. Rybka · A. Pudov · A. Abyzov · S. Sokolov · L. Davydov · S. Melnikov · G. Kholomyeyev · K. Kutny · I. Kolodiy
National Science Center "Kharkov Institute of Physics and Technology", Kharkov, Ukraine
e-mail: kutny@kipt.kharkov.ua

R. Austin
Florida Polytechnic University, Lakeland, FL, USA

U. Rohatgi
Stony Brook University of New York, Stony Brook, NY, USA

C. Palestini (ed.), *Advanced Technologies for Security Applications*, NATO Science for Peace and Security Series B: Physics and Biophysics,
https://doi.org/10.1007/978-94-024-2021-0_21

Keywords Xenon · HPXe · Ionization chamber · Charge carrier lifetime measurements · Modeling

21.1 Introduction

The spectrometers of γ-radiation based on high-pressure high-purity xenon gas (HPXe) have excellent detecting properties and found numerous applications. Their energy resolution has reached 1.7% (at 662 eV), and the possibility of their operation with the high acoustic noise background has been demonstrated [1]. Such spectrometers have been used in the studies of the space gamma radiation. Their applicability in a number of other areas has been demonstrated: characterization of spent fuel bundles [2]; characterization of radioactive waste [3]. The use of HPXe detector bundles in port-of-entry security portals is studied. Among the advantages of the HPXe detectors are their high energy resolution, high sensitivity, large detecting volume and its scalability, ability of the detector to operate at elevated temperatures [4].

The Ukraine's first HPXe spectrometric breadboard detector model was made at National Science Center "Kharkov Institute of Physics and Technology" (NSC KIPT) within SPS project #G4655 (2014–2017). The primary goal of the current project (SPS project #5373) is the design and construction of a prototype hand-held gamma detector with improved characteristics based on HPXe for the detection, identification and characterization of radioactive materials. We hope to improve the detector registration efficiency by a factor of 5–10 (up to ~70% for the 100 keV γ-quanta) by reducing the detector wall thickness down to ~500 μm and by using novel carbon-fiber-based reinforcement layers for structural strength. Such improvement would result in the increased detecting range of the device, therefore extending the areas of its application. The new detector design will use its own custom high-pressure components, the Xe purity level will be increased. The development work should lead to the production of highly reliable portable γ-radiation detectors.

In the project, NSC KIPT is responsible for the development of the hand-held HPXe spectrometric detector model and the Xe purification system, Florida Polytechnic University, USA, is responsible for the work related to the 3D printing of the outer layers of thin-wall detectors, Stony Brook University of New York, USA, provides the scientific guidance and the financial management of the project.

The detector is intended for use in security applications to detect whether radioactive materials surrounded by benign cargo are being transported in the normal stream of commerce. Some cargo contains radioactive materials for peaceful purposes (naturally occurring radioactive isotopes, medical and industrial isotopes etc.); other cargo may contain special radioactive materials like weapon grade plutonium intended for malevolent use. Spectrometric identification and characterization of these materials is necessary to determine what radioisotope is present and make a decision on whether it is intended for malevolent use. Detectors with spectrometric capabilities can resolve the γ-spectra of radioactive materials and accurately identify them. Gamma detectors based on HPXe, having an energy resolution of 2.0–2.5%, are such detectors. In addition, from the experience gained with this detector, we see a wider range of applications of HPXe detectors in diagnostics, nuclear power engineering and other applications.

This particular article reports on the aspect of improving the measurements of the Xe purity, by measuring the respective charge carrier lifetime. The requirement for the high purity of Xe gas arises due to the detrimental role of electro-negative impurities that cause the charge carriers, created in the events of the registration of ionizing particles in the chamber of the detector, to get trapped before reaching the collecting electrode. In practice, the usability of Xe or its mixtures for spectroscopic applications can be determined through the measurement of the electron lifetime of gas. There is a number of different lifetime determination methods and measurement chamber designs reported in the literature. Typically, they use a separate dedicated ionization chamber in the Xe gas transfer installation.

In the present paper, we report on proving the possibility and conducting the measurements of the electron lifetime in the detector (a cylindrical ionization chamber), developed and built by us together with an installation for Xe purification and transfer. For the data analysis, we developed and used an analytical model for charge transport inside an ionization chamber of cylindrical configuration. To compare our results to the more typical parallel plate measurement configuration, we created the respective purity control ionization chamber with the parallel-plate measurement anode and cathode, which will also be used in the lifetime determination studies.

21.2 Experimental

The cylindrical ionization chamber of the HPXe γ-detector prototype was designed and built [5, 6] (Fig. 21.1). In this study, it was used for the measurements of the electron lifetime, which is indicative of Xe gas purity. A second ionization chamber was designed and built [7] (Fig. 21.2) for Xe purity control via pulse measurements in the parallel-electrode configuration. The parameters of the chambers are listed in Table 21.1.

The chambers described above were attached, one at a time, to the Xe purification and transfer installation, already designed and built by us [5]. The installation can work with high vacuum (2×10^{-9} mbar), high pressure (<60 bar), high-purity gas mixtures, possesses the adaptability to attach new gas purification units and other units, has the function of the cryogenic Xe transfer and purification. The installation is equipped with a heating system – the flexible resistive heating elements and a controller, which allows us to heat seven independent zones of the installation up to 450 °C. Such heating is used for degassing of the inside walls of all chambers, cylinders and pipes of the installation prior to pure Xe gas transfer procedures.

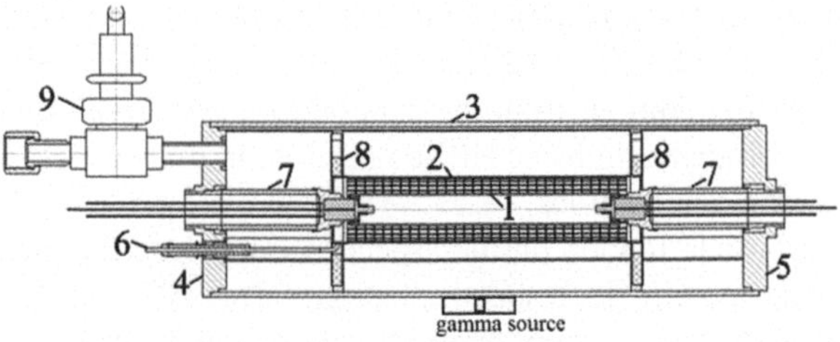

Fig. 21.1 A drawing of the created HPXe detector ionization chamber: *1* – hollow Ti anode; *2* – Frisch grid; *3* – SS cathode (the body); *4, 5* – left and right flanges; *6* – Frisch grid feedthrough; *7* – anode feedthroughs; *8* – ceramic grid supports; *9* – high-pressure vent

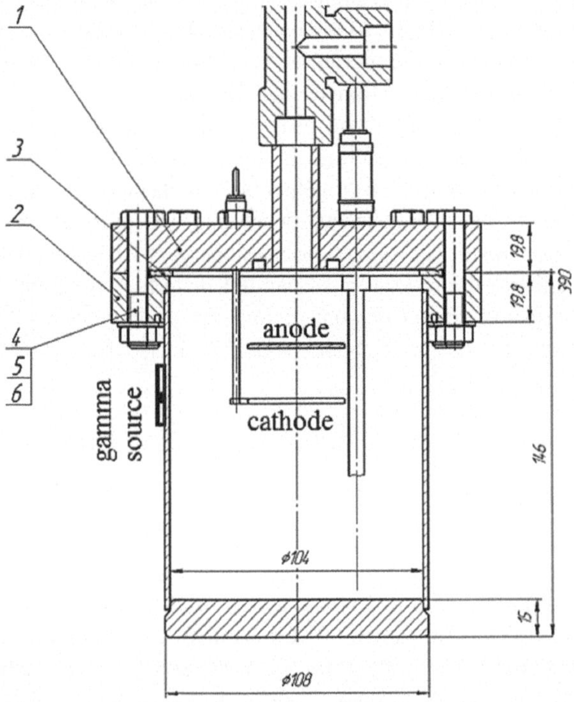

Fig. 21.2 Purification and purity control chamber

The typical degassing (heating to ~250 °C for several days) of the installation with chambers was carried out in order to prepare it for work with pure gases. The detector chamber was subsequently filled with Xe+0.25%H_2 mixture from 5 to 50 bar. The purity control chamber was filled with pure Xe from 10 to 40 bar. At each pressure, the gas electronic characteristics were evaluated. In this work, we used Xe gas of 99.9999% purity doped with 0.25% H_2, as well as undoped Xe gas of 99.9999% purity. The impurity analysis of the xenon was carried out. The presence of such impurities as Kr, N_2, O_2, C_nH_m, CO_2, CF_4 (fluorine-containing),

Table 21.1 The characteristics of both ionization chambers

Detector characteristic	Value, units	Purity control chamber characteristic	Value, units
Working anode length	180 mm	Inner diameter	104 mm
Anode outer diameter	20 mm	Inner volume height	131 mm
Frisch grid length	180 mm	Inner volume	1.11 l
Frisch grid outer diameter	40 mm	Measurement electrode diameter	5 cm
Cathode inner diameter	100 mm	Measurement electrode spacing	2 cm
Cathode wall thickness	3 mm	Chamber wall thickness	2 mm
Working gas volume	≈1.41 l		
Frisch grid transparency	96%		

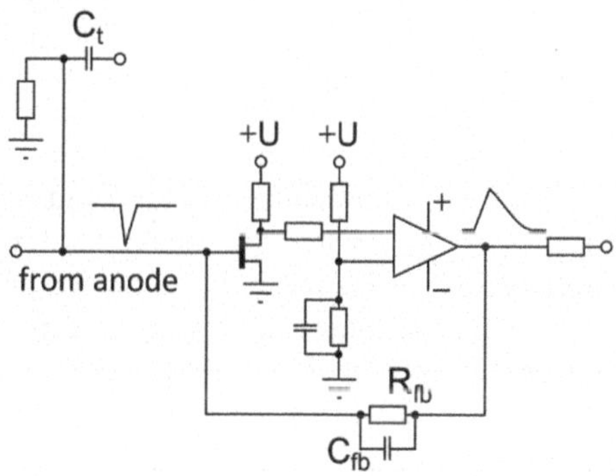

Fig. 21.3 A custom preamplifier electronic diagram

H_2O by gas chromatography, IR-spectrometry, measuring the humidity of gases, and by other mass spectroscopy methods, was determined.

Cosmic muons, as in [8], were used as ionization particles for the detector lifetime measurements. For both detector ionization chamber and the purity control chamber, ^{137}Cs standard gamma source was also used. The placement of the γ-source during the measurements is indicated in Figs. 21.1 and 21.2.

The spectroscopic measurement channel was developed and used for acquisition and analysis of individual pulses from the γ-radiation source and muons. It consists of a Spellman SL30 high-voltage supply, an Amptek A250CF CoolFET charge-sensitive preamplifier, an Ortec 927 Multichannel Analyzer, an Ortec 672 Spectroscopy Amplifier, a Tektronics MDO 3034 oscilloscope. The Amptek preamplifier was used for the control chamber measurements. A custom preamplifier (Fig. 21.3) was used for the detector measurements.

The carrier lifetime (τ) evaluation was performed by measuring and analyzing individual current pulses produced by ionizing radiation in the ionization chamber. The actual sensitive part of the volume of the detector chamber was between the anode and the Frisch grid, and it was between the two measurement electrodes for the purity control chamber. The applied voltage between the electrodes was as low as possible, typically 50 V or higher if the application of 50 V did not result in any signal.

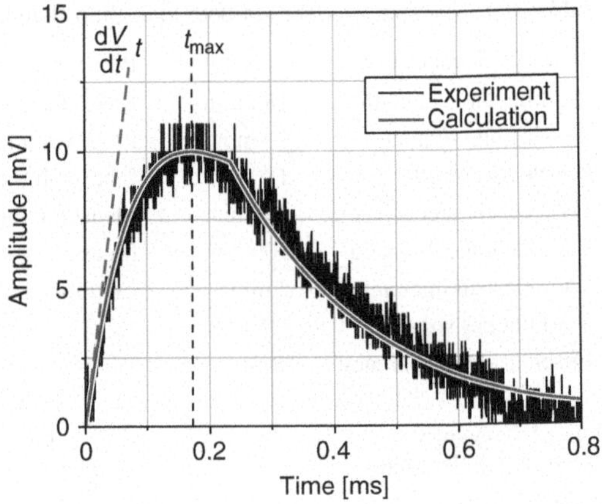

Fig. 21.4 A measured pulse (thin line) from a single cosmic muon. A calculated pulse (thick line)

Figure 21.4 shows an example of a measured pulse. According to literature (see e.g [10].), it is possible to evaluate the low limit of τ by determining time of the rising portion of the pulse (t_{max}). Such argument can be made from the Eq. (21.1), where Q is the total charge generated in Xe by an ionizing particle, ΔQ is the collected charge, t_C is the charge collection time. The implied basic conclusion of this equation is that the charge carrier lifetime is equal to or higher than the collection time.

$$\frac{\Delta Q}{Q} = 1 - \exp\left(-\frac{t_C}{\tau}\right) \approx \frac{t_C}{\tau}, \quad \tau = t_C \frac{Q}{\Delta Q}, \qquad \tau \geq t_C \tag{21.1}$$

The actual value of the electron lifetime is most likely higher than the measured pulse rise time as the method implies.

We have developed an analytical model for the charge transport in the parallel-plate and the cylindrical ionization chambers and analyzed the obtained lifetime measurement results. The next section describes this model. It will demonstrate, that simple logic of Eq. (21.1) can be improved.

21.3 The Analytical Model of the Charge Transport in the HPXe Chamber

21.3.1 The Equation of State for Xenon

First of all, the calculation of the equation of state of the gas was performed in order to be able to obtain the values of the gas density at certain gas pressure. For the calculation of dependence of Xe density on pressure, we will apply a slightly modified equation of state [9]

$$p = \frac{RT(1-\epsilon)}{V^2}(V+B) - \frac{A}{V^2}, \tag{21.2}$$

$$A = A_0\left(1 - \frac{a}{V}\right), \quad B = B_0\left(1 - \frac{b}{V}\right), \quad \epsilon = \frac{c}{VT^3}. \tag{21.3}$$

$$V = \frac{M}{\rho}. \tag{21.4}$$

Here, pressure p is given in bar, V is the molar volume in cm^3, T is the temperature in Kelvins, ρ is the density in g/cm^3, and the following parameter values were used:

$$R = 83.1451, A_0 = 4.8 \cdot 10^6, a = 34, \quad B_0 = 71, b = 2.2, c = 2 \cdot 10^8,$$
$$M = 131.293 \text{ g/mol}, \tag{21.5}$$

which gave a good agreement to the measurement results in the pressure range of interest $p = 0 \ldots 60$ atm.

The value of the Xe atomic concentration is determined from the expression:

$$N = \rho \frac{N_A}{M}. \tag{21.6}$$

Figure 21.5 shows the simulated Xe atomic concentration (N) dependence on pressure for temperatures 20, 30, 40, and 50 °C.

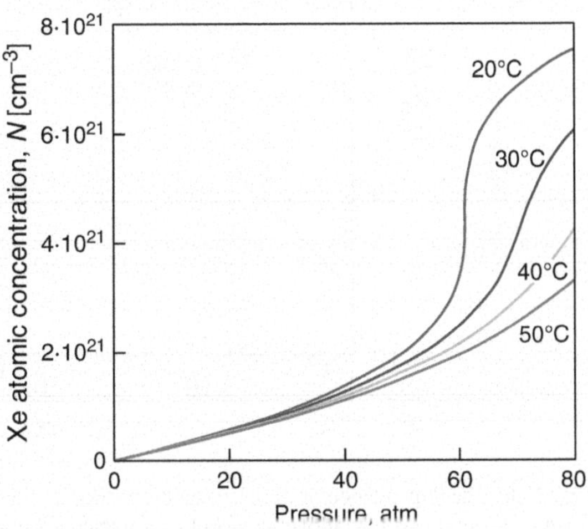

Fig. 21.5 Xe atomic concentration (N) dependence on pressure for different temperatures

21.3.2 Mobility and Drift Velocity of Electrons in Xenon

The next step is to calculate the dependence of the electron drift velocity as a function of xenon concentration.

Electric field strength between the cathode and the anode in the cylindrical configuration (detector – det.) and flat electrode configuration (purity control chamber – PCC) is determined by the expressions:

$$E = \frac{U}{r \ln \frac{R_2}{R_1}}, \text{(det.)}; \qquad E = \text{const}\,(PCC) \tag{21.7}$$

where R_1, R_2 are the anode and cathode radii, respectively, and r ($R_1 \leq r \leq R_2$) is the radial distance from the cylinder axis. For the detector measurements, the electrodes' (the anode and the grid, in our case) radii are $R_1 = 1$ cm, $R_2 = 2$ cm, and the field strength from the grid to the anode grows in the radial direction, as in Eq. 21.7, by a factor of ~ 2.

To determine the electron drift velocity W for pure Xe (measurement in PCC), we use the data from [10] (Fig. 21.6). This data at low ratios of E/N are well approximated by the analytical expression:

$$W = 2.433 \cdot 10^5 \frac{E}{N}\left(1 + 32.5 \cdot 10^{17} \frac{E}{N}\right) \tag{21.8}$$

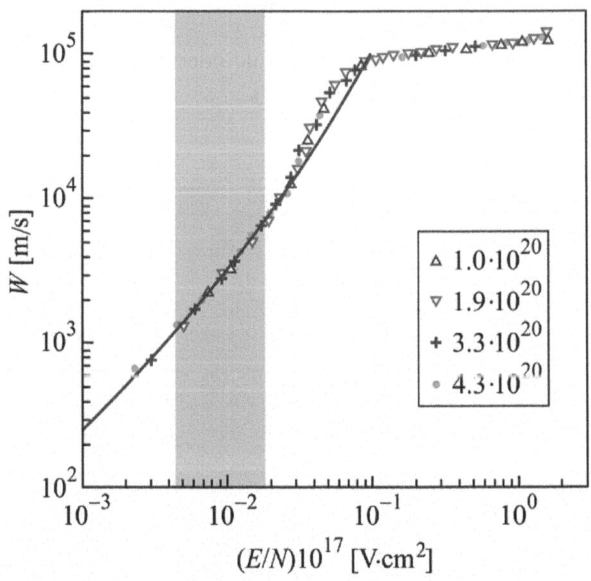

Fig. 21.6 The dependence of the electron drift velocity on the E/N for different concentration of hydrogen from [12]. The gray area span E/N values used in the measurements

Similarly, taking into account the influence of the 0.25% H_2 addition (measurements in the detector).

$$\mu N = \begin{cases} \mu N_0, & \text{if } \mu N_0 > 4 \cdot 10^{22} \\ 4 \cdot 10^{22}, & \text{if } \mu N_0 \le 4 \cdot 10^{22} \end{cases}, \tag{21.9}$$

$$\mu N_0 = 10^{20} f\left(-\frac{E/N - d}{\Delta}\right) + 9.07 \cdot 10^9 \left(\frac{E}{N}\right)^{-0.82} f\left(\frac{\frac{E}{N} - d}{\Delta}\right)$$

$$f(x) = \frac{1 + \tanh x}{2}, d = 10^{-18}, \Delta = 0.6d. \tag{21.10}$$

From these expressions, we get the values for W:

$$W = \frac{E}{N} \mu N \left(\frac{E}{N}\right), \tag{21.11}$$

The curves calculated in this approximation are presented in Fig. 21.7, and they describe the experimental data of W vs. electric field strength in the interval $E = 1.5 \ldots 10$ kV [11] well.

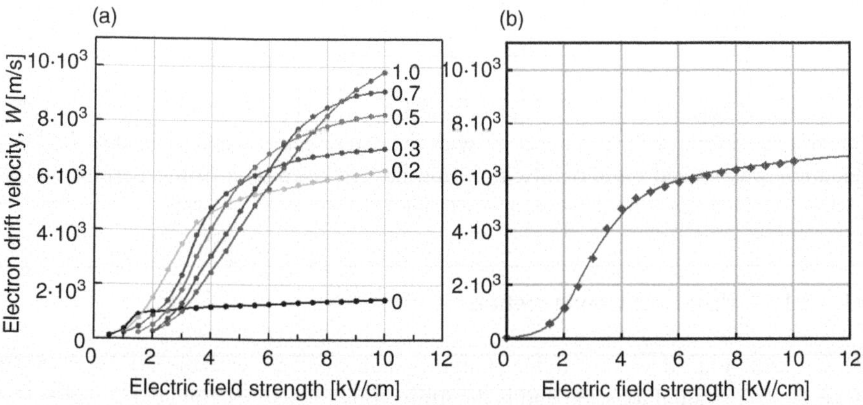

Fig. 21.7 (**a**) The dependence of the electron drift velocity on the electric field strength for different concentration of hydrogen from [13], (**b**) The line is calculated from Eqs. (21.9, 21.10, and 21.11), the points are the data from [13] calculated for the 0.25% H_2 concentration

21.3.3 The Calculation of the Pulse Shape

As mentioned in the literature, in [10] for example, the duration of the rise portion of the registered pulse was used to assess the low limit of the electron lifetime. However, to perform a more precise electron lifetime determination, it is possible to fit the measured pulses with the mathematically modeled curves and obtain the sought electron lifetime that way. We have created the analytical model for the calculation of pulses for the parallel-plate and cylindrical chamber geometries so that the measured pulses could be fitted.

21.3.3.1 Parallel Plate Configuration

The current induced by the moving charge in the chamber is determined by the expressions:

$$j(t) = \frac{q_0}{L} W \exp\left(-\frac{t}{\tau}\right) \tag{21.12}$$

where q_0 is the generated charge, τ is electron lifetime, L is the anode-cathode distance.

The voltage in the output of the charge-sensitive preamplifier, u, is determined by differential equation

$$\frac{du}{dt} = \frac{j(t)}{C} - \frac{u}{\tau_{RC}}. \tag{21.13}$$

where $j(t)$ is determined by Eq. (21.12), and $\tau_{RC} = RC$ is the time constant of the integrating circuit of the preamplifier with the capacitance C and resistance R. This equation was solved numerically, the lifetime was used as a fitting parameter for getting the best fit of the dependence $u(t)$ and the measured pulses.

21.3.3.2 Cylindrical Configuration

The current induced by the moving charge q_0 in the detector with the axial symmetry depends on its radial location and is determined by the expression [12]:

$$j(r) = \frac{q_0 E(r)}{U} W(r) \exp\left\{-\frac{t(r)}{\tau}\right\}. \tag{21.14}$$

When solving this equation together with the charge movement equation

$$dr = W(r)dt, \tag{21.15}$$

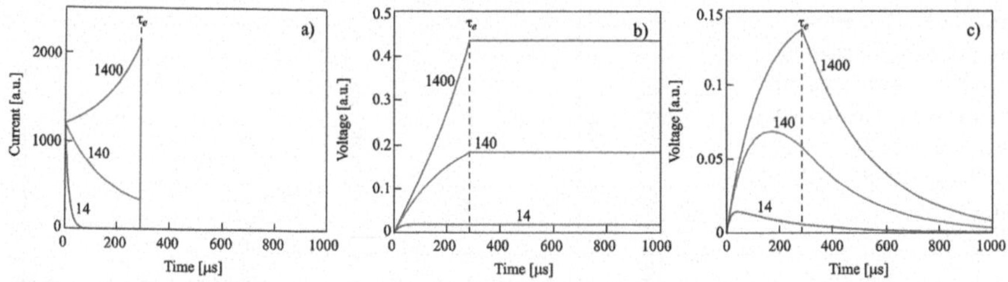

Fig. 21.8 Dependences on time of: (**a**) anode current Eq. (21.11); (**b**) preamplifier voltage for $\tau_{RC} \to \infty$; (**c**) preamplifier voltage for $\tau_{RC} = 250$ μs Eq. (21.13). The example calculations use the following parameters: $r = 1.5$ cm, $\tau = 14$, 140 and 1400 μs, pressure $p = 20$ atm

it is easy to obtain the anode current dependence on time $j(t)$, and then with Eq. (21.13) the voltage in the output of the charge-sensitive preamplifier.

For the case of the cylindrical configuration, Fig. 21.8 presents examples of the dependences $j(t)$ and $u(t)$, calculated by using the above model, for three different electron lifetime values; the value of the integrating circuit time is constant.

21.4 Results

As mentioned, a significant number of pulses from cosmic muons and ^{137}Cs γ-quanta were measured in the ionization chambers filled with Xe and $Xe + H_2(0.25\%)$ up to several pressure values. The general shape of the measured pulses was typical as in Fig. 21.4, though sometimes different pulse shapes were observed.

21.4.1 The Parallel Plate Configuration Measurements in Xe

The values of t_{max}, that is, time from the beginning of the pulse to its maximum, were determined for all pulses measured in the parallel plate configuration of the purity control chamber for pure Xe, and they are shown in Fig. 21.9 as a function of Xe pressure. At each pressure, there is a significant spread in the values, but it is obvious that t_{max} grows with pressure up to 20 bar and then decreases. This indicates that, up to 20 bar, t_{max} depends on (practically equal to) the electron collection time t_c. Above this pressure, the collection time becomes longer in comparison to the charge carrier (electron) lifetime, and so, now, the latter limits t_{max} [13]. The figure also shows several calculated dependences $t_{max}(p)$ as a function of the fitting parameter τ(20 bar) (electron lifetime at 20 bar).

In this work, we assume that the electron lifetime depends inversely on gas pressure:

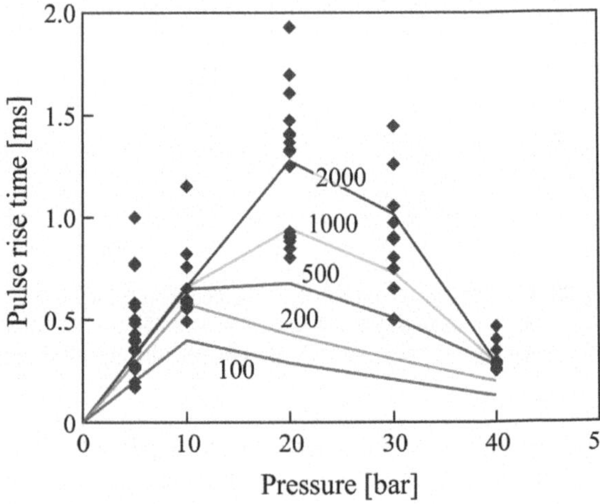

Fig. 21.9 Rise time (t_{max}) vs. Xe pressure: measured (symbols) and calculated (lines) data. The lines are labelled with the values of model parameter τ(20 bar) (ns)

$$\tau(p) = \tau(p_0)\frac{N(p_0)}{N(p)}, \tag{21.16}$$

where $\tau(p)$ is the lifetime at any given pressure p, $\tau(p_0)$ is the lifetime at a reference pressure p_0, and $N(p_0)/N(p)$ is the fraction of atomic concentrations at those pressures. Therefore, knowing τ(20 bar), we can determine the $\tau(p)$ for other pressure values.

From the figure, it is seen that a precise numerical fitting, by using our model, could not be performed. Even though, the fitting gave the order-of-magnitude estimation of the electron lifetime vs. pressure (with τ at 20 bar equal ~2 ms), yet we attempted to obtain improved results.

In order to do this, each ionization pulse was fitted by the preamplifier voltage Eq. (21.15). An example of such fits is shown in Fig. 21.4.

The results of the determination of the electron lifetime in pure Xe as a function of pressure, performed by fitting individual pulses, are presented in Fig. 21.10. The symbols represent the lifetime as determined from the individual pulse fits. The solid line is the best fit of Eq. 21.16 to the data. The obtained dependence τ(p) has greater precision than in the approach of Fig. 21.9.

21.4.2 The Cylindrical Configuration Measurements in Xe+0.25%H$_2$ Mixture

For the electron lifetime determination in Xe+0.25%H_2 mixture in the cylindrical chamber of the detector, pulse measurements of cosmic muons and the ^{137}Cs γ-quanta were also carried out. The values of the pulse rise time t_{max}, calculated for all measured pulses at different pressures, are presented in Fig. 21.11a.

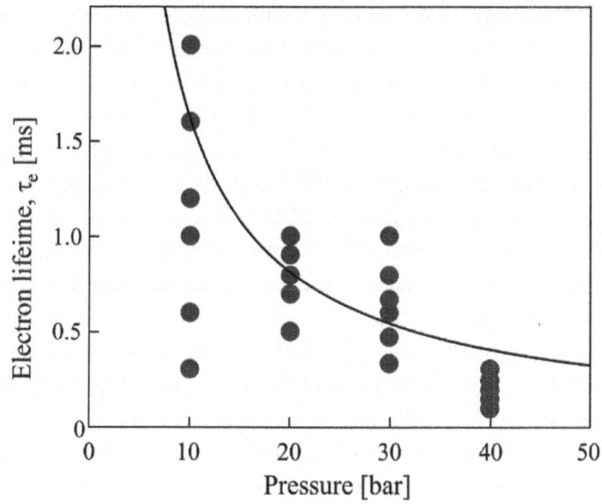

Fig. 21.10 The electron lifetime, determined by fitting individual pulses (symbols), and the respective Eq. (21.16) fit (solid line)

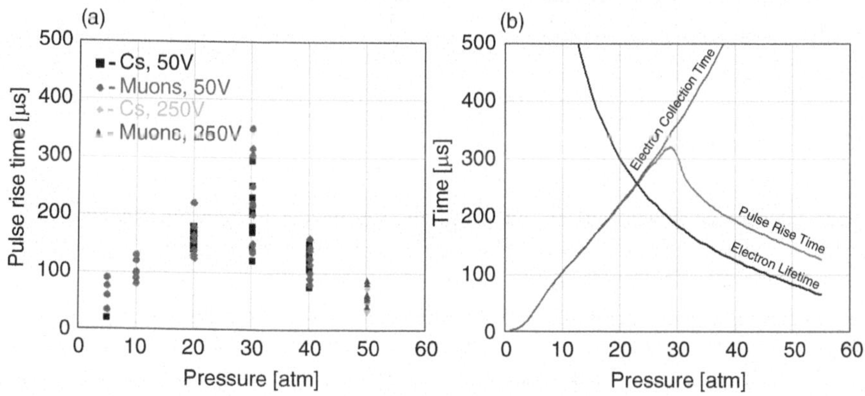

Fig. 21.11 (**a**) The pulse rise time dependence on pressure (measurements); (**b**) The rise time dependence on pressure, as well as the collection time and electron lifetime dependences, obtained from the analytical model

Similarly to the pure-Xe measurements in the parallel plate configuration, at each pressure value and applied voltage, a spread of the determined rise time values was observed, though this data set looked somewhat cleaner than that for pure Xe. This may have resulted from the ionizing particles getting registered in different locations of the detection volume, and, thus, the electrons drifting different distances (from 0 to 1 cm) from the place of gas ionization to the anode. The longest rise times at each pressure should originate from the charge generated near the negative electrode (~1 cm electron drift distance), and it was practical to take only those points for determining the lifetime. The data points for cosmic muons and the data points for ^{137}Cs overlap, so there is a good agreement of the results from the two different radiation sources.

Similarly to the pure-Xe data, the Xe-H_2 rise time data vs. pressure also has a maximum, only at ~30 atm. The probable reason, as in the pure Xe case described earlier, is that t_{max} is limited by t_c at lower pressure and by τ at higher pressure.

By varying parameters, such as the electron lifetime, the charge generation coordinate, and others, in our analytical pulse shape model, we were able to calculate the pulse rise time dependence on pressure (Fig. 21.11b), which well agrees with the experimental one. The respective fitted dependences of the electron collection time and the electron lifetime on pressure are also shown. The given electron lifetime curve is the sought result. That is, it presents the best estimate of the electron lifetime values as a function of pressure for our gas. Thus, at low pressures, the electron lifetime should be close to one millisecond, and it is progressively smaller: ~180, 120, 80 μs for 30, 40, and 50 atm, respectively. Note that from our modeling it follows that if we estimated the lifetime by the pulse rise-time values only, we would have underestimated it by orders of magnitude for low pressure; for high pressure we would have gotten relatively close estimate (within a factor of 1.5–2). By implementing the modeling of the pulse rise time on pressure, we obtained much more precise lifetime values.

21.5 Conclusions

The presented work was carried out within the framework of creation of a hand-held HPXe γ-spectrometer, which could be used at ports of entry to search for potentially transported malevolent radioactive substances.

We have reported on the improved methods of determining the electron lifetime in pure Xe, which is used as the detecting medium in the HPXe spectrometers. The high Xe purity is critical when creating HPXe γ-spectrometes with the energy resolution values, which are sufficient to ensure the identification of different radioactive isotopes. It was shown that, by using the analytical model, developed by us, it is possible to determine the electron lifetime, which is indicative of Xe purity, with a good accuracy. This is done by fitting the individual ionization pulses or by fitting the obtained dependence of the pulse rise time on gas pressure. It was demonstrated that using the parallel plate measurement configuration and the cylindrical configuration is equally possible, and the good results can be obtained, when the analytical model is applied. The measured lifetime values on the order of 1–2 ms are in the range to be sufficient in the spectroscopic applications.

Thus, Xe purity (charge carrier lifetime) control aspect of the work was presented in this study. The following are the next steps in building of the hand-held HPXe-based spectrometer: finding an optimized spectrometer design; reducing the detector body thickness and reinforcing it with 3D-printed low-absorbing armor materials in order to obtain better detector sensitivity.

Acknowledgements The authors gratefully acknowledge the funding from the NATO's Science for Peace and Security Program (SPS project #5373).

References

1. Novikov AS, Chernysheva IV, Dmitrenko VV, Grachev VM, Shustov AE, Ulin SE, Uteshev ZM, Vlasik KF, Ulina IV (2013) Xenon gamma-ray detector's electrical signals digital processing technique. In: Proceedings of XXIV International Symposium on Nuclear Electronics & Computing, September 9–16, 2013, Varna, Bulgaria, pp 192–197
2. Ghosh VJ, Bolotnikov A, Rohatgi US (2012) A feasibility study exploring the use of high-pressure xenon (HPXe) detectors for the characterization of spent fuel bundles. The Brookhaven National Laboratory report, BNL-98177-2012-IR, June 2012
3. Pyae SN, Grachev VM, Dmitrenko VV, Ulin SE, Vlasik KF, Uteshev ZM, Shustov AE, Novikov AS, Petrenko DV, Chernysheva IV (2015) Xenon gamma-detector applicability for identification and characterization of radioactive waste. Phys Procedia 74:352–356
4. Beddingfield DH, Beyerle A, Russo PA, Ianakiev K, Vo DT, Dmitrenko V (2003) High-pressure xenon ion chambers for gamma-ray spectroscopy in nuclear safeguards. Nucl Inst Methods Phys Res A 505:474–477
5. Kutny VE, Rybka AV, Davydov LN, Pudov AO, Sokolov SA, Kholomeyev GA, Melnikov SI, Turchin AA (2017) Gamma-ray detector based on high pressure xenon for radiation and environmental safety. Probl At Sci Technol 2(108):103–108
6. Pudov AO, Abyzov AS, Sokolov SA, Davydov LN, Rybka AV, Kutny VE, Melnikov SI, Kholomyeyev GA, Leonov SA, Turchin AA (2018) Measurements and modeling of charge carrier lifetime in compressed xenon. Nucl Inst Methods Phys Res A 892:98–105
7. Pudov A, Abyzov A, Sokolov S, Davydov L, Rybka A, Kutny V (2020) Determination of electron lifetime in compressed Xe gas for gamma-spectroscopy. Nucl Inst Methods Phys Res A 951:163037
8. Bolotnikov A, Ramsey B (1996) Purification techniques and purity and density measurements of high-pressure Xe. Nucl Instrum Methods Phys Res A 383:619–623
9. Beattie JA, Barriault RJ, Brierley JS (1951) The compressibility of gaseous Xenon I an equation of state for xenon and the weight of a liter of xenon. J Chem Phys 19:1219
10. Dmitrenko VV, Romanyuk AS, Suchkov SI, Uteshev ZM (1983) Electron mobility in dense xenon gas. Sov Phys Tech Phys 28:1440
11. Ulin SY (1999) Gamma spectrometers based on compressed xenon (development, characterization, applications). PhD thesis, Moscow State Institute of Engineering and Physics, Moscow
12. Poenaru DN (1967) Collection time of electron-hole pairs in a coaxial Ge(Li) radiation detector. IEEE Trans Nucl Sci 14:1–7
13. Kalashnikova VI, Kozodayev MS (1966) Detectors of Elementary Particles, "Nauka" Moscow [in Russian]

Part IV
Unmanned and Autonomous Systems

Chapter 22
Agile Tyre Mobility: Observation and Control in Severe Terrain Environments

V. Vantsevich, D. Gorsich, A. Lozynskyy, L. Demkiv, T. Borovets, and S. Klos

Abstract This research study develops fundamentals for a new ground vehicle technology to radically improve and protect off-road vehicle mobility by providing agile (fast, exact and pre-emptive) responses and advanced mobility controls in severe terrain conditions.

The current framework of terrain vehicle mobility that estimates a vehicle capability "to go through" or "not to go through" the given terrain conditions cannot provide an analytical basis for novel system design solutions. Indeed, modern traction control and other mobility related electronic control systems possess control response time within the range of 100–120 milliseconds and greater. With this response time, the actual control occurs after the vehicle has reached a critical motion situation, e.g., a wheel(s) is/are spinning and the vehicle is already losing its mobility. In this study, the developed methods allowed for estimating tyre mobility and controlling tyre motion before the tyre starts spinning. As shown in the conducted analysis, the response time, which occurs within the longitudinal tyre relaxation time constant of 40–60 ms, is sufficient for a tyre to avoid spinning and to maintain its required mobility.

Most common traditional approaches to observation of data supplied by virtual sensors were simulated and improved by means of machine learning algorithms. Computational simulations of an one-wheel-locomotion module driven by an electric driveline system demonstrated a sufficient performance of the proposed observation method to estimate mobility margins of the module in real time.

V. Vantsevich
University of Alabama at Birmingham, Birmingham, AL, U.S.A

D. Gorsich
U.S. Army CCDC Ground Vehicle Systems Center, Warren, MI, USA

A. Lozynskyy · L. Demkiv (✉) · T. Borovets · S. Klos
Lviv Polytechnic National University, Lviv, Ukraine
e-mail: lozynsky@lp.edu.ua

C. Palestini (ed.), *Advanced Technologies for Security Applications*, NATO Science for Peace and Security Series B: Physics and Biophysics,
https://doi.org/10.1007/978-94-024-2021-0_22

A hybrid intelligent control algorithm was designed, in which reinforcement learning was used to fine-tune the parameters of a fuzzy logic controller. A new wheel mobility index was utilized as a cost function to guarantee a designed behavior of the locomotion module. A fuzzy corrector was additionally designed to take into account both the dynamic state of the system and the dynamics of the tyre-terrain interaction. The fuzzy corrector supports upper level controls of autonomous vehicle dynamics by decreasing tyre slippage on severe terrains.

Computer simulations testified both stability of the controller (due to utilization of fuzzy logic polynomial control) and its desired performance (due to application of reinforcement learning). The fine-tuned controller requires minimal online computations.

This paper provides an extended summary of the above-listed research studies. Further details can be found in publications referenced in the paper.

Keywords Agile terrain mobility · Real-time observer and controller

22.1 Introduction

Further improvements of terrain mobility of vehicles can be achieved by reducing the response time of the modern vehicle electronic systems, including Traction Control and Torque Vectoring systems. Indeed, by using real-time controls being capable to operate within a short period while tyre is still developing its slippage would enhance vehicle mobility.

When developing a controller, a cost-function that provides the desired system behavior has to be utilized. Such function based on the wheel mobility index was introduced in Vantsevich et al. [1]. This index was further used to synthesize an intelligent control of the open-link locomotion module Vantsevich et al. [2, 6]. Two approaches were used for this purpose: one approach utilized reinforcement learning and the other one used hybrid intelligent control, in which reinforcement learning was used to fine-tune the parameters of fuzzy logic controller. The second approach testified both stability of the obtained controller (due to utilization of fuzzy logic polynomial control) and the desired performance (due to application of reinforcement learning). The utilization of the fine-tuned fuzzy controller for real-time agile control became possible due to minor online computations required by this type of controller. The designed intelligent control significantly decreased the response time in various severe terrain conditions. The controller demonstrated its robustness and stability.

Additionally, based on a comprehensive analysis of both traditional and modern control methods was done in Vantsevich et al. [3], a fuzzy corrector was introduced in Vantsevich et al. [4]. The fuzzy corrector main goal was to support either a driver or to contribute the system level control of an autonomous vehicle to decrease tyre slippage.

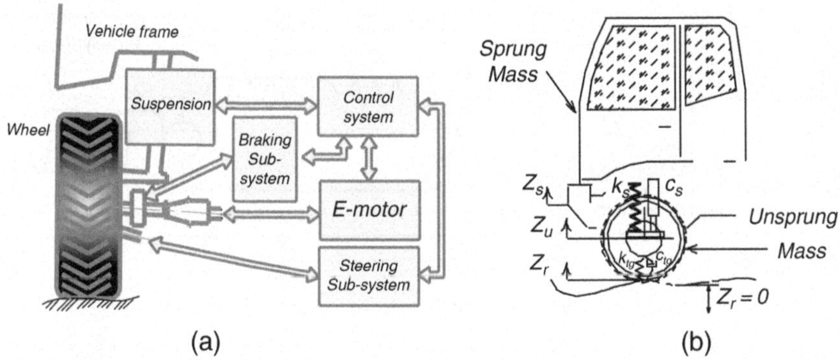

Fig. 22.1 Diagram of the open-link locomotion module: (**a**) rotational subsystem (**b**) vertical subsystem

22.2 Open-Link Locomotion Module Model

The open-link locomotion module model is a mathematical representation of a vehicle module (see Fig. 22.1a) that combines a wheel driven by an electric motor, a gear set, a brake sub-system, steering, and suspension [10, 11]. The model is represented by Eqs. (22.1) and (22.2), where Eq. (22.1) describes the rotational dynamics and Eq. (22.2) corresponds to the normal dynamics (see Fig. 22.1b).

$$\begin{cases} \frac{di_m}{dt} = \frac{1}{L_a}\left(U - R_a i_m - e\right), \\ U = \frac{k_{bat} u}{u_{\max}}, \\ J_{eq} = J_m + \sum_{l=1}^{k} \frac{J_l}{i_l^2}, \\ J_{eq}\dot{\omega}_m = T_m - k_{eq}\left(\varphi_m - i\varphi_w\right) - c_{eq}\left(\omega_m - i\omega_w\right) - T_{fm}, \\ T_{fm}\left(\omega_m\right) = \alpha_{0m}\operatorname{sign}\left(\omega_m\right) + \alpha_{1m}\exp\left(-\alpha_{2m}\left|\omega_m\right|\right)\operatorname{sign}\left(\omega_m\right), \\ i = \frac{\omega_m}{\omega_w}, \\ T_m = k_t i_m, \\ T_s = k_{eq}\left(\varphi_m - i\varphi_w\right) + c_{eq}\left(\omega_m - i\omega_w\right), \\ \varphi_m = \varphi_1 - \varphi_2 \\ \varphi_w = \varphi_3 - \varphi_4, \\ J_w\dot{\omega}_w = ik_{eq}\left(\varphi_m - i\varphi_w\right) + ic_{eq}\left(\omega_m - i\omega_w\right) - T_{wl} - T_{fw}, \\ T_{fw}\left(\omega_w\right) = \alpha_{0w}\operatorname{sign}\left(\omega_w\right) + \alpha_{1w}\exp\left(-\alpha_{2w}\left|\omega_w\right|\right)\operatorname{sign}\left(\omega_w\right). \end{cases} \tag{22.1}$$

here, J_{eq} is a rotational inertia of the equivalent mass; J_m and J_l are the rotational inertias of the rotor and an l-gear of the gear set, $l = 1, k$ correspondingly (see Fig. 22.2a); i_l is the velocity ratio between the l-gear of the gear set and the rotor, i is the total velocity ratio of the gear set between the motor and the wheel; ω_m, and ω_w are the angular velocities of the rotor and the wheel; k_{eq} is the equivalent torsional damping; c_{eq} is the stiffness shafts; T_s is the internal elastic-damping torque in the

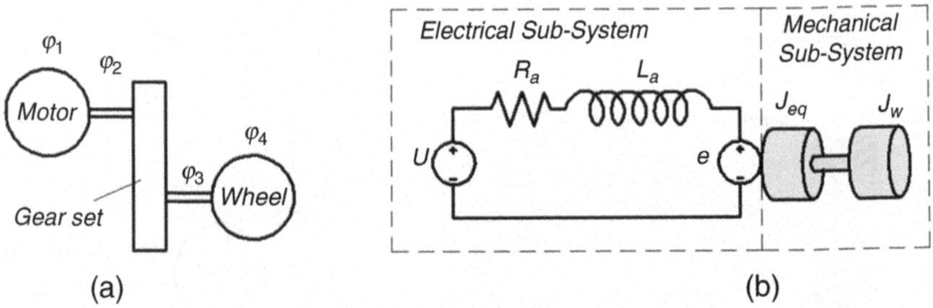

Fig. 22.2 Kinematic (**a**) and mechatronic (**b**) diagram of the drive sub-system

gear set; φ_m and $\varphi_w$are the instant revolution angles of the e-motor shaft and the wheel shaft; R_a is the armature electric resistance; L_a is the armature inductance; U in Fig. 22.2b stands for the armature voltage; e is the back electromotive force, which is the product of the back EMF constant, k_{emf}, and the angular velocity of the e-motor, ω_{m}; T_m is e-motor torque; i_m is electric current; k_t is the e-motor constant; u is a control voltage; k_{bat} is the maximum voltage output of the *Pulse Width Modulation* battery; u_{max} is the maximum control voltage; T_{wl} is the load torque caused by terrain; T_{fm} (ω_m) is the mechanical friction torque; α_{0m} is a constant that represents the Coulomb friction torque; α_{1m} is a constant that represents the difference between the Coulomb friction torque and the static friction torque; α_{2m} is a time constant; T_{fw} is the mechanical friction torque in the wheel bearings; α_{0w} is the Coulomb friction; α_{1w} is the static friction torque; α_{2w} is a time constant

$$\begin{cases} R_z = \mathrm{W}_w \cos\theta_n + k_{tg}\left(z_r - z_u\right) + c_{tg}\left(\dot{z}_r - \dot{z}_u\right) \\ m_s \ddot{z}_s = k_s\left(z_u - z_s\right) + c_s\left(\dot{z}_u - \dot{z}_s\right), \\ m_u \ddot{z}_u = k_{tg}\left(z_r - z_u\right) + c_{tg}\left(\dot{z}_r - \dot{z}_u\right) - k_s\left(z_u - z_s\right) - c_s\left(\dot{z}_u - \dot{z}_s\right). \end{cases} \tag{22.2}$$

where, R_z is the dynamic normal reaction; θ_n is the slope of the surface of motion; W_w is the static wheel load caused by the sprung mass and the unsprung mass; k_{tg} is the tire-soil normal stiffness; c_{tg} is the tire-soil damping factor; z_s and z_u are the displacements of the sprung and unsprung masses, m_s and m_u; z_r is the height of the terrain profile; k_s and c_s are the reduced stiffness and damping of the suspension.

In Eq. (22.1), φ_1, φ_2, φ_3, φ_4 represent instant revolution angles of the ends of the shafts between motor and gear set, wheel and gear set (Fig. 22.2a). Due to the elastic properties of material of the shafts and gears, these angles are different.

22.3 Agile Control Response Time

As pointed out in [5], the agility of the tyre control means the ability of a control algorithm to respond within a predefined time, which is small enough to implement the control before the tyre develop an extensive slippage. The aim of the small response time is to establish a new traction force between the wheel and the terrain that would eliminate the extensive slippage. The electric motor torque cannot be instantly converted in the traction force due to the longitudinal deflections of tyre and soil, which are referred as the longitudinal relaxation time constant:

$$\tau_{rl}(t)\frac{dF_x(t)}{dt} + F_x(t) = F_{ssx} \qquad (22.3)$$

here, $\tau_{rl}(t)$ is the longitudinal tire relaxation time constant, $F_x(t)$ is the dynamic circumferential wheel force, F_{ssx} is the steady-state circumferential wheel force.

As shown in [6], the longitudinal relaxation time constant is a measure of the tyre reaction to variations of terrain and operational conditions. A control algorithm that is to be deployed on the vehicle has to function in response to the conditions within the time interval that is the longitudinal relaxation time constant. Taking into account considerations research outcomes of [6], the response time of the control has to be within 40–60 ms. This will allow for providing agile real-time control of tyre mobility.

22.4 Virtual Sensor Design

The design of virtual sensors means the usage of virtual observers that can provide necessary and sufficient information on the system states using a few physical sensors, i.e., the observers serve as virtual sensors. A detailed overview of existing observers can be found in [7]. Figure 22.3 and Fig. 22.4 illustrate applications of some algorithms to observe the wheel normal reaction and the elastic damping torque of the module. An analysis of these simulation results allowed for concluding on utilizing the observers for in real-time applications (see [8]).

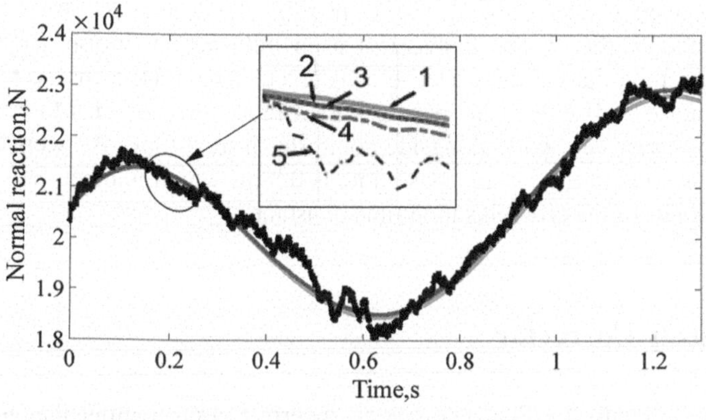

Fig. 22.3 Estimation of the wheel normal reaction by Extended Kalman Filter (EKF), Unscented Kalman filter (UKF), Particle filter (PF), Luenberger observer (LO)

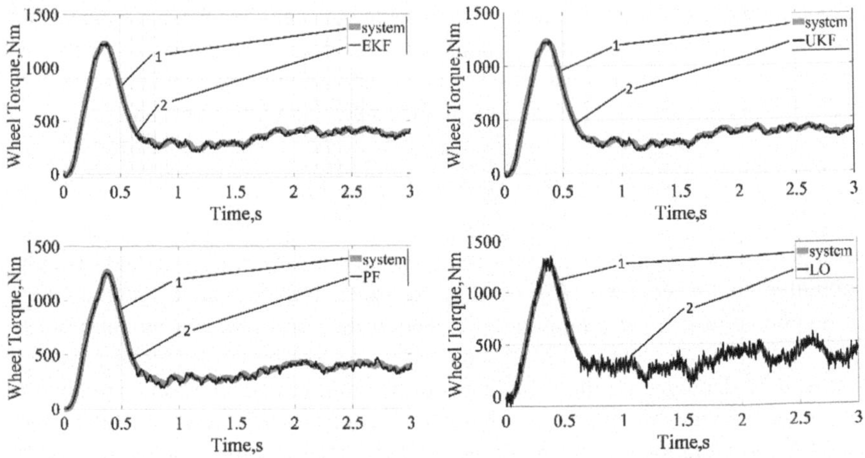

Fig. 22.4 The estimation of the elastic-damping torque by EKF, UKF, PF, and LO: black lines are the computed states; cyan lines are the estimated states

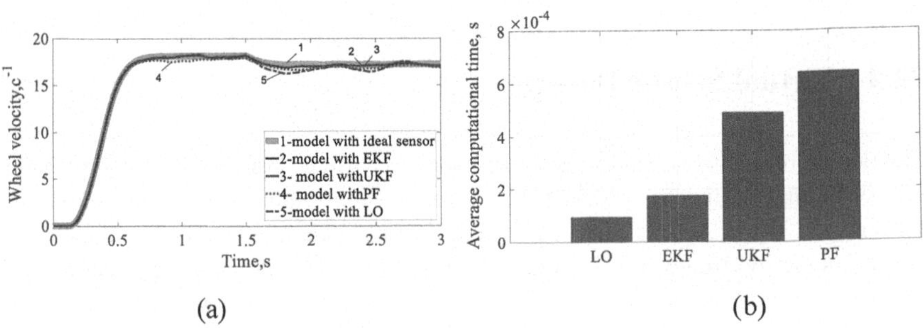

Fig. 22.5 (**a**) The dynamics of the system (1)–(2) with ideal sensors and the system with EKF, UKF, PF and LO; (**b**) Average time of observation using four algorithms

The obtained results show that it is possible to design an observer that is accurate enough, robust, and convergent even influences to noise. Figure 22.5a demonstrates the values of the wheel velocity calculated with the observer subjected to sensor noise and with ideal sensors (i.e. without any noise). Moreover, each designed observer can work alongside controller in real-time. As seen from Fig. 22.5b, the maximum time of response time is 0.8 ms which is much less than 40–60 milliseconds of the tyre relaxation time constant.

22.5 Fuzzy Corrector

A fuzzy logic controller that modifies an incorrect control input that may come either from a driver or from an autonomous control system of the electric motor. The incorrect input can be smoothened and a new, modified input can

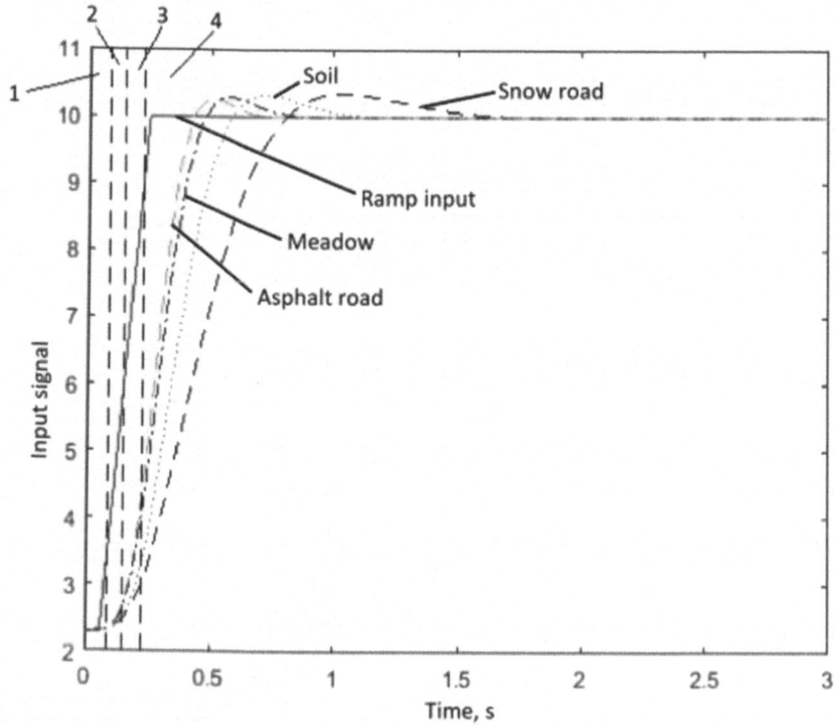

Fig. 22.6 Incorrect input signal illustrated by the ramp input and smoothened input signals on different terrains

be introduced. Details on this technical approach are explained in Vantsevich et al. [4].

Assuming an incorrect input signal varies from 0 to u_{max}, the maximal value that may be generated by the Pulse Width Modulation signal is also u_{max}. Figure 22.6 illustrates the input signal that is considered as an incorrect. This signal is further smoothened by the proposed fuzzy corrector. The smoothened input with a longer response time can reduce extra tyre slippage. The subdomains, labeled as 1, 2, 3 and 4 correspond to different levels of activation of the fuzzy logic rules from the rule base. The corrector acts swiftly and in a short time frame, however its performance has significant influence on the behavior of the locomotion module.

As an example of computational results, Fig. 22.7 demonstrates characteri-stics of the module for the ramp input and the smoothed input on a snow road.

The fuzzy corrector reduces the tyre slippage and provides the desired rotational velocity of the wheel at the set point. This was obtained by smoother changes of the electric current, the torque, and the rotational velocity in area 1. An increased rate of the controller's signal in the middle of the wheel acceleration process (areas 2 and 3) allows for keeping up with a response time of the locomotion module; a steady

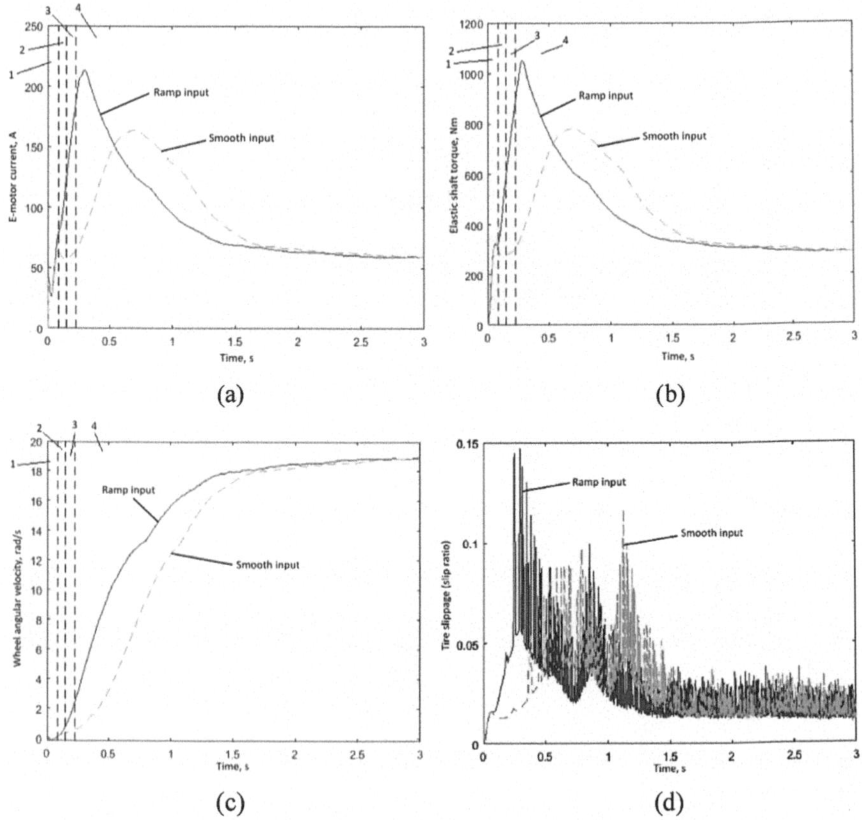

Fig. 22.7 Comparison of system characteristics on a snow road: (**a**) comparison of motor current values; (**b**) comparison of motor shaft torque values; (**c**) comparison of wheel angular velocities; (**d**) comparison of tyre slippages

state phase is achieved after 1.5 sec. Area 4 shows the inputs (i.e., the rotational velocity) smoothened to the set point.

22.6 Fuzzy Controller

As shown in [9, 12], a utilization of an unstable subsystem may significantly increase efficiency of a controller. The unstable controller acts only in the region of the large errors, which means that a system that is close to the steady state will be stable, and there is a transient region where both stable and unstable subsystems are active.

Figure 22.8a represents the hodograph of the system with unstable subsystem. Traditionally, the pass of the roots is along the convex curve. However, other trajec-

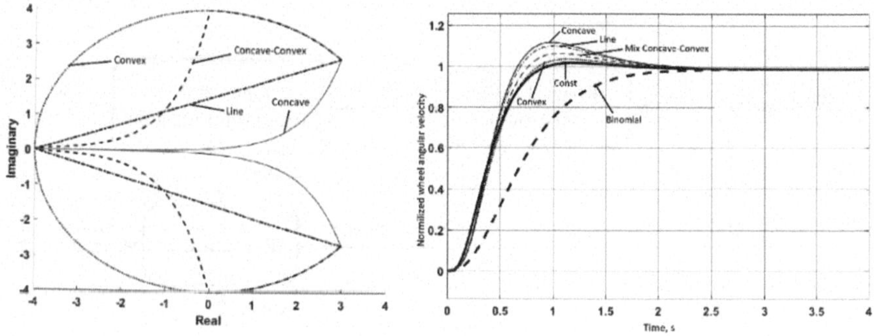

(a) (b)

Fig. 22.8 (**a**) The hodograph of the roots in the case of a system with a fuzzy controller; (**b**) Trajectories of the output signal of the system in the case of shown hodographs

tories can be possibly utilized: concave, linear, etc. Moreover, these trajectories can be obtained by calculating the weight function of the fuzzy rules that would allow for obtaining the desired trajectories.

The results of the simulation are presented on Fig. 22.8b. As one can see, the transients of the system with non-standard trajectories are better than in the case of convex trajectory. However, computational efforts are significant. Therefore, to ensure real-time computation, a convex trajectory of the root has to be used.

22.7 Fuzzy Reinforcement Learning Controller

As shown in the previous sections of this paper, the utilization of the fuzzy controller and the fuzzy corrector adds significant improvements to the behavior of the system. However, these controller and corrector require a manual tuning. In this study, reinforcement learning was applied to calculate the output signal of the controller depending in the state of the system and disturbances (including terrain conditions). Hence, reinforcement learning was applied to fine-tune the parameters of the fuzzy membership function. This approach allowed for obtaining an optimal behavior of the control system and, what is even more important, for achieving stability of the system. As this study confirmed, with the use of the fuzzy reinforcement learning approach, all subsystems were stable.

The schematics of the above-proposed control is presented in Fig. 22.9. The output of the fuzzy controller is the input to the open-link locomotion module model. At the same time, the reinforcement learning part is used to provide the optimal parameters of the fuzzy controller. The reward and tuning process of the fuzzy reinforcement learning controller is presented in [6] in detail.

The results of the simulations are given in Fig. 22.10. The results testify that the application of fuzzy logic itself may cause fluctuations or longer transients of the rotational velocity if the locomotion module's wheel. This is due to the fact that

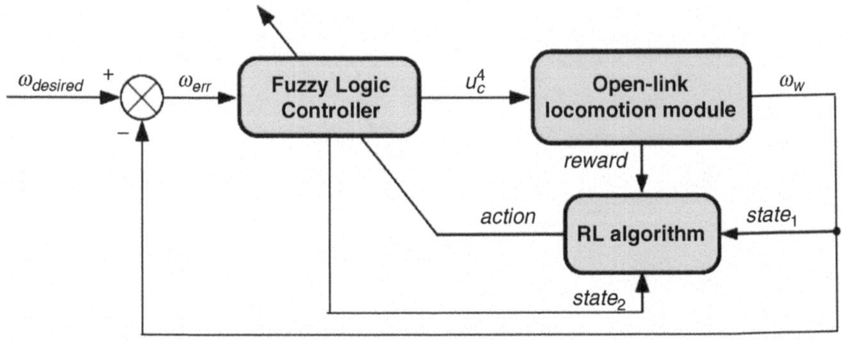

Fig. 22.9 Block diagram of the closed loop control system of the locomotion module

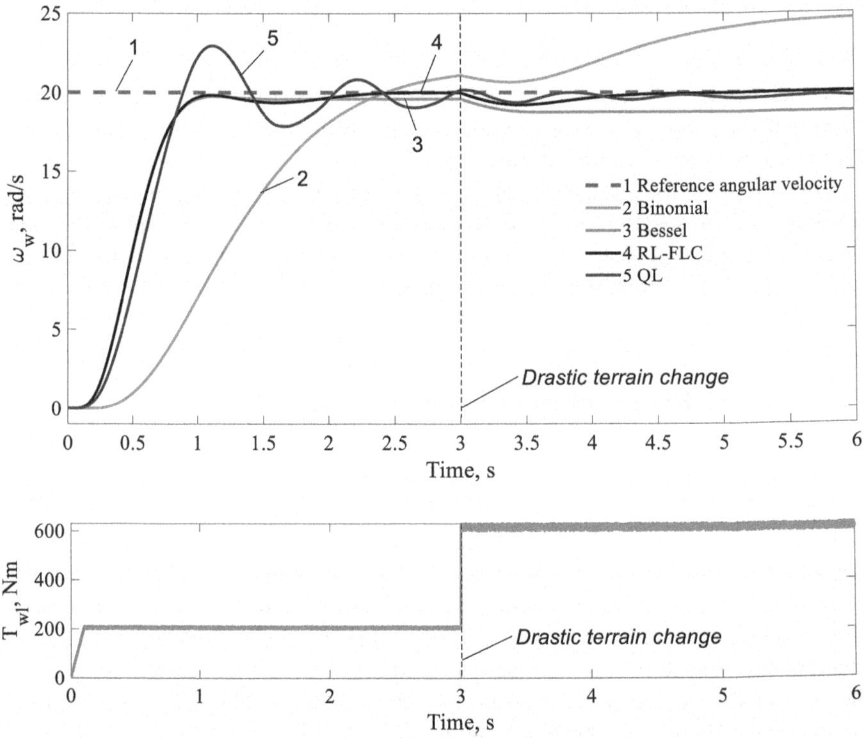

Fig. 22.10 Simulation results of wheel torque load and the wheel angular velocity control by different control methods (*1* – reference angular velocity, *2*, *3* – actual angular velocities controlled by single Binomial and single Bessel; *4* – actual angular velocity controlled by RL-FLC; *5* – actual angular velocity controlled by Q-learning)

the final conditions may differ (due to, for instance, additional external disturbances, etc.) from the conditions at which the system has been trained. At time of t = 3 s, a drastic terrain change occurs, which influences the wheel torque (see Fig. 22.10). However, the designed fuzzy reinforcement controller is robust and agile enough to overcome this dynamic change without significant oscillations of the rotational velocity of the wheel.

An analysis presented in [6] proved that the computational time of the control algorithm allows for running the control in real-time computations. Thus, the system of the designed state observer, the fuzzy reinforcement learning controller, and the fuzzy corrector is fast to operate within the time interval of the longitudinal relaxation time constant.

22.8 Conclusions

The technical problem of the improving of tyre mobility in severe terrain conditions has been solved by designing an agile tyre mobility control system that comprises a state observer, a fuzzy reinforcement controller, and a fuzzy corrector. The control system was designed for the locomotion module and computational simulations were conducted. The simulations proved the ability of the control system to operate in real-time within the longitudinal relaxation time constant when the tyre and soil are gaining the longitudinal deflections and an extended slippage did not occur yet. In future research plan, the designed control will be extended to a hybrid computer simulation of a 4×4 truck when one of the wheels in simulation is substituted with a physical wheel that simultaneously runs on the MTS FlatTrac LTR.

Acknowledgements This study has been supported by a grant of the NATO Science for Peace and Security Programme: MYP SPS G5176 "Agile Tyre Mobility for Severe Terrain Environments".

References

1. Vantsevich V, Gorsich D, Lozynskyy A, Demkiv L, Borovets T, Klos S (2019b) Reference torque history for sustainable mobility and energy efficiency under target velocity profiles. In: Proceedings of the ISTVS 15 th European-african regional conference, Prague, Czech Republic, September 9–11, 2019 (in print)
2. Vantsevich VV, Lozynskyy A, Demkiv L, Holovach I (2017) Fuzzy logic control of agile dynamics of a wheel locomotion module, Dynamics of vehicles on roads and tracks Vol 1: Proceedings of the 25th International Symposium on Dynamics of Vehicles on Roads and Tracks (IAVSD 2017), 14–18 August 2017, Rockhampton, Queensland, Australia
3. Vantsevich VV, Lozynskyy A, Demkiv L, Klos S (2017a) A foundation for real-time tire mobility estimation and control. In: Joint 19th Iiternational & 14th european-african regional conference, Budapest. ISBN:978–1–942112-49-5
4. Vantsevich V, Lozynskyy A, Demkiv L (2017b) A wheel rotational velocity control strategy for an open-Link locomotion module. In: 19th international & 14th european-african regional conference of the ISTVS
5. Vantsevich VV, Demkiv LI, Klos SR (2018) Analysis of tire relaxation constants for modeling vehicle traction performance and Handling. In ASME 2018 dynamic systems and control conference. American Society of Mechanical Engineers

6. Vantsevich V, Gorsich D, Lozynskyy A, Demkiv L, Klos S (2019d) A reinforcement-learning algorithm with relaxation time constant-based boundaries for real-time off-road traction control. The 26th iavsd international symposium on dynamics of vehicles on roads and tracks (IAVSD), 12–16 August 2019, Gothenburg, Sweden
7. Vantsevich V, Gorsich D, Lozynskyy A, Demkiv L, Borovets T (2019a) State observers for terrain mobility controls: a technical analysis. In: IFToMM world congress on mechanism and machine science. Springer, Cham, pp 3681–3690
8. Lozynskyy AO, Demkiv LI, Vantsevich VV, Borovets TV, Gorsich DJ (2019) An estimation accuracy of state observers under uncertain initial conditions. Math. Model. Comput. 6(2):320–332
9. Lozynskyy A, Demkiv L (2016) Application of dynamic systems family for synthesis of fuzzy control with account of non-linearities. Adv Electr Electron Eng 14(5):543–550
10. Paldan JR, Gray JP, Vantsevich VV (2015) Sensor signal limitations in wheel rotational kinematics estimation model. In: ASME 2015 dynamic systems and control conference. American Society of Mechanical Engineers, pp V003T45A001–V003T45A001
11. Andreev AF, Kabanau V, Vantsevich V (2010) Driveline systems of ground vehicles: theory and design. CRC Press, Boca Raton
12. Lozynskyy A, Vantsevitch V, Demkiv L (2018) Enhancement of dynamical characteristics of a fuzzy control system by using unstable subsystem. IEEE International Conference on Fuzzy Systems

Chapter 23
Development of a Novel Wheel Speed Sensor for Enhanced Close-Loop Feedback Control of Vehicle Mobility

Samuel R. Misko, Vladimir Vantsevich, and Lee Moradi

Abstract The maximization of vehicle mobility in off-road environments requires implementation of real-time control systems that govern individual wheel rotational velocity based on closed-loop feedback from wheel rotation sensors. A paradigm shift in the type of feedback sensor is required to facilitate faster vehicle control loop speeds such that a subsequent paradigm shift can be realized in vehicle mobility control. This anticipated shift in vehicle mobility is facilitated by a transition from the traditional reactive control paradigm (response time of 100–150 ms) to an agile control paradigm (response time of 40–60 ms); wherein adjustments can be made within the tire relaxation time.

A novel wheel speed sensor (WSS) was developed to demonstrate possible improvements over the two main types of conventional OEM wheel speed sensors that use toothed reluctor rings; passive variable reluctance sensors, and active magneto resistive sensors. The number of teeth on a given reluctor ring fundamentally limits the resolution these OEM sensors can provide to the mobility control systems of a vehicle (e.g., traction control, anti-lock braking, wheel torque vectoring) – thereby limiting vehicle mobility. The output of these OEM sensors are in the style of incremental encoders that produce a discrete pulse train signal where the real-time control system must measure the duration of a full pulse or period to obtain each measurement of wheel speed. This technique of measurement introduces: (1) latency, in that the control system must wait for next signal transition before measurement is made, and (2) error, in that the wheel speed measurement is an unweighted average over the whole pulse/period duration. To eliminate these measurement limitations, a paradigm shift was required such that the output of the feedback sensor is instead a continuous signal.

The top level design of the novel WSS built upon work done previously to theorize the configuration and behavior of such a continuous output sensor. The design was then reduced to practice through an iterative series of design with multi-physics

S. R. Misko · V. Vantsevich (✉) · L. Moradi
University of Alabama at Birmingham, Birmingham, AL, USA
e-mail: miskosr@uab.edu; vantsevi@uab.edu; moradi@uab.edu

C. Palestini (ed.), *Advanced Technologies for Security Applications*, NATO Science for Peace and Security Series B: Physics and Biophysics,
https://doi.org/10.1007/978-94-024-2021-0_23

simulations and laboratory validation at the component and subsystem levels. The novel WSS was then implemented alongside two OEM sensors for experimental test and evaluation on an MTS Flat-Trac LTR Tire Test System during dynamic testing of an off-road truck tire (Continental MPT 81365/80 R20). In these tests, the novel WSS sensor demonstrated a significant improvement to measurement latency during low speed acceleration of more than 200 ms, and provided data for use in the design and evaluation of new vehicle control systems to determine comparative measurement error and alternative techniques for measurement of high wheel speeds. This demonstrated improvement to measurement latency from more than 200 ms down to <5 ms, is more than sufficient to facilitate the desired subsequent paradigm shift in overall vehicle mobility control.

Keywords Wheel speed · Feedback sensor · Vehicle mobility

23.1 Introduction

The maximization of vehicle mobility in off-road environments requires high speed iterative intervention by a vehicle's mobility control systems **during** the evolution of a critical motion situation. This stands in contrast to existing modern electronic vehicle control systems such as traction control, wheel torque vectoring systems, and anti-lock braking which are largely designed to take affect **after** the vehicle has begun to experience loss in mobility due to a critical motion situation. In the current "reactive control" paradigm the response time is typically within the range of 100–200 ms. In the proposed "agile control" paradigm, the response time is required to be less than the vehicle's tire relaxation time (<<60 ms). This new paradigm seeks to facilitate iterative control system adjustments **during** the mechanical response of the tire due to interactions between the wheel and terrain.

To facilitate the development of this new vehicle mobility paradigm, a similar paradigm shift in wheel speed sensor (WSS) technology is required to provide feedback to the vehicle's control system with an unprecedented level of temporal and spatial accuracy regardless of the instantaneous wheel speed. Existing OEM WSSs provide a discrete pulse train signal that is derived from either toothed reluctor rings or magnetic encoder hub assemblies. Toothed reluctor rings are still the most common OEM solution and typically have no more than 48 teeth. Magnetic encoder technology has been adopted into some vehicles to improve the discrete angular resolution by embedding alternating magnetic poles in the hub bearing assembly itself, but this approach is still fundamentally constrained by the number of discrete positions. This discrete approach introduces latency and error into the control system by requiring detection of at least two discrete signal transitions to provide an unweighted estimation of the average speed over that duration of time. Discrete approaches can be further said to generally have lower temporal accuracy at low wheel speeds, and lower spatial accuracy for tires with larger diameters.

Despite these drawbacks, some forward thinking research performed with ultra-high resolution magnetic encoders has demonstrated promise not only in the area of improved instantaneous speed estimation, but also in the area of real-time road condition estimation based on "frequency analysis of the speed variation components extracted from the rotational pulse signals" [1].

The goal of this research is to begin to explore the implications of shifting away from discrete WSS signals towards continuous WSS signals, which seeks to impose no independent electronic restrictions to latency and error outside of those inherent in the control system's own data acquisition limitations. By the very nature of the continuous approach, the limitations on temporal and spatial accuracy should be largely independent of instantaneous wheel speed and tire diameter. This independence should also facilitate a significant reduction if not total removal of dependence on vehicle speed for real-time estimation of terrain conditions based on frequency analysis of speed variation in future research.

23.2 Design

In previous work [2–4] it was theorized that a magnetic shunt could be used to incrementally interrupt the flow of magnetic flux from a rare earth magnet to a Hall Effect sensor that is positioned across an air gap of some arbitrary distance (Fig. 23.1) such that a continuous output signal could be obtained to estimate angular velocity of a wheel. This work used modeling that assumed the magnitude of magnetic flux density passing through the Hall Effect sensor to be directly proportional to the total unobstructed area of the magnet from the perspective of the Hall Effect sensor. This approach predicted a highly linear response could be obtained from such a sensor for approximately 340–350° of rotation. As theorized, a spiral disc vane could be used to provide this type of incremental occlusion of the magnet, but such an implementation would have to account for the sharp change in signal at the discontinuity point where the radius of the spiral must jump from its smallest value to largest. In previous work, a method for mitigation of the effects of this discontinuity were modeled as a software based solution that would adjust the signal to provide an instantaneous signal (Fig. 23.2).

23.2.1 Initial Investigation & Selection of Hall Effect Sensor, Magnet, and Shunt

To initiate the design process of the novel WSS, a series of tightly controlled laboratory experiments were performed to thoroughly define the nature of the interaction between a single axis SS495 Hall Effect sensor, a number of different shapes and thicknesses of low-carbon steel shunts, and a number of different shapes

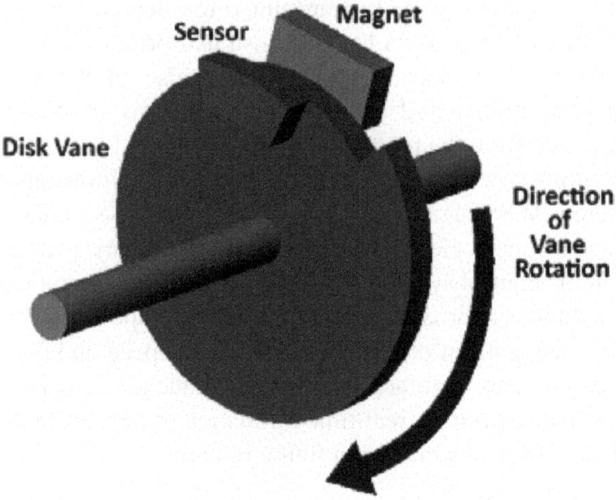

Fig. 23.1 Conceptual Design of Novel WSS [2]

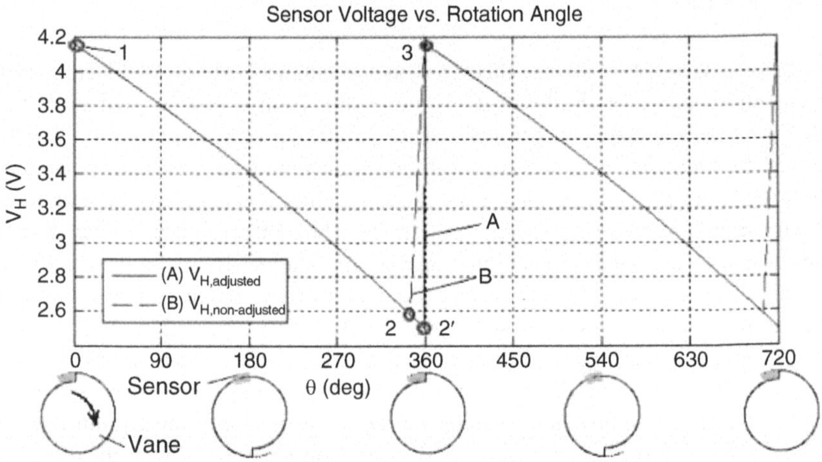

Fig. 23.2 Estimation of Novel WSS Output [2]

and sizes of neodymium rare earth magnets. To characterize the interaction between the Hall Effect sensor and the magnet, a 3D printer's modified extruder stage was used to precisely position the sensor in 3D space with respect to series of magnets that were mounted to the print bed. The resulting data provided a series of 3D "heat maps" that were used to determine that the thickness and overall surface area of the magnet to be the primary drivers for the magnitude of magnetic flux density and field shape. This test also allowed for strategic selection of the air gap between the sensor and the magnet such that the maximum range of the Hall Effect sensor could be utilized.

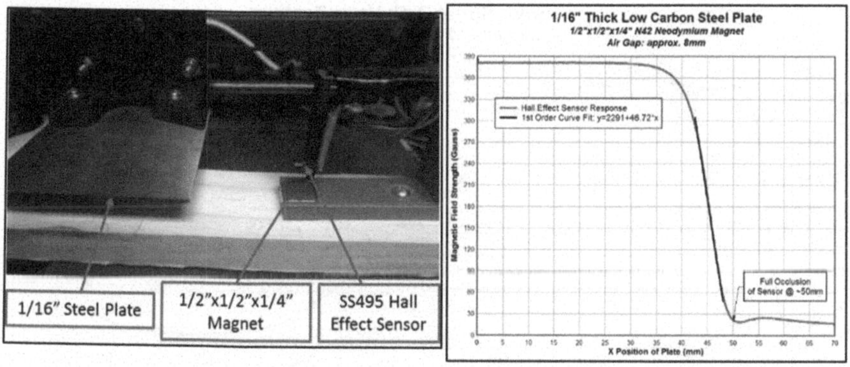

Figs. 23.3 & 23.4 Test Setup and Data Obtained from Shunt Introduction Experiment

Fig. 23.5 Novel WSS Proof of Concept Test Configuration

A second test was performed to examine the effect of introducing a magnetic shunt into the air gap between the sensor and magnet. Figures 23.3 & 23.4 show the setup and data obtained from shunt introduction testing, where a 1/16″ thick low carbon steel plate is incrementally (0.1 mm step resolution) introduced through the middle of the ~7 mm air gap between a ½″×½″×¼″ N42 Neodymium magnet and the SS495a Hall Effect sensor. The data obtained successfully demonstrates the desired linear shunting effect, where there is a near-complete shunting of the magnetic field away from the SS495 as the steel plate approaches full occlusion of the sensor's line of sight to the magnet.

A rudimentary proof of concept design for the novel WSS sensor's spiral disk vane was developed based on the linear range of response observed in Fig. 23.4, such that a spiral disk was fabricated with an evenly distributed change in radius of ~5 mm over the 360-degree spiral curve. This disk was fabricated from a 1/16″ thick plate of low carbon steel with a 2-axis CNC, and was integrated into a

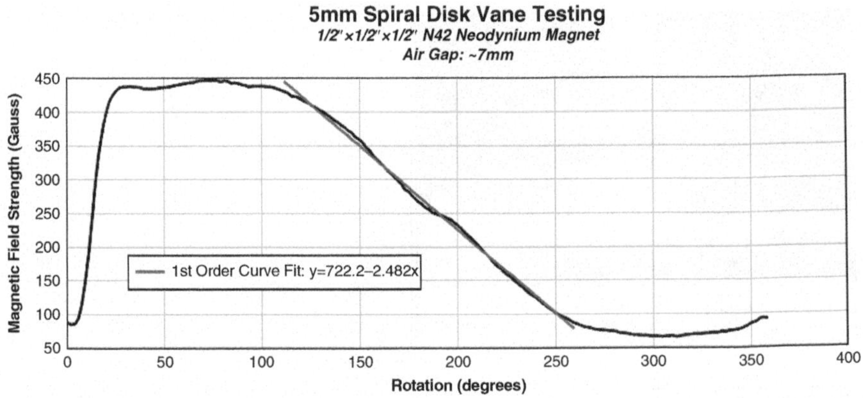

Fig. 23.6 Experimental Results from Novel WSS Proof of Concept Experimentation

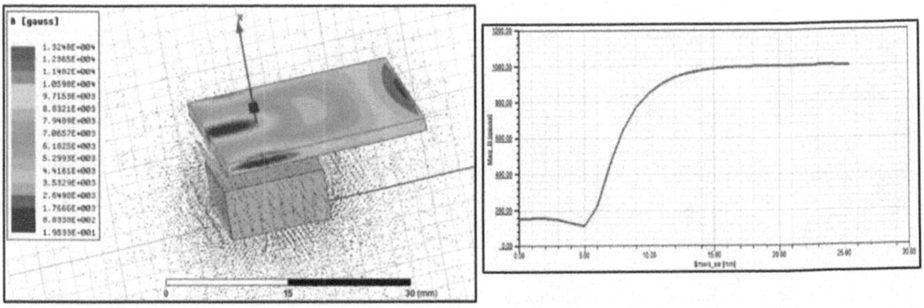

Figs. 23.7 & 23.8 Simulation Results for Validation of Magnetic Shunt Behavior

test fixture that integrated a NEMA23 stepper motor and fixtures for mounting of a ½″×½″×¼″ N42 Neodymium magnet and SS495a Hall Effect sensor with a ~7 mm air gap (Fig. 23.5). The spiral disk vane was mounted with its axis of rotation aligned with that of the stepper motor and positioned in the air gap such that at maximum radius it would fully occlude the Hall Effect sensor's line of sight to the magnet. A LabVIEW program was developed to rotate the stepper motor in single step rotations (1.8° degree per step), where the motor's movement is allowed to "settle" for ~2 s before an NI 9224 analog input module is used to record 3 s of data from the SS495a at a sample rate of 1 kHz. The arithmetic mean of these data points was then computed and saved to a file for each step as the motor performs one full rotation. Figure 23.6 shows that data obtained from the of the proof of concept experiment, where it was concluded that further experimentation and iterative development would be required to discover the implementation details required to ensure that the linear range of response covers upwards of 340–350° of rotation.

23.2.2 Multi-Physics Modeling and Simulation for Selection of Novel WSS Implementaton Details

To expedite and optimize this iterative design effort, 3D magneto-static models of the novel WSS design were developed with ANSYS's Maxwell 3D multi-physics software. The simulation environment and models were setup incrementally in an effort to ensure that the experimental results obtained to date are reproducible in this analytical environment. First, the ferrous metal shunt introduction experiment was recreated by modeling representative geometries of the magnet, metal plate, and hall effect sensing element (Fig. 23.7). The resulting data (Fig. 23.8) matched the data obtained experimentally (Fig. 23.4) with a high degree of accuracy, thereby validating the multi-physics environment for modeling of ferrous shunt – NdFeB magnet interaction behavior.

The spiral disk vane design component geometries where then modelled, parameterized, and iteratively simulated such that the shape, thickness, and orientation of all components could be varied to discover optimal design for each (Fig. 23.9). The simulation was setup to converge the solver based on the total magnetic flux flowing in a normal direction through a 1 mm square plane – which approximated the effective sensing area of the SS495a Hall effect sensor.

The plot in Fig. 23.10, shows a analytical method developed to optimize the novel WSS design through analysis of the slope of the representative sensing element's response leading up to and leaving the vane's discontinuity by applying 1st Order Curve fits to each section. This analysis yielded the ideal translational position of the vane within the air gap by identifying the point, at which the slopes of both curve fits are approximately equal – thereby providing a linear response over the largest possible range of rotational positions leading up to and away from the vane's discontinuity.

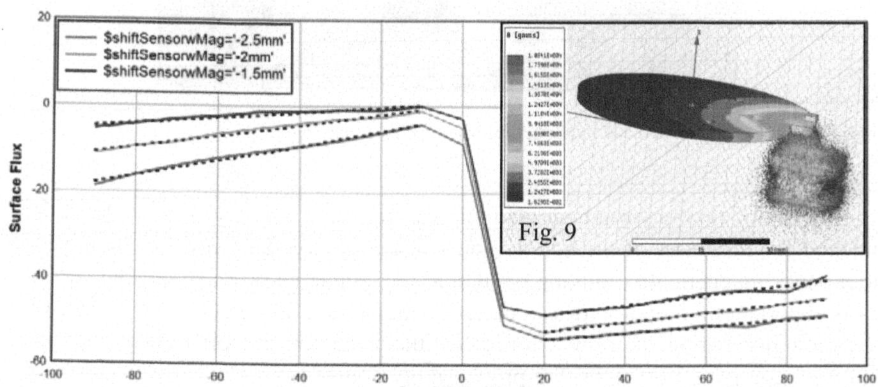

Figs. 23.9 & 23.10 Example of Iterative Spiral Disk Simulations to Drive Sensor Design

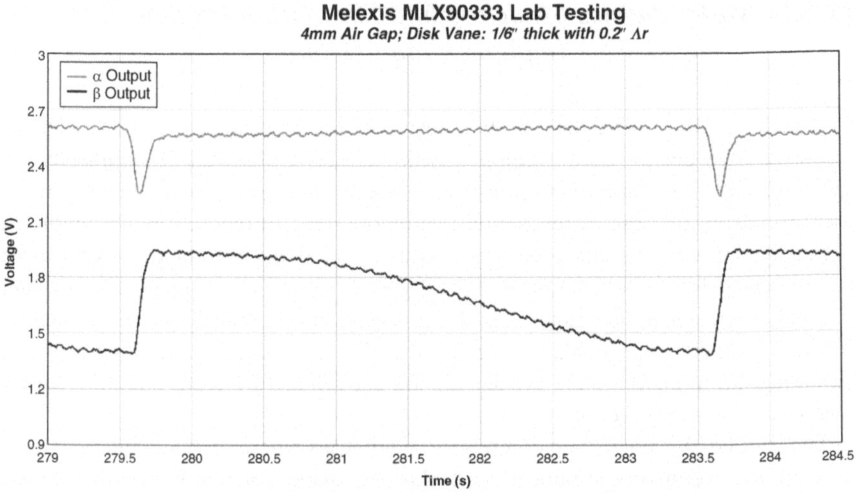

Fig. 23.11 Experimental Test Results for Stepper Motor Test Bed with MLX90333

Fig. 23.12 MTS FlatTrac LTR System with MPT 81 Tire Installed

This simulation environment also allowed for the discovery of an independent method for detection of the spiral vane's discontinuity. This involves being able to monitor magnetic flux moving parallel to the simulated primary sensing surface. This important finding provides an independent method for detection of the sensor's discontinuity region, thereby eliminating the need for the previously theorized software compensation algorithm proposed previously [1].

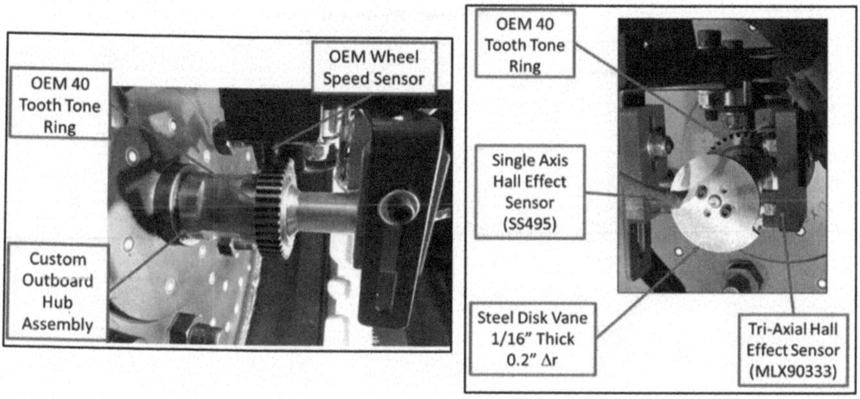

Figs. 23.13 & 23.14 OEM and Novel WSS Implementations on MTS FlatTrac LTR System

23.3 Laboratory Validation Experiments

With the geometry and orientation of all three sensor components (the hall effect sensor, the magnet, and the spiral disk vane) identified from simulation based design iterations, a MLX90333 tri-axial Melexis Hall Effect sensor was selected to provide the multi-axis sensing of magnetic flux required. It was theorized that this sensor could provide two analog output signals representative of both the angular position of the WSS as well as independent detection of the disk vane discontinuity. This sensor was integrated into the aforementioned stepper motor actuated test bed with a 4 mm air gap between it and a small Neodymium N35 magnet (0.25″ × 0.25″ × 00.18″). Four disk vanes where manufactured with an industrial CNC to provide two different changes in total radius (0.1″ and 0.2″) for two different thicknesses of low carbon steel (1/16th, 1/8th). Laboratory test results validated the anticipated behavior of a significant change in out of plane flux during transition of the discontinuity through the air gap (Fig. 23.11, "α Output"). The angular position signal, however, exhibited unexplained non-linearities that were not predicted in the multi-physics simulation, but are currently thought to be within the range of usability (Fig. 23.11, "β Output").

23.3.1 *Wheel Speed Sensor Performance Comparison During Dynamic Tire Tests*

The MTS FlatTrac® LTR Tire Test System at the National Tire Research Center (NTRC) in Alton, VA was used for our dynamic tire testing with a Continental 365/80 R20 MPT 81 tire (Fig. 23.12). The Flat-Trac LTR system is a larger, higher force version of Flat-Trac technology, specially designed for dynamic tests on light

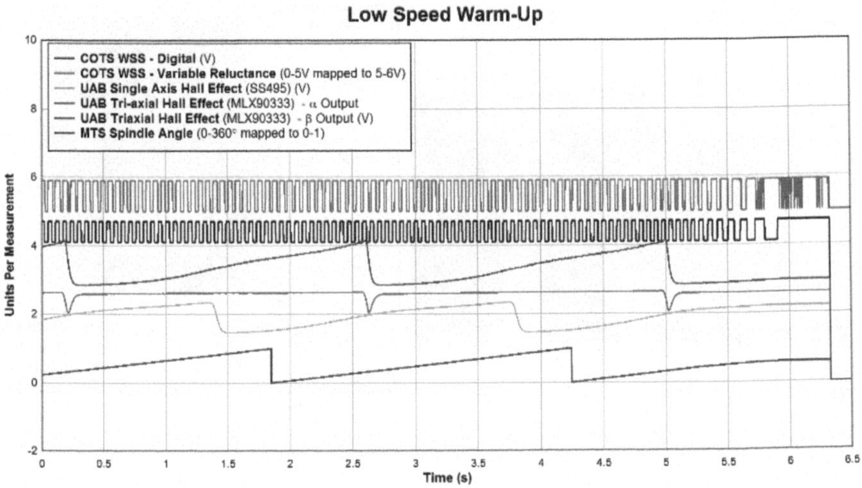

Fig. 23.15 Sensor Output Comparison During Low Speed

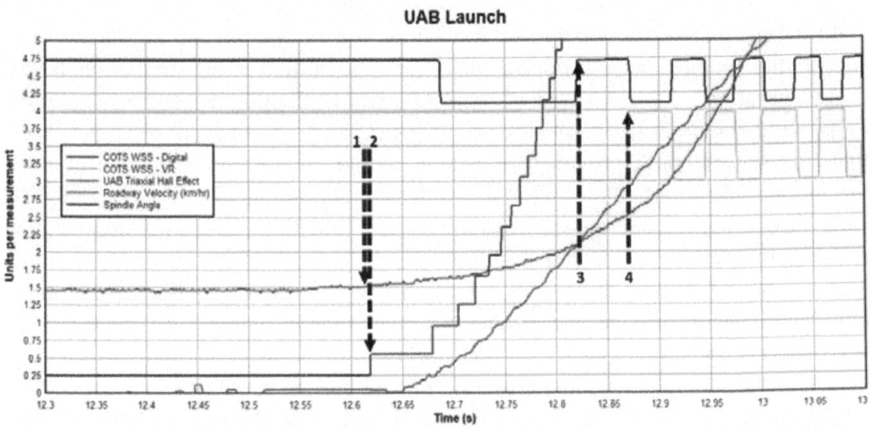

Fig. 23.16 Sensor Output Comparison During High Acceleration Launch

truck and racing tires. NTRC provided a custom built 20// wheel and hub adapter that provided a robust coupling to the FlatTrac's drive spindle. NTRC also fabricated a custom wheel speed sensor bracket and hub assembly to facilitate the attachment and alignment of two OEM wheel speed sensors to a OEM tone wheel, as well as two different Hall Effect sensors for our novel wheel speed sensor's spiral disk vane (Figs. 23.13 & 23.14). The wheel speed sensor assembly was implemented on the outboard side of the wheel-spindle adapter assembly to allow for clear viewing of the sensors' operation and for simplifying the custom design of the sensor's bracket and the hub assembly.

The MTS FlatTrac LTR data acquisition system was used to capture the data from each of the wheel speed sensors to ensure proper temporal synchronization with other FlatTrac system measurements. The system had all proper analog signal conditioning including anti-aliasing filters on all analog signal input lines. For simplicity and for highest accuracy, all wheel speed sensors (analog and digital) where attached to individual, isolated 14-bit analog input channels with 1 kHz sampling rates. In summary, there were 5 channels of wheel speed data collected as follows: OEM digital sensor, OEM variable reluctance sensor, single axis SS495 hall effect sensor, and two channels from the MLX90333 tri-axial Hall Effect sensor. The sensor output from each of these sensors, as well as the ultra-high accuracy encoder used on the MTS drive spindle are shown below in Fig. 23.15 (where the y-axis is unitless so that linearity and performance can be better compared side by side). Both implementations of the novel WSS exhibit some degree of non-linearity when compared to the pre-processed spindle angle signal, but are considered in the range of usability. Furthermore, the ability of the MLX90333 to independently detect the vane discontinuity is again verified. Also noticeable here is the instability of the variable reluctance sensor as the speed approaches zero.

Of primary interest however is the demonstration of our novel WSS's ability to provide unprecedented levels of accuracy for the detection of subtle tire movements. This is best demonstrated during a Launch test which simulates a zero-to-65mph rapid acceleration event (Fig. 23.16). First discernable detection of wheel speed increase proceeded as follows: (1) novel WSS @ t = 0, (2) high resolution MTS encoder @ t = 3.9 ms, (3) digital OEM @ t = 205.8 ms, (4) variable reluctance OEM @ t = 255.6 ms.

23.4 Conclusion

The development of a novel wheel speed sensor for enhanced closed loop feedback control of vehicle mobility has been conceived, designed, and validated in laboratory environments. The sensor has shown a dramatic improvement to wheel speed estimation latency at low speeds over other OEM technology solutions, and even demonstrated latency improvements over ultra-high accuracy magnetic encoder solutions. This solution represents the desired paradigm shift away from fundamentally limited discrete wheel rotation sensors toward a new generation of continuous output wheel rotation sensors that can facilitate the development of the next generation of agile vehicle control systems.

Acknowledgements This study has been supported by a grant of the NATO Science for Peace and Security Programme: MYP SPS G5176 "Agile Tyre Mobility for Severe Terrain Environments".

References

1. Nishikawa K, Takahashi T, Duret C (2013) Hub bearing with an integrated high-resolution rotation sensor, NTN-SNR Roulements. Technical Article No. 81
2. Paldan JR (2015) Sensor signal limitations in Wheel rotational kinematics estimation model. In: Proceedings of the ASME 2015 dynamic systems and control conference (DSCC2015–9769), Columbus, Ohio, USA
3. Gray J, Paldan J, Vantsevich V Multi-domain modelling of a new wheel rotational kinematics. In: Proceedings of the 24th symposium of the international association for vehicle system dynamics (IAVSD 2015). CRC Press, Graz, pp 253–262
4. Gray JP, Paldan JR, Vantsevich VV (2015) An agile tire slippage estimation based on new tire and wheel rolling characteristics. In: 13th European conference of the ISTVS 2015, at Rome, Italy

Chapter 24
ThreatDetect: An Autonomous Platform to secure Marine Infrastructures

Paolo Casari, Lutz Lampe, Stephane Martinez, and Roee Diamant

Abstract The NATO SPS multi-year project ThreatDetect investigates an autonomous platform to secure marine infrastructures by reliably detecting divers and mines in real time. Our system combines acoustic remote detection with verification using pattern recognition on underwater imagery. For diver detection, we rely on active acoustics from a single transceiver, and analyze the acoustic reflections to detect and localize a target that fits the pattern of a diver. For mine detection, we segment sonar images from an autonomous underwater vehicle (AUV) to differentiate between background, highlight, and shadow. In case of detection, we steer the AUV's trajectory so as to closely observe the target, and transmit segmented sonar images to a surface station via underwater acoustic communications. At the time of writing, the project is performing final technology tuning and integrated sea experiments.

Keywords Underwater threat detection · Infrastructure security · Diver detection and tracking · AUV · Submerged mine detection · Underwater acoustic communications · Sea experiments

P. Casari (✉)
Department of Information Engineering and Computer Science, University of Trento, Trento, Italy
e-mail: paolo.casari@unitn.it

L. Lampe
University of British Columbia, Vancouver, BC, Canada
e-mail: lampe@ece.ubc.ca

S. Martinez
Department of Marine Biology, Morris Kahn Marine Research Station, Sdot Yam, Israel
Leon H. Charney School of Marine Sciences, University of Haifa, Haifa, Israel
e-mail: smartinez@univ.haifa.ac.il

R. Diamant
University of Haifa, Haifa, Israel
e-mail: roee.d@univ.haifa.ac.il

C. Palestini (ed.), *Advanced Technologies for Security Applications*, NATO Science for Peace and Security Series B: Physics and Biophysics,
https://doi.org/10.1007/978-94-024-2021-0_24

24.1 Outline of the Project

The global marine industry represents a multi-billion-dollar-per-year business, whose sensitive infrastructure may include oil and gas facilities, offshore rigs, submerged pipes and cables, as well as harbor installations. Such strategic infrastructure must be protected against hostile intruders: once an unwanted access event is detected, surface control units must be promptly and securely alerted. Currently, these tasks are performed manually, requiring considerable manpower investment (for example, about 100 people for a single gas rig in Israel), restricting real-time response capabilities.

In particular, no holistic solution is currently available to secure facilities against direct diver attacks or against the deployment of submerged mines. While there exist active acoustic diver detection and autonomous underwater vehicle (AUV)-based mine detection systems, they often suffer from significant false alarm rates and require human interaction. The main reason is that most algorithms to identify acoustic reflections from a target remain suboptimal in the presence of strong reverberation, without a full characterization of the target's reflection pattern.

In this project, we design a novel autonomous early detection and identification system for divers and mines, that combines active acoustic remote detection with verification via an approaching AUV (see the system's structure in Fig. 24.1). Verified detections are communicated acoustically to a control station. We rely on a single acoustic transceiver, which greatly improves the applicability of our solution to small platforms, boats and buoys.[1]

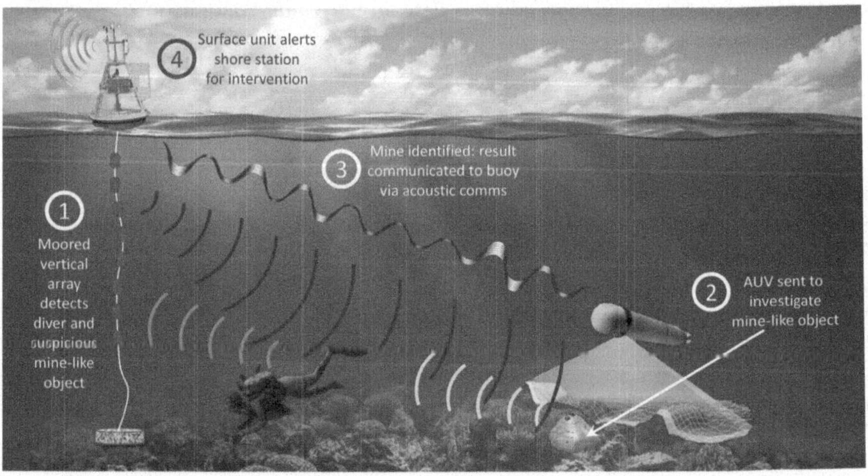

Fig. 24.1 Concept of our solution for the identification of intruding divers and mines

[1] Visual project overview: https://tinyurl.com/NATO-SPS-ThreatDetect-video.

The acoustic detection component processes reflected acoustic signals over time via a track-before-detect approach, and extracts patterns via dynamic programming. This method distinguishes a moving target (e.g., a scuba diver) even when its acoustic footprint is deeply buried in noise or outweighed by strong static reflections. To localize the target using a single transceiver, we match its range and bearing with a model-based reproduction of acoustic propagation under water.

The detection of submerged mines from sonar images is based on a combination of detection and segmentation algorithms, that offer a good tradeoff between true positive and false negative detection rates. All the above systems are integrated into an automatic solution, where an AUV scans the area around some marine infrastructure through a sonar, and automatically adapts its course to approach detected targets and scan them more accurately. Relevant sonar segments including detections are compressed and transmitted acoustically from the AUV to a surface station. Such communications are robustified against anthropogenic and natural interference via interference suppression algorithms that leverage the characterization of shipping noise and other simultaneous acoustic transmissions.

At the current stage of the project, we have developed all required subsystems and tested them in multiple sea experiments, both in the Mediterranean sea and in the Red sea. These experiments include detections of real scuba divers from boats and buoys, and detection of submerged mines from our AUV and other surface vehicles. The AUV detection, course adaptation and image communication chain has also been fully integrated. In the following, we provide a high level description of the system's components.

24.2 Detection and Localization

24.2.1 Acoustic Detection, Classification and Tracking

Our method to detect scuba divers addresses the limitations of currently existing solutions. These limitations include the need to setup a rigid array of multiple receiving elements, the false alarms caused by mistaking marine fauna as scuba divers, and the use of high-power acoustic transmissions that may harm marine animals. Instead, we rely on a single transceiver, small enough to be deployed from any platform, and transmit at lower source levels by leveraging wideband acoustic signals. A full description of the system is available in [9].

Since our transmission level is low, we accumulate reflections from multiple acoustic transmissions and arrange them into a time-distance matrix, where each row contains the reverberation pattern corresponding to one transmission (see, e.g., the data in Fig. 24.2a, collected near the Haifa shore, Israel). Detections are then

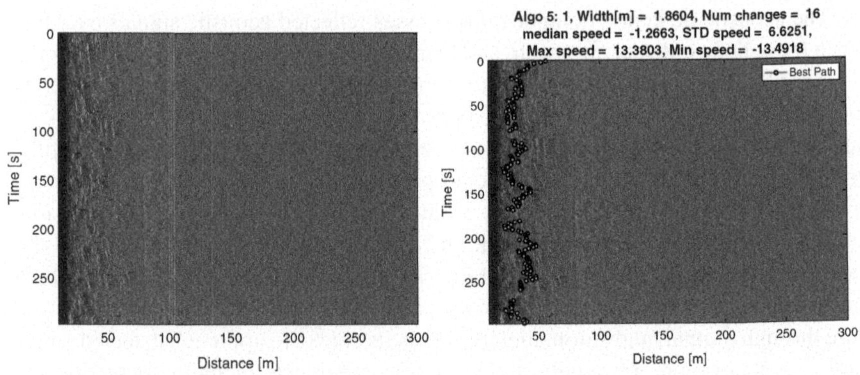

a. Time-distance matrix b. Detected track

Fig. 24.2 A detected track (**b**) over a time-distance matrix with reflections from a diver (**a**)

achieved based on pattern recognition from all reflections (track-before-detect approach). Specifically, we assume that clutter is random, whereas reflections from a real target are stationary. We single out the latter from the former through probabilistic analysis, based on a modified Viterbi algorithm informed with the maximum speed and orientation change rate that are compatible with diver motion.

To reject static reflections from such objects as rocks, weights, and chains, we embed a clustering step based on expectation-maximization (EM). This scheme separates static reflections (whose energy arrives from the same location over time) from non-static ones, by operating column-wise on the time-distance matrix. The application of this method to Fig. 24.2a results in detecting the diver trajectory highlighted in Fig. 24.2b.

24.2.2 Localization Methods

We developed two different localization methods: one aimed at low-complexity systems with a single receiver, and a second one targeting systems that embed a vertical array. In the former case, we rely on the diversity of the bathymetry around a fixed receiver. This induces a different multipath propagation profile depending on where the source of an acoustic signal: therefore, knowing the bathymetry around the receiver and the sound speed profile allows us to infer the location of the source from the multipath distortion.

We proceed as follows. For an arbitrarily dense grid of possible source locations, we predict the channel impulse response (CIR) that acoustic signals would be subject to using a numerical acoustic propagation model. Every time we receive an acoustic signal we estimate the CIR, and correlate it with our CIR database to extract a set of candidate source locations. We then collect multiple transmissions

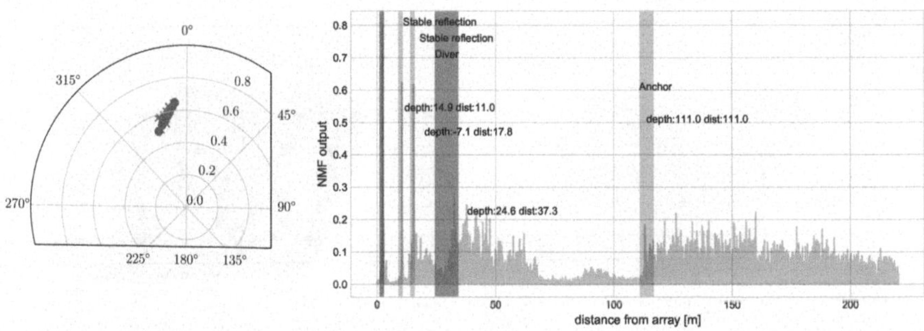

a. Path detection b. Angle-of-arrival estimation

Fig. 24.3 Path detection using a single receiver (**a**); angle of arrival estimation for different targets and a diver using a vertical array (**b**)

and fuse the corresponding source location estimates using a trellis search algorithm. This yields the most likely source path with a lower complexity than the Viterbi algorithm. Figure 24.3a shows an accurate trajectory estimate for a moving source using environmental data from the San Diego bay area, USA. The algorithm correctly selected a sequence of locations from a grid composed of about 4 million points.

For localization via a vertical array, we employ wideband acoustic beamforming helped by side information. We observe that underwater equipment often imposes mounting constraints that may impede to fully control the shape of an acoustic array. For systems working at centimeter wavelengths λ, it may be even impossible to guarantee that the array preserve the $\lambda/2$ spacing. We solve these challenges through a processing chain that involves matched filtering, clustering of match filtering peaks, elimination of stationary arrivals, and wideband direction of arrival estimation. The latter is helped by time-difference-of-arrival information in order to bound the angle of arrival search and remove spatial ambiguity. This system was tested in a sea experiment involving two divers. The results in Fig. 24.3b show the output signal recorded by one of the array's hydrophones, where relevant peaks are tagged with range and depth estimates. We observe that our scheme correctly identifies the diver (brown) at about 25 m of depth and 37 m from the array.

24.2.3 *Segmentation of Sonar Images from AUVs*

We employ a multi-stage chain for the detection of submerged mines from sonar images. We start with a rough detection of regions of interest (ROIs) within the sonar image, and then segment the ROI to separate background, highlight, and shadow regions. We report a detection based on prior knowledge about the target (e.g., minimum number of pixels, height above the bottom, and so forth). Specifically, we use likelihood ratio combining of highlight and shadow identifications. The former

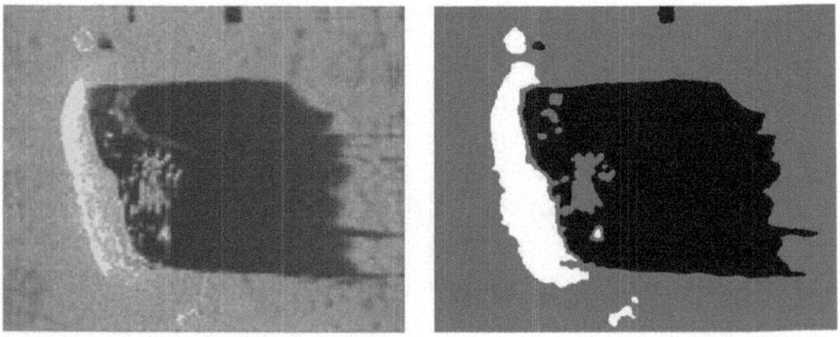

Fig. 24.4 A region of interest including a target (**a**), and segmentation results (**b**)

is based on segmenting a higher moment of the sonar image, whereas the latter is based on a blind classification using a support vector machine (SVM).

The segmentation of the highlights relies on the assumption that acoustic reflections from the target have homogeneous characteristics, whereas for shadows we assume that the distribution of the shadow pixels is the same throughout the image, hence the shadow's intensity level in the ROI can be learned from other sonar image sections. Figure 24.4a shows an example of ROI. A full description of the segmentation process is available in [3].

We perform fine-grained target detection by clustering the ROI via two strategies. The first is based on clustering the ROI using an EM algorithm for a mixture of alpha distributions, which offers sufficient flexibility. For better performance, we allowed hard decisions to be made during EM iterations: this way, we can exploit the expected dependencies among the image pixels related to the target. For full details, we refer the reader to [1]. The above solution is accurate but computationally demanding. Therefore, we designed a second segmentation strategy based on fuzzy logic. Here, we manage the inhomogeneity within the sonar image through two fuzzy terms that reflect the location of the segmented pixels within the image [2]. For both solutions, we designed a de-noising filter that smooths the image before segmentation. An example for a segmented image using our approach is shown in Fig. 24.4b.

We tested the above solutions via a specifically designed simulator that generates sonar images with targets over different types of seabed (e.g., sand ripples, grass, or rocks). Furthermore, using our own AUV, we collected a database of more than 1000 sonar images, and hand-labeled them for targets. Since these images were taken from different sea environments (in France, as well as in southern, central, and northern Israel), the database enables a sufficiently robust statistical exploration of ROI detection performance.

24.2.4 Automatic Scan by an AUV

A distinctive feature of our system is a fully automatic behavior without humans in the loop. In this vein, we have designed a protocol that allows the AUV to change its predefined course in real time upon detection events. Such capability is essential to detect submerged mines, which often can be identified as such only after viewing them from different points of view. An automatic procedure is paramount in this case, as navigation errors may prevent an operator to successfully steer the AUV towards the suspected target, and would imply significant delays for surfacing and re-diving the AUV.

We implement the automatic adaptation of the AUV's mission by acting on its backseat driver. For each sonar image collected, the AUV detects and segments ROIs in real time. It then determines the geographic location of the ROI, and correspondingly sets a new mission, which includes back-tracking to observe the object in the ROI from the opposite direction, and returning to the previous location to resume the original mission. During this last step, the AUV also hovers above the ROI, so as to profile the target via optical or sub-bottom scans. Finally, the AUV fuses all collected information, compresses the best ROI image, and sends it to an operator.

Figure 24.5a shows an example of the above procedure operated by the AUV in Fig. 24.5b, in southern Israel in June 2019. Blue lines represent the AUV's pre-determined mission, whereas the yellow arrows shows the actual position of the AUV over time. Once it reaches the bottom-right corner of the trajectory, the AUV drops its pre-defined mission and loops around a suspect location, in real time without any intervention by an operator.

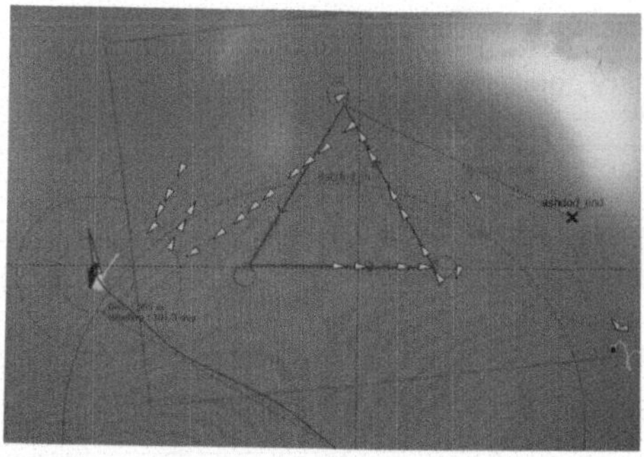

a. Image of the track record

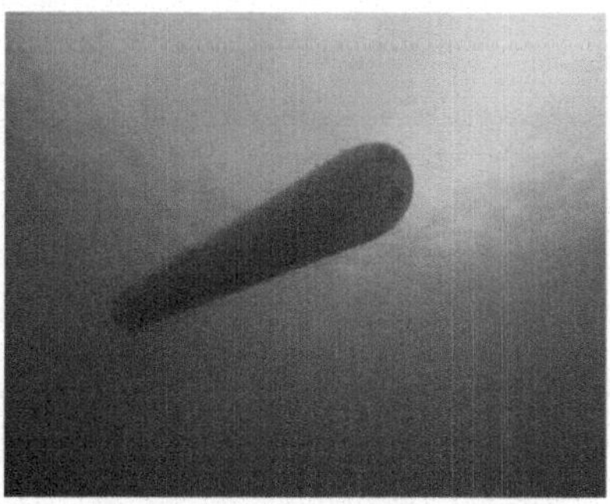

b. The AUV performing the mission

Fig. 24.5 AUV (**a**) track record from a sea experiment (**b**). Yellow arrows show the actual track of the vehicle. Blue lines show the pre-determined mission track

24.3 Noise Characterization and Interference Cancellation

As we foresee our system to operate in the proximity of underwater installations or harbors, we investigate methods to make communications robust against noise generated by shipping activities as well as co-existing signals, which we expect to constitute major impediments in our scenarios of interest. To model shipping noise, we evaluated acoustic recordings from the Ocean Networks Canada (ONC) database available at [11], which includes a large number of measurements from hydrophone arrays, some of which are located close to busy shipping routes. As a starting point for our modeling attempt, we considered the impulsiveness of shipping noise as also reported in previous work such as [7, 10], and processed data to fit a Gaussian mixture model via the EM algorithm. Figure 24.6 compares measured noise and noise according to a two-term Gaussian mixture model trained with the EM algorithm. We observe that shipping noise could be modelled with a two-term mixture having a 15 dB higher variance for the impulse noise component.

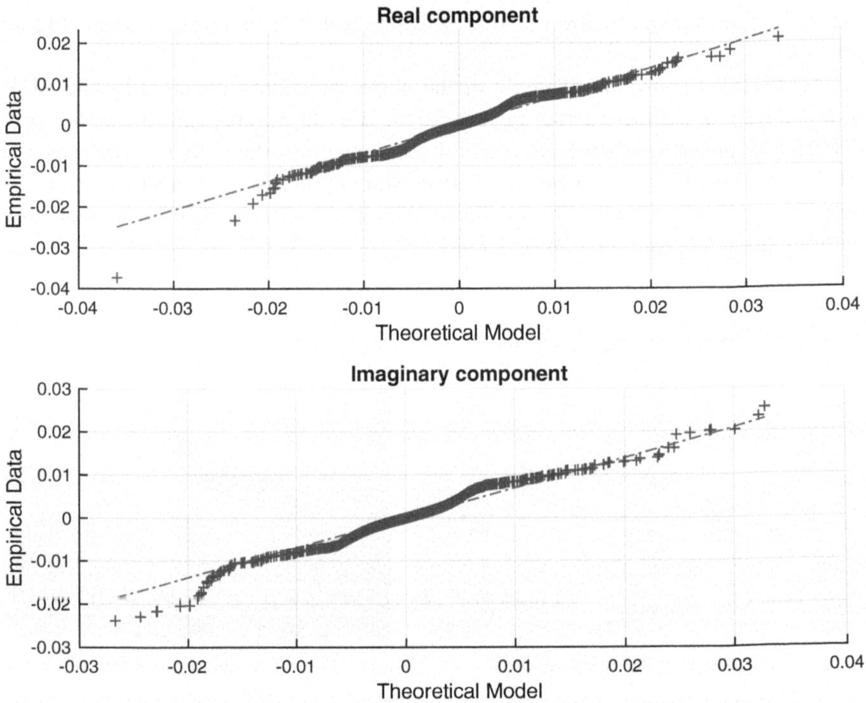

Fig. 24.6 QQ-plot of empirical quantiles from a complex baseband noise signal in the 8–16 kHz band versus theoretical quantiles from a Gaussian mixture random variable

24.3.1 *Interference Cancellation*

24.3.1.1 Shipping Noise

The structure inherent to shipping noise can be exploited for noise mitigation or cancellation. We consider the latter, making use of null-subcarriers in a multicarrier underwater acoustic communication signal, which enables noise (or interference) estimation. For this purpose, the fact that the impulse component of the noise is sparse can be exploited via compressed-sensing based signal recovery [6]. This can be extended by also considering the noise distribution or by learning a sparse representation [4]. Using 192 ship noise recordings obtained from the ONC database and considering 25% of null-subcarriers, we conclude that communication systems operating from 0.1 to 16 kHz can achieve relative cancellation gains of about 1 to 3 dB [5].

24.3.1.2 Co-signal Interference

Co-existing acoustic transmissions constitute one of the strongest interference sources for underwater acoustic communications. Such signals can originate from acoustic systems (e.g., echo sounders for depth measurement, ADCP water current meters) or from transmissions by nearby modems. When these transmissions intersect in the same bandwidth of the desired signals, the signal-to-interference ratio (SIR) may be too low to correctly receive the message. Considering this challenge, we have designed an interfering cancellation filter that specifically handles the case of low SIR.

Our algorithm modifies the traditional noise cancellation filter in order to remove a reference signal from the signal to be cleaned. The solution first detects the strong interference and finds its analytic form (i.e., its time-frequency pattern). Then, we employ an adaptive filter to equalize the channel from the interference source to the receiver. The result is subtracted from the received signal, and is fed back into the adaptive filter. To discriminate between the two channels (i.e., from the interference and from the designed source), we lock onto the channel taps of the interferer and zero-force all other taps. This way, the energy of the desired signal is not effected. The full details of this approach are available in [8].

We have tested our system in several sea experiments including two interferers and a common receiver. An example of the performance of our algorithm is provided in Fig. 24.7, where we show the spectrogram of the received signal before and after interference cancellation for a sea experiment conducted in Ashdod, Israel. We observe that the strong interference is almost completely removed from the signal.

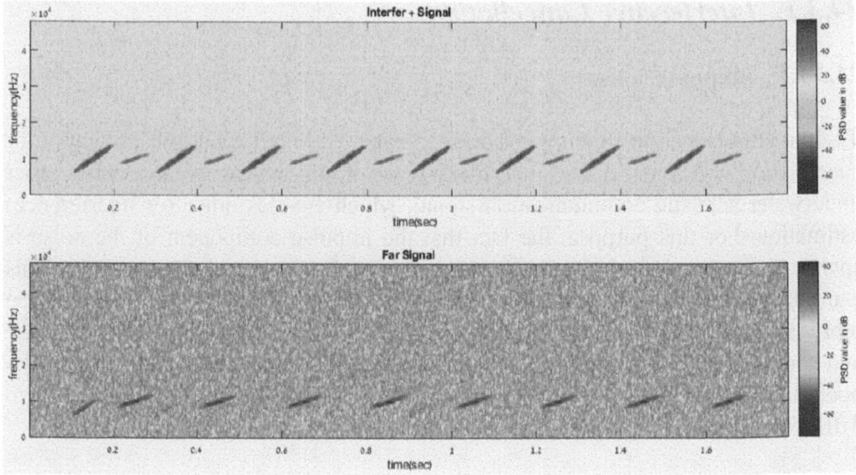

Fig. 24.7 A spectrogram showing two signals before (upper panel) and after (lower panel) interference cancellation

24.4 Conclusions

The ThreatDetect project targets the protection of sensitive and strategic marine infrastructures from intruding divers and submerged mines. In this paper, we reported on the progress of our research activities in the context of diver detection, threat localization, fully automatic detection of mines by an AUV, as well as noise and interference cancellation for robust communications. A number of integrated sea trials demonstrate the feasibility of our solution.

Acknowledgments This research has been sponsored in part by the NATO Science for Peace and Security Programme under grant G5293.

References

1. Abu A, Diamant R (2018) Unsupervised local spatial mixture segmentation of underwater objects in sonar images. IEEE J Ocean Eng 44:1179–1197
2. Abu A, Diamant R (2019) Enhanced fuzzy-based local information algorithm for sonar image segmentation. IEEE Trans Image Process 29:445–460
3. Abu A, Diamant R (2019) A statistically-based method for the detection of underwater objects in sonar imagery. IEEE Sensors J 19:6858–6871
4. Aharon M, Elad M, Bruckstein A, et al (2006) K-SVD: an algorithm for designing overcomplete dictionaries for sparse representation. IEEE Trans Signal Process 54:4311
5. Atanackovic L, Zhang R, Lampe L, Diamant R (2019) Statistical shipping noise characterization and mitigation for underwater acoustic communications. In: Proceedings on MTS/IEEE OCEANS

6. Caire G, Al-Naffouri TY, Narayanan AK (2008) Impulse noise cancellation in OFDM: an application of compressed sensing. In: IEEE international symposium on information theory, pp 1293–1297
7. Chen P, Rong Y, Nordholm S, He Z, Duncan AJ (2017) Joint channel estimation and impulsive noise mitigation in underwater acoustic OFDM communication systems. IEEE Trans Wirel Commun 16:6165–6178
8. Diamant R (2018) Robust interference cancellation of chirp and CW signals for underwater acoustics applications. IEEE Access 6:4405–4415
9. Diamant R, Kipnis D, Bigal E, Scheinin A, Tchernov D, Pinchasi A (2019) An active acoustic track-before-detect approach for finding underwater mobile targets. IEEE J Sel Top Signal Process 13:104–119
10. Mahmood A, Chitre M (2016) Robust communication in bursty impulsive noise and Rayleigh block fading. In: 11th ACM international conference on underwater networks & systems, p 13
11. ONC (2019) Search Hydrophone Data. https://data.oceannetworks.ca/SearchHydrophoneData

Chapter 25
BalSAR: A Stratospheric Balloon-Borne SAR System

Marco Martorella and Elias Aboutanios

Abstract Surveillance systems are continuously employed for both military and civilian applications, including homeland security and border protection, which are two main concerns to NATO and in particular to the Science for Peace and Security (SPS) programme. Several platforms and systems, developed in past years, have turned into surveillance systems that are currently used in such scenarios. This paper describes a high-altitude balloon-borne synthetic aperture radar (BALSAR) system, which is currently under development as part of a NATO funded project within the SPS programme. Such a system will be able to perform surveillance tasks by acquiring radar data, forming SAR images and using them to extract valuable information.

Keywords High-altitude platform · Stratospheric balloon · Airborne surveillance · Synthetic aperture radar

25.1 Introduction

Military and civilian information gathering is an essential part of maintaining security and significant effort and money are spent on systems to enable these functions. Current technologies, which mainly employ satellites, aircraft (both manned and unmanned – UAVs), and drones suffer from a number of shortcomings. Space-borne systems operate from a large distance and, provided careful constellation design, are

M. Martorella (✉)
Department of Information Engineering, University of Pisa, Pisa, Italy

Radar and Surveillance Systems National Lab, CNIT, Pisa, Italy
e-mail: m.martorella@iet.unipi.it

E. Aboutanios
School of Electrical Engineering and Telecommunications, UNSW Sydney, Sydney, NSW, Australia

C. Palestini (ed.), *Advanced Technologies for Security Applications*, NATO Science for Peace and Security Series B: Physics and Biophysics,
https://doi.org/10.1007/978-94-024-2021-0_25

able to cover almost all areas on the surface of the earth. However, space missions are expensive and their use depends on the presence of a satellite over the designated area which can only happen at particular times that are determined by the orbit. Also, they do not offer a rapid and timely response as the ability to repeat measurements over a given area is constrained by the satellite orbit, which may impose intervals of several hours or even days between revisits. Airborne systems operate at much lower altitudes and piloted aircraft missions put the lives of pilots at risk as they are inherently vulnerable to attacks. Their size, and therefore radar cross section, and flight altitude limits make them easier to detect and target from the ground or the air. Moreover, the cost of such aircraft is quite high, justifying the launch of an intelligent missile against them. While UAVs take the pilot out of the equation, they still suffer from all of the other problems associated with piloted aircraft including cost and vulnerability. In fact, all of these systems are not expendable and must be protected.

High altitude platforms (HAPs) [1] have the potential to complement the above two systems and address many of their shortcomings. HAPs have been proposed in military applications for the gathering of surveillance data and are poised to play a key role in the area of national security. The USA, China, Japan and European Union have had military HAPs projects. These platforms enjoy a number of unique advantages with respect to both aircraft and satellites. HAPs operate at altitudes exceeding 20 km, and include certain aircraft, airships and balloons [1]. Their high altitude gives them a higher degree of immunity against attack as compared to aircraft while providing them a wider field of view [2]. On the other hand, they are a cheaper alternative to traditional satellite systems as their development and deployment costs are much lower than those of spacecraft. Their comparatively low altitude, with respect to spacecraft, makes them more versatile and recoverable meaning that they can be maintained and even upgraded. Also for remote sensing applications, they do not suffer from long revisit times that are a drawback of satellite systems.

While some systems employing HAPs exist [1–3] or have been proposed, most tend to be either airships or UAVs, and are large and very expensive. The Zephyr [4] is a solar powered UAV that is developed by Airbus. It is described as a High Altitude Pseudo-satellite (HAPS) as it is designed to hover for an extended period of time over a designated region. Among the intended uses of the Zephyr, Airbus lists maritime and border surveillance, environmental surveillance, missile detection, navigation, and continuous imagery. Lockheed Martin's ISIS [1] and Raytheons Radar Aerostat [5] are two US projects developing high altitude airships. All three systems are manoeuvrable and intended to provide long mission durations (on the order of days, months or even years at a time). Consequently, they are quite expensive and require significant infrastructure for deployment. As an example, the UK Ministry of Defence was reported to have an order for two Zephyr worth USD18 million [1].

Stratospheric balloons have been used for decades for scientific experiments and remote sensing and both NASA [6] and JAXA [7] conduct high altitude balloon missions. These free floating balloons are very large and carry payloads weighing many hundreds of kilograms or even a few tonnes. Stratospheric balloon constellations with trajectory control have also been proposed for communications [8, 9], and scientific missions [10]. Small (sounding) balloons are extremely cheap even in comparison to the large HAPs discussed above. These are mainly used for weather sensing [11], education [12, 13] and amateur activities.

We propose a new high-altitude (in excess of 20 km) balloon-borne Synthetic Aperture Radar (SAR) system that overcomes the shortcomings of existing systems without compromising performance, such as resolution, signal-to-noise ratio and hence target detection and recognition. The new system has a number of significant advantages over the aforementioned solutions. It is very low cost and hence expendable, rapidly deployable, has low probability of intercept (LPI) characteristics and is practically immune to attacks, requiring a very expensive guided missile to intercept it. There are a number of scenarios where airborne/space-borne radar surveillance is needed. Such scenarios include border protection, battlefield surveillance and, in more general terms, critical area surveillance where ground systems cannot be deployed either because the area is far from any logistic support or because of the unacceptably high risks that would be involved in deploying the system. In this paper we describe the proposed architecture and detail various subsystems.

25.2 Proposed Solution

The proposed solution employs a combination of a very light SAR system and a high altitude balloon platform to provide a balloon-borne SAR (BalSAR) system. The BalSAR system would provide the means to fly at a very high altitude, 20 km and above, and therefore operate at a safe distance from the surveillance area. In addition to the BALSAR system, the project will also produce the support systems including the flight prediction and mission planning software, launch (balloon inflation and release) system, telemetry tracking and command (TT&C) system, and HAP recovery system. A high altitude balloon borne SAR has in fact been recently reported in [14], which demonstrates the feasibility of the concept. Our system, however, aims to achieve a number of specific aims including low cost, rapid deployment and standardization of the bus.

The BALSAR system is shown in Fig. 25.1. It comprises a balloon capable of lifting a High-altitude Platform (HAP) weighing around 20 kg, a flight termination system, and a parachute to facilitate the HAP recovery. Capitalizing on the experience in building the UNSW cubesat UNSW-EC0 [15, 16], which formed part of the European QB50 constellation [17, 18], the HAP itself is modeled on the cubesat standard [19, 20]. Platform standardization will enable the use of commercial-off-the-shelf (COTS) components, directly leading to greater flexibility, as well as lower costs and reduced risk. Additionally, as COTS components improve being driven by market competition and pressures, improvements in the HAP will result. Finally a standard bus will greatly facilitate the opportunities of payloads other than the SAR system to be carried by the HAP. This enhances the utility of the proposed HAP

Fig. 25.1 Illustration of the BALSAR system. Note that the radome is not drawn in order to show the antennas

system. Therefore, the standardisation of the HAP is one of the core goals of this project.

The project involves significant challenges both in the SAR and HAP systems. In order to deliver a low-cost and rapidly deployable system, the overall size and complexity must be kept low. The proposed BALSAR system rides on the stratospheric winds and does not include any trajectory control. Therefore, the mission goals are ensured through an innovative architecture that includes the subsystems and support systems described in what follows.

25.3 The Flight Subsystem

The flight subsystem consists of the balloon, termination mechanism and parachute. The balloon is helium-filled and is rated to carry a payload weighing 20 kg up to an altitude of 30 km. As the balloon's trajectory is not controlled, a flight termination subsystem is included in order to terminate the mission in case it diverges significantly from the set flight plan. The termination subsystem is positioned between the parachute and balloon and receives the termination commands from ground control. Flight termination is achieved by burning a nichrome wire to separate the parachute and HAP from the balloon. The termination subsystem can also be used if a desired landing site is specified. The mission is then terminated at the point that results in the HAP landing at the designated site. Also note that automatic flight monitoring may be programmed into the on-board computer (OBC) which can then command the termination mechanism to activate if a pre-specified mission envelope is exceeded.

The parachute ensures that the HAP descends at a rate that prevents damage if it is to be recovered. Thus the parachute is located below the termination device and is sized to provide the desired descent rate, which is typically between 5 and 10 m/s. During the descent phase of the flight, the parachute automatically opens when the atmospheric density becomes high enough (typically at altitudes higher than 15 km).

25.4 The HAP Subsystem

The proposed HAP architecture is shown in Fig. 25.2. As alluded to earlier, it is modelled on a spacecraft bus, and in particular on a cubesat design. Therefore, the HAP comprises an OBC, an electrical power subsystem (EPS) and associated batteries, a communications module (comms), inertial measurements units (IMUs), and various positioning units to provide tracking of the platform. The HAP subsystem also houses the payload, that is the SAR subsystem. Micro-controller Units (MCUs) are employed as intermediaries between the OBC and a number of other subsystems in order to perform specific tasks associated with these subsystems. This philosophy ensures that critical tasks are handled by their decidated MCUs which ensure the reliable operation of the HAP. The main subsystems of the proposed HAP architecture are described below:

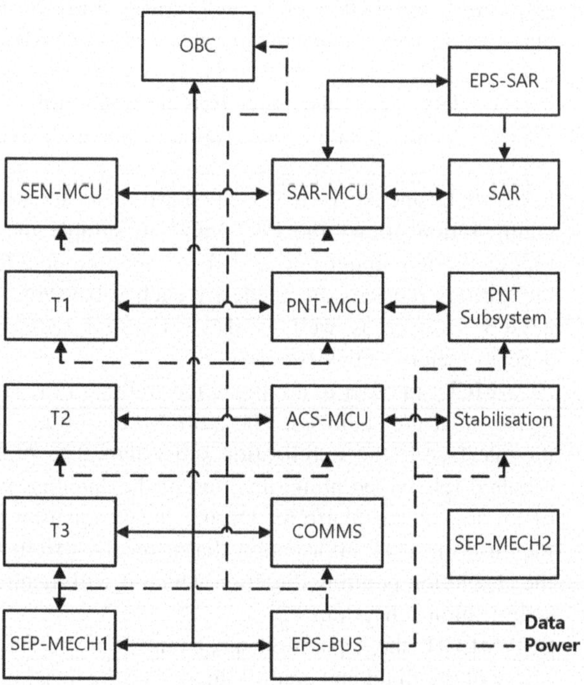

Fig. 25.2 Architecture of the HAP bus showing the data and power interconnections

- OBC: the OBC manages the flight, executes the schedule, and monitors the various subsystems. Upon powering the bus, the OBC will first initialise the other subsystems and load the flight parameters to the SAR and stabilization microcontroller units (SAR-MCU and S-MCU respectively). The OBC will then periodically query the other subsystems to check the health of the HAP. Deviations from the nominal flight envelope will lead to termination of the flight. Finally, if requested by ground control, the OBC will gather relevant information on the HAP subsystems and communicate it to the ground via the comms subsystem.
- EPS: the HAP carries two electrical power subsystems, one for the bus (EPS-BUS in Fig. 25.2) and another for the SAR payload (EPS-SAR in Fig. 25.2). Each EPS includes its own battery pack and is fully controllable allowing various rails to be switched on and off. The use of a separate EPS for the payload is dictated by the power requirement of the radar.
- COMMS: the COMMS module operates at UHF in the amateur band. It provides communications with ground control in order to monitor the HAP health and track it. It also permits critical commands, such as flight termination to be uplinked to the HAP. Note that the SAR data is stored onboard and is not downlinked to the ground during the flight.
- SAR-MCU: This microcontroller interfaces the SAR payload to the HAP bus and performs two primary functions: firstly, it controls the operation of the radar by first turning it on at the right point of the flight, instructing it to start the acquisition and then stopping and turning it off. Secondly, it stores the position tags for the radar snapshots. To this end, the SAR-MCU receives an interrupt from the radar every time the latter acquires a snapshot and then fetches the position data, tags it and stores it in the positioning file.
- PNT-MCU: the position, navigation and timing MCU logs the data from the PNT subsystem and then services requests for tracking data from the OBC, SAR-MCU, and ACS-MCU (which drives the antenna stablization system). This configuration allows the PNT-MCU to sample the high precision navigation unit at the maximum rate and then accommodate the different rates at which the various requests are made by each subsystem. The PNT system includes a GPS receiver as well as IMUs (inertial measurements units) comprising accelerometers and gyroscopes.
- ACS-MCU: In order to minimize the weight of the HAP, only the antenna arrays of the radar are stablized. The attitude control and stabilization MCU provides the interface to the stabilization subsystem. The ACS-MCU is present with the required relative pointing direction of the antennas with respect to the direction of motion of the platform. During the acquisition phase, the ACS-MCU will then use the PNT information that it receives from the PNT-MCU to calculate the absolution pointing direction which it will then relay to the controller of the stabilization subsystem.
- SEP-MECH: the separation mechanisms are included primarily to ensure the safety of the flight and control the risk of the mission. The OBC will continually monitor the flight parameters and verify that they are within the acceptable

mission envelope. Should the flight move outside this envelope, the separation mechanisms are activated to terminate the flight. Two independent mechanisms are included to provide redundancy. One of these mechanisms, SEP-MECH2, is an independent system that was mentioned in the previous section. The other, namely SEP-MECH1, is connected to the OBC and forms part of the HAP bus. In addition to the PNT system described above, multiple tracking subsystems, denoted as T1, T2 and T3 in Fig. 25.2, will be used ensure adequate tracking system redundancy at all stages of the flight.

- SEN-MCU: the sensor suite MCU provides the functionality for various HAP environmental sensors, such as temperature and pressure, and other flight information sensors, such as flight dynamics sensors, to be. These are logged by the SEN-MCU and stored on a dedicated SD card. The MCU can also be queried by the OBC if the relevant data is required.

In addition to the subsystems detailed above, the HAP provides environmental protection to the bus and payload. During the flight, the atmospheric temperature can drop as low as $-70°$ and the pressure decreases to approximately 1% of its value at ground level. Foam insulation, combined with the heat generated by the various subsystems, will ensure that the internal temperature of the HAP remains above $-20°$. Furthermore, thermal and vacuum testing will be employed prior to the flight to verify the system performance under the expected environmental conditions.

25.5 The SAR Payload

As the HAP is limited to around 20 kg in total weight, the project requires a miniaturized SAR system. The SAR system is restricted to a total of 10 kg, with the electronics weighing around 7 kg and the antennas 3 kg. Furthermore the SAR draws its power from the HAP, which then places a power consumption requirement on it. The limits on weight and power consumption are particular challenges for the SAR design.

25.5.1 Radar Front-End

The radar front-end is made up of a single board in standard ITX format (17×17 cm) that is stacked and interconnected according to the requirements to other boards (Embedded Digital processor and power supply subsystem) and an external power amplifier. The front end implements an X-band direct-conversion Linear-FMCW radar transceiver architecture. The transmitted waveform is generated by a PLL-based frequency synthesizer. This approach focuses on a compact, low cost and low power consumption solution that allows for the generation of large bandwidth and high chirp rate Linear-FMCW waveforms. The front-end mainly consists of the following sub-sections:

- Waveform generation and transmitter: this consists of a PLL based, X-band, programmable signal generator (phased detector, VCO and loop filter), a RF pre-amplifier, a splitter for the generation of the OL signal, a RF medium power amplifier (MPA) and a digital control interface. The generator is locked to the same low noise reference source that feeds the ADC stages, in order to perform coherent data processing.
- RF receiver: the receiver includes an input limiter, a RF band-pass filter, a low-noise amplifier and a quadrature demodulator. The demodulator is fed by OL signals generated by the aforementioned synthesizer.
- Base-band signal conditioning: down-converted signals are sent to a base-band signal conditioning stage that employs a programmable attenuator and an active band-pass filter. This allows to tune the overall gain and adapt the signal to the input dynamics of the AD converters.
- Power amplifier: the output of the transmitter section is sent to a solid state GaN based linear high power amplifier (HPA), with a maximum output power of 40 dBm.
- Power down-conversion subsystem: this is a multiple output down-conversion subsystem that employs mixed linear and switch-mode topologies in order to fulfill the power requirements of all the previous described sections. Such subsystem down-converts the voltage(s) provided from the HAP and converts them to the rails required by the various SAR subsystems.

The radar front-end specifications are summarized in Table 25.1, and picture of a test board of the radar front-end is shown Fig. 25.3a.

25.5.2 *Power Amplifier*

The power amplifier is a COTS device by Keylink Microwave. This model is a GaN based high power amplifier operating between 9.1 and 10.1 GHz and offers a wide dynamic Range with 10 W of output power. It has long term reliability and

Table 25.1 Radar front-end specifications

Parameter	Value
Waveform type	Linear-FMCW
Frequency range	9.3 to 9.9 GHz
Chirp rate	up to 1 THz/s
Output power	40 dBm (max)
TX attenuator	0–30 dB (1 dB step)
Minimum detectable signal	−130 dBm
Noise figure	6 dB
IF bandwidth	from 10 to 40 MHz
RX attenuators	0–40 dB (0.5 dB step)

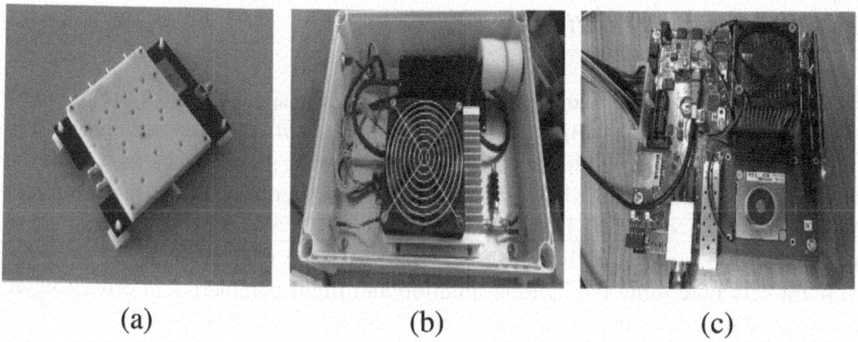

Fig. 25.3 SAR Payload – (a) front-end, (b) power amplifier, (c) embedded digital radar processor

high efficiency and it is ideal for X-Band linear applications. A picture of the same component used for another system is shown in Fig. 25.3b.

25.5.3 Embedded Digital Radar Processor

The Embedded Digital Radar Processor has been designed around the Trenz Electronic TEBF0808 carrier board which is a baseboard for the Xilinx Zynq Ultrascale+ MPSoC modules TE0808 and TE0803. A picture of the developed system is shown in Fig. 25.3c.

The main sub-systems are:

1. Carrier board
2. System on a Module
3. Acquisition board
4. Mass memory storage

25.6 Mission Support Systems

In order to enable the mission execution, a number of supporting subsystems are being developed. These include the flight prediction and planning, inflation and release system, and telemetry tracking and command system. These are described in what follows.

25.6.1 *Flight Prediction and Planning*

Flight prediction and mission planning are important aspects of the system. Given a launch location, date and time, the flight prediction software uses weather data to determine the expected flight trajectory. This allows the ground track as well as the landing site to be determined. Mission planning, however, can require that a number of waypoints be observed by the balloon. Therefore, when provided with these waypoints, the mission optimization software uses the mission planning program to iteratively determine the balloon inflation and flight parameters in order to give the optimal trajectory that is as close to the desired waypoints as possible.

25.6.2 *Inflation and Release System*

The balloon will have a pre-launch diameter of approximately 10 m and will display a large area to any light breeze. Therefore, it needs to be anchored to the ground and handled properly during inflation. This requires an inflation and release rig that is being developed as part of the project. The rig is modular to facilitate its transport to the launch site. It is also easy to set up and is able to provide measurement of the neck lift of the balloon to ensure the correct inflation is achieved.

25.6.3 *Telemetry Tracking and Command System*

Although the SAR data will be stored onboard the HAP and will not be downlinked to the ground during the flight, a TT&C system is being developed to continually track the HAP and monitor its state. The TT&C system, which operates at UHF in the amateur band, comprises a mobile ground station that will enable simple commands to be uplinked to the HAP and health check data to be received from it. Additionally, the TT&C system receives the position data from the HAP and updates the estimated flight path and landing position. This permits the mission to be monitored and decisions to be made on the termination of the flight if required.

25.7 Conclusions

This paper describes a novel balloon-borne synthetic aperture radar that is under development as part of a NATO funded project. The full system realisation is predicted to be completed by 2020 and results in terms of SAR imagery should appear soon after the system completion. The BALSAR system is intended to be low-cost and rapidly deployable in order to provide enhanced surveillance capability in hostile environments. Furthermore, the BALSAR system will find applications in remote sensing and monitoring applications, such as border protection and disaster monitoring.

References

1. d'Oliveira FA, de Melo FCL, Devezas TC (2016) High-altitude platforms present situation and technology trends. J Aerosp Technol Manag 8:249–262
2. Wang W, Shao H (2014) High altitude platform multichannel SAR for wide-area and staring imaging. IEEE Aerosp Electron Syst Mag 29:12–17
3. Yang H, Li Z, Wu J, Huang Y, Yang J, Yang X (2013) Near-Space slow SAR high-resolution and Wide-Swath imaging concepts. In: 2013 IEEE radar conference (RadarCon13), pp 1–5
4. Airbus Zephyr High Altitude Pseudo-satellite (HAPS). https://www.airbus.com/defence/uav/zephyr.html. [Online], last Accessed May 2020
5. Raytheon Aerostat. http://www.raytheon.com/capabilities/products/jlens/. [Online], last Accessed May 2020
6. NASA's Scientific Balloon Program. http://asd.gsfc.nasa.gov/balloon/. [Online], last Accessed May 2020
7. JAXA high altitude balloon projects. http://global.jaxa.jp/projects/sas/balloon/topics.html. [Online], last Accessed May 2020
8. Google Loon Project. https://loon.com/. [Online], last Accessed May 2020
9. Facebook High Altitude Platform project. https://code.fb.com/connectivity/high-altitude-connectivity-the-next-chapter/, [Online], last Accessed May 2020
10. Nock K, Heun M, Aaron K (2002) Global stratospheric balloon constellations. Adv Space Res 30:1233–1238
11. NOAA NWS Radiosonde Observations. https://www.weather.gov/upperair/factsheet. [Online], last Accessed May 2020
12. Stratoflights Classroom on the Edge of Space. https://www.stratoflights.com/en/education/info/. [Online], last Accessed May 2020
13. UNSW's BLUEsat High Altitude Balloon Team. https://bluesat.com.au/teams/high-altitude-ballooning/. [Online], last Accessed June May 2020
14. Zaugg EC, Margulis A, Bradley JP, Kozak AH, Roehrich WK (2019) SAR imaging from stratospheric balloons: first results. In: IEEE radar conference
15. Osborne B, Aboutanios E, Dempster A, Cetin E, Heiser G, Glennon E (2013) UNSW EC0 cubesat design: experiments in radiation tolerance critical systems, GNSS remote observation and 3-D printed satellite structures. In: 5th European cubesat symposium, p 41
16. Cheong JW, Southwell BJ, Andrew W, Aboutanios E, Lam C, Croston T, Li L, Green S, Kroh A, Glennon EP, Bultitude J, Broadbent T, Guo TBQ, Aligno JG, Dempster AG, Osborne B (2020) A robust framework for low-cost Cubesat scientific missions. Space Sci Rev 216(1):8
17. Muylaert J, Reinhard R, Asma C, Buchlin J, Rambaud P, Vetrano M (2009) QB50: an international network of 50 cubesats for multi-point, in-situ measurements in the lower thermosphere and for re-entry research. In: ESA atmospheric science conference, Barcelona, pp 7–11
18. The QB50 project. https://www.qb50.eu. [Online], last Accessed May 2020
19. Puig-Suari J, Turner C, Ahlgren W (2001) Development of the standard CubeSat deployer and a cubesat class picosatellite. In: 2001 IEEE aerospace conference proceedings (Cat. No.01TH8542), vol 1, pp 1/347–1/353

20. Spangelo SC, Kaslow D, Delp C, Cole B, Anderson L, Fosse E, Gilbert BS, Hartman L, Kahn T, Cutler J (2012) Applying model based systems engineering (MBSE) to a standard CubeSat. In: 2012 IEEE aerospace conference, pp 1–20
21. The UNSW-EC0 QB50 CubeSat. http://www.acser.unsw.edu.au/QB50. [Online], last Accessed May 2020

Chapter 26
A Dynamic and Flexible Architecture Based on UAVS for Border Security and Safety

Fabrizio Granelli, Claudio Sacchi, Riccardo Bassoli, Reuven Cohen, and Itzik Ashkenazi

Abstract NATO SPS G5428 DAVOSS project focuses on merging those two concepts (5G SBA and UAVs) to build a communication and service architecture capable of providing coverage for monitoring, border security and safety applications. Indeed, DAVOSS project aims to develop a multi-layer virtualised system in which all the technologies listed above work together to guarantee efficient and effective borders and ports surveillance.

Keywords UAVs · Border surveillance · 5G · LoRA

26.1 Introduction

With the advent of 5G, networks are expected to include an un-precedented functionality: fast deployment. Indeed, 5G requirements include the possibility of deploying a functional next-generation wireless networks in less than 90 min, where for LTE this would have required days. Such requirement would require the usage of a software-based and configurable network architecture as well as the availability of proper network nodes capable of moving in the territory. The 5G Service Based Architecture defined in 2019 provides an important step forward on the softwarization of the telco infrastructure by enabling the deployment and orchestration of Virtual Network Functions on the network infrastructure. On the other hand, Unmanned Aerial Vehicles have recently gained the attention of the

F. Granelli (✉) · C. Sacchi
DISI – University of Trento, Trento, Italy
e-mail: fabrizio.granelli@unitn.it

R. Bassoli
Technical University of Dresden, Dresden, Germany

R. Cohen · I. Ashkenazi
Technion, Haifa, Israel

C. Palestini (ed.), *Advanced Technologies for Security Applications*, NATO Science for Peace and Security Series B: Physics and Biophysics,
https://doi.org/10.1007/978-94-024-2021-0_26

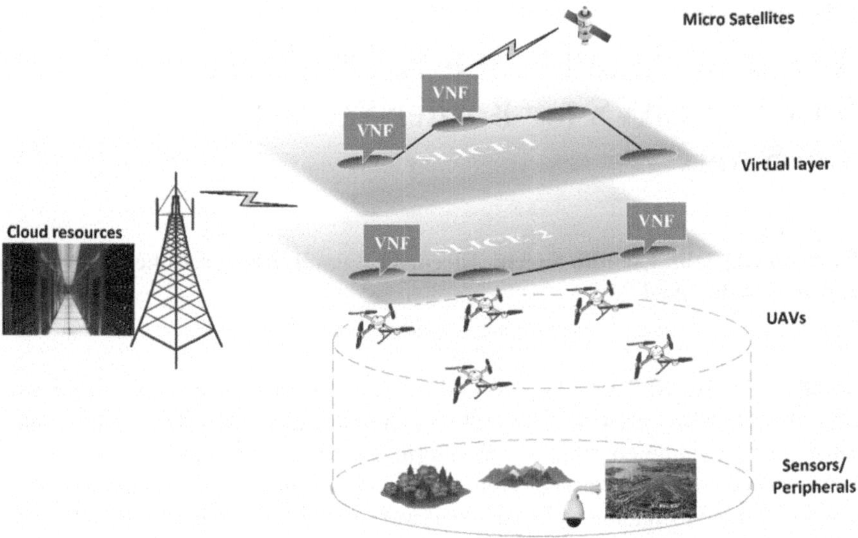

Fig. 26.1 Structure of the proposed system to guarantee efficient and effective border control and security against terrorist threats. There are four layers. The virtual layer dynamically assigns virtual network functions (VNFs)

communication community as a potential technology to deploy movable and agile network nodes for fast deployment of network nodes.

NATO SPS G5428 DAVOSS project focuses on merging those two concepts (5G SBA and UAVs) to build a communication and service architecture capable of providing coverage for monitoring, border security and safety applications. Indeed, DAVOSS project aims to develop a multi-layer virtualised system in which all the technologies listed above work together to guarantee efficient and effective borders and ports surveillance. Figure 26.1 depicts the proposed network and architecture. The proposal identifies four main system layers: the one constituted by sensors (Layer 1), the one composed by UAVs (Layer 2), the one deploying virtualisation (Layer 3) and the one including micro satellites (Layer 4).

The project is approaching the end of the research activities, leading to the selection of the technologies to actually deploy and test in the project testbed. Agreements are being prepared in order to run experiments at the Trento Firemen (Vigili del Fuoco) training facilities for providing a suitable framework for testing DAVOSS results.

Next sections of the paper introduce the main research activities in the different layers of the DAVOSS project architecture.

26.2 Layer 1 – UAVs and Sensors

Layer 1 of the DAVOSS project consists of a sensor network with a drone-based mobile gateway. Basically, the drone flies above the sensors and communicates directly with each of them to collect their data. Therefore, designing an appropriate path is important for several reasons. First, such a path allows to reduce the energy consumption of the drone, which is translated to increasing the area it can cover and minimizing the time needed for collecting urgent information. Second, an efficient path allows the drone to hover very close to each sensor, which is translated to increasing the wireless throughput of each sensor and of the whole network.

The project is testing path planning algorithms using the following hardware components:

- **Drone:** RE470 Quadcopter equipped with Pixhawk PIX PX4 2.4.8 Flight Controller and GPS.
- **LoRaWAN Gateway:** Raspberry Pi-3 with RAK-831 Multi-channel LoRaWAN Gateway that uses Semtech sx1272 LoRa RF transceiver.
- **LoRaWAN Sensors:** The sensors combines STM Nucleo L073RZ base board with Semtech sx1272 LoRa RF transceiver.

The algorithm consists of three stages. The first stage is initial sensors data collection. In this stage, the drone is configured with the GPS coordinates (waypoints) of each sensor, using the mission planner tool. The drone is then dispatched using the "Auto" flight mode to collect the data from each LoRaWAN sensor. In this mode, the drone simply flies from one waypoint to another. In the second stage, after the drone returns from its mission, its LoRaWAN gateway is connected to a LoRa server in order to extract the collected sensors data, as well as the metadata related to the communication with each sensor. In the third stage of the project, our Path Planning algorithm receives the metadata related to each sensor and computes an optimized path for the next drone flight.

The proposed algorithm consists of two parts. In the first part, the algorithm calculates the waypoints that should be traversed. These points are referred to as key waypoints. To this end, the algorithm takes into consideration the obstacles, as shown in the example in Fig. 26.2. After determining the key waypoints, the Path Planning algorithm determines the exact path that traverses all the key waypoints, while minimizing the length of the route and not entering the obstacle boxes (Fig. 26.3).

26.3 Layer 2 – UAV Networking

Layer 2 of DAVOSS architecture consists of drones which can communicate either among each other or through ultra-light vehicles and balloons, which is used for backhauling. The drones have mechanical parts, which are responsible for flight

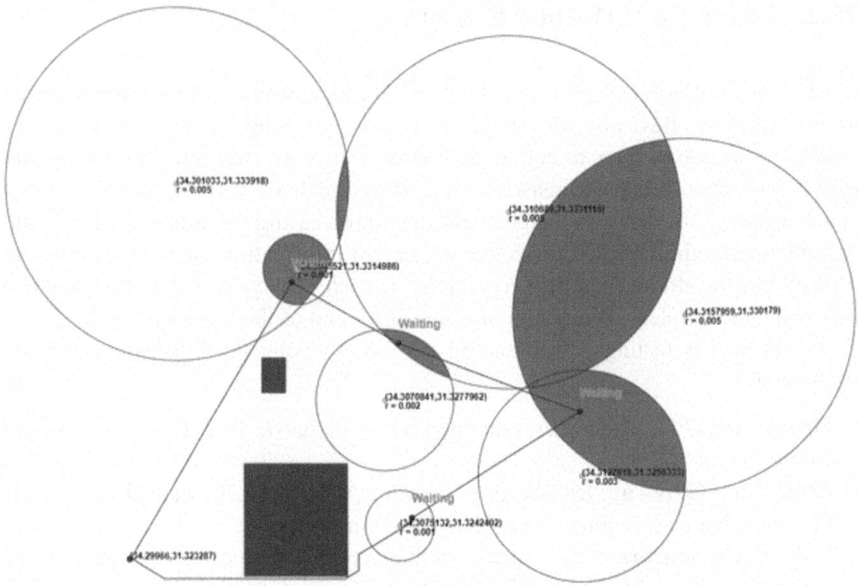

Fig. 26.2 Path planning example in presence of an obstacle

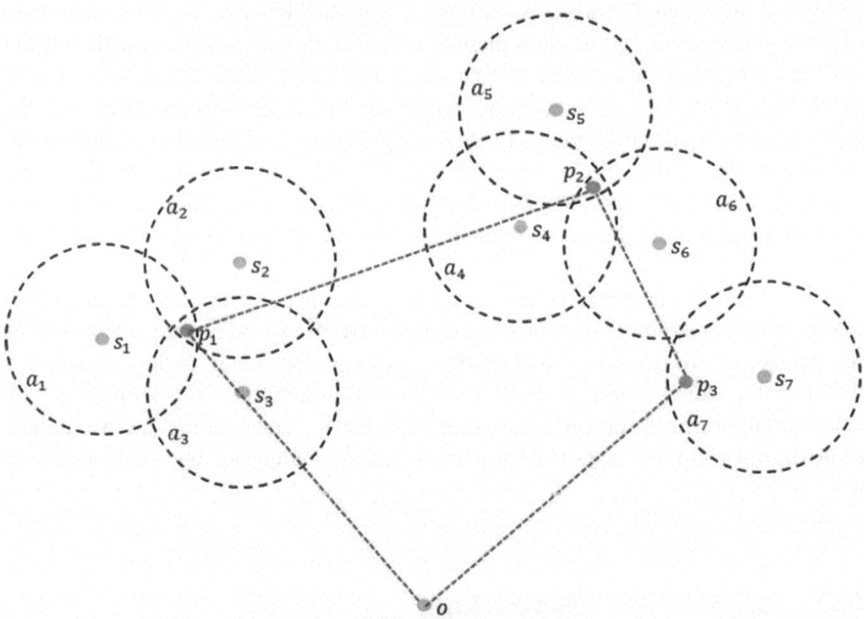

Fig. 26.3 Path planning example (without obstacles)

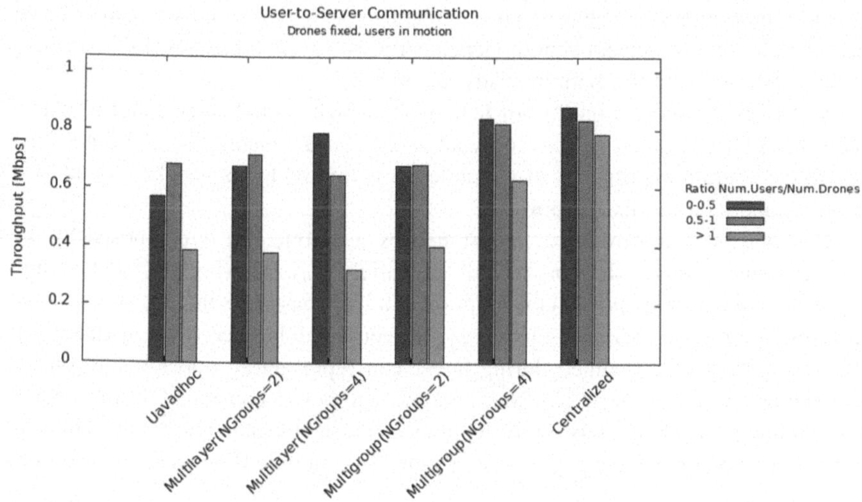

Fig. 26.4 Communication performance for different UAVs network configurations: ad-hoc, centralized, multi-layer and multi-group

operations and a battery for power supply of carried equipment. Furthermore, the unmanned aerial vehicles (UAVs) of our system are considered as mobile base stations (BSs) thus they carry radio equipment (supporting LTE/4G standard communications to the ground peripherals) and hardware for baseband processing, called baseband unit (BBU).

Several configurations of the interconnections among UAVs and other platforms (aerial or terrestrial) are being studied using ns-3 network simulator (see Fig. 26.4).

26.4 Layer 3 - Virtualization

The idea behind DAVOSS' design of Layer 3 is the virtualization of most of the network operations in order to avoid loading the battery of the drone and the drone itself. That becomes fundamental to increase drones' battery life and capabilities. A preliminary analysis of this aspect was presented in [1]. This represents a novel study in the research panorama because it seems the first work considering BBU impact on mobile BSs.

The mobile BSs collect data from the peripherals and transmit them to Layer 2. Without loss of generality, we considered the drones hovering and active (i.e. transmitting/receiving data from peripherals). For a correct theoretical analysis, we decided to use stochastic geometry, which represents the current most reliable mathematical model for radio access network of existing 4G cellular networks. As DAVOSS objective is to provide connectivity in complex scenarios and borders (areas without any existing reliable network infrastructure), we can reasonably consider mobile BSs and terrestrial peripherals (belonging to Layer 1) distributed according to two-dimensional homogeneous Poisson point processes (PPPs) respectively called Φ_{bs} and Φ_s, with intensity λ_{bs} and λ_m.

A UAV is normally a multirotor helicopter, which should carry either a remote radio head (RRH) and a BBU or only an RRH. As just mentioned, the flight time t_{fl} and the operational time t_{op} of a mobile BS is limited by its weight, its battery's capacity and its transmission power.

The total average power p_{tot}, consumed by a mobile BS, is composed by the average power consumed during takeoff (p_{to}), flight (p_{fl}), hovering (p_{ho}) and landing (p_{la}). Next, the average power consumption of a UAV-based BS includes the average transmission p_{tr} and processing power p_{pr}. In general, it has been demonstrated that the average power consumed during hover can represent an upper bound on the average power during flight [2]. Next, the average power consumed during takeoff and landing is approximately equivalent to the power consumed hovering. Then, an upper bound on total average power, consumed by a mobile BS, can be expressed as

$$p_{tot} = 4p_{ho} + p_{tr} + p_{pr}$$

Mobile BSs were assumed to be comparable to pico-BSs in terms of power consumption and coverage [3]. So, reasonable values of power consumption can be considered 1.9 W for power amplifier, 1 W for radio frequency hardware and 3 W for BBU. In the analysis we considered the general virtualization of all BBU operations thus, the deployment of all BBU on the pico-satellites (CubeSats). In this way, we obtained a sort of lower bound in terms of power consumption, which can give an upper bound on how much power we can save at the drones via virtualization. That is because this calculation still neglected the differentiation of power consumption for the various schemes of BBU's sub-function virtualization. Moreover, it did not considered the power consumption required by hardware for backhaul/fronthaul transmissions, which permit the communications between drones and satellites. Figure 26.5 depicts the upper bound on the power gain according to the weight of the drone.

26.5 Layer 4 – Aerial Backhaul (Microsatellites, Blimps)

26.5.1 CUBESAT Link Requirements

The virtualization of BBU network function imposes some requirements in terms of capacity to the link connecting the two network nodes involved in the RRH-BBU splitting operations, in our case the LTE RRH installed on board of the UAV and the BBU mounted on the cubesat rack. Such requirements depend on the splitting configurations, shown in Fig. 26.6.

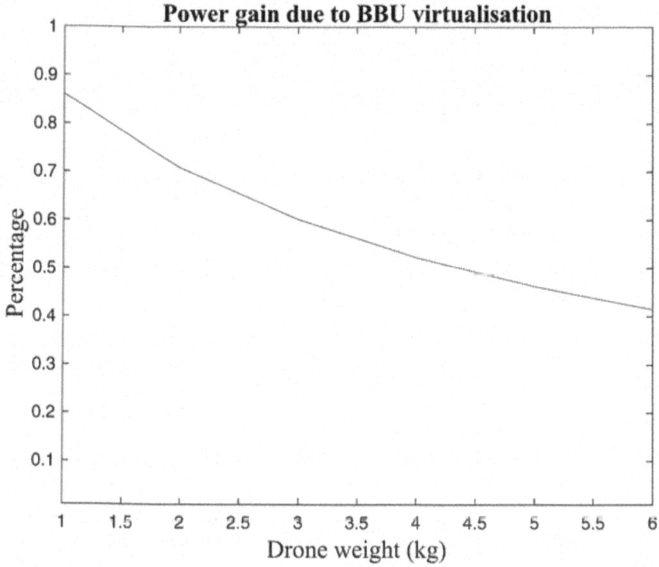

Fig. 26.5 Power gain at the mobile BSs (drones) when BBU is virtualized. Obviously, by increasing the load of the drone, the impact of the weight of the BBU decreases

The baseline splitting configuration considered in DAVOSS is the Split D of Fig. 26.7, where L2H and L3 are relocated to the cubesat BBU unit, leaving to the RRH the functions mostly related to PHY-layer management (RF, L1L, L1H, L2H). The link requirements for the baseline configuration are given as follows:

- link capacity $\geq$180 Mb/s (achieved with a quasi-zero bit-error-rate,
- say: $\leq 10^{-12}$);
- link delay not exceeding 4 ms.

The aforesaid target value of capacity looks quite ambitious for small cubesats operating in the usual L and S bands, while the delay may be affordable. In the following, we shall analyze the transmission techniques and the cubesat equipment that would allow to reach the expected performance.

26.5.2 CUBESAT-Based RRH-BBU Connection Setup

In Fig. 26.8, the architecture of the virtualized monitoring system exploiting cubesats is depicted. A broadband sensor (e.g. a 3D scanner) transmits data from the border area to the drone upon the LTE standard. The data are turbo-encoded by the LTE UE and transmitted to the drone. The RRH mounted on the drone perform the demodulation of the encoded data and forward them to the BBU installed on

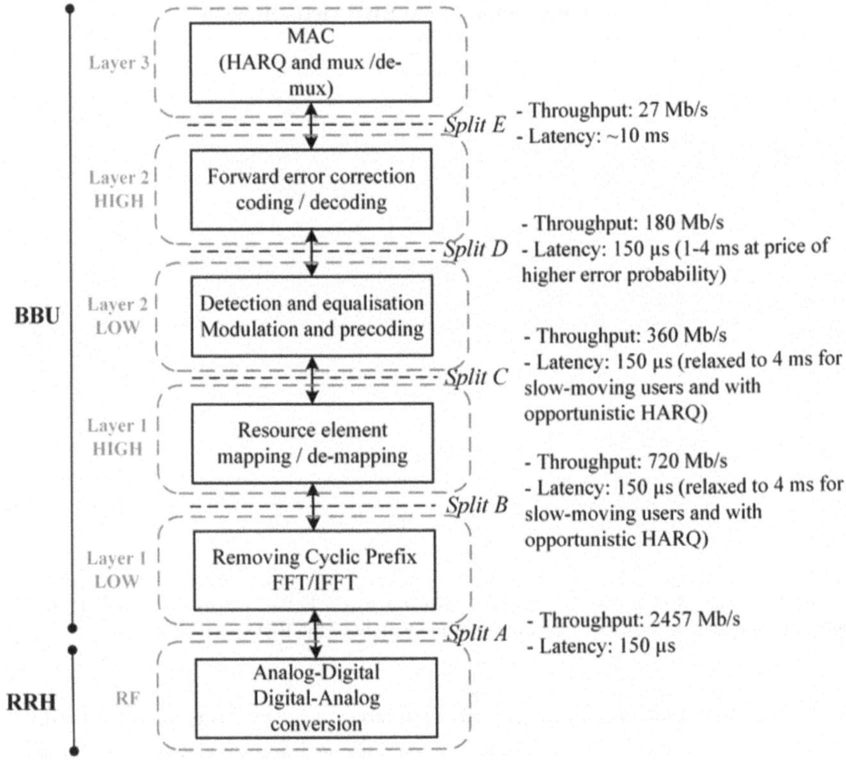

Fig. 26.6 Basic requirements of different RRH-BBU splitting configurations

Fig. 26.7 Two possible Cloud RAN slicing configurations

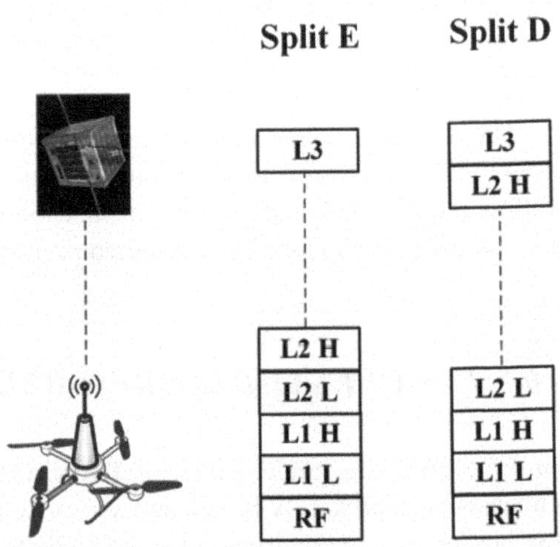

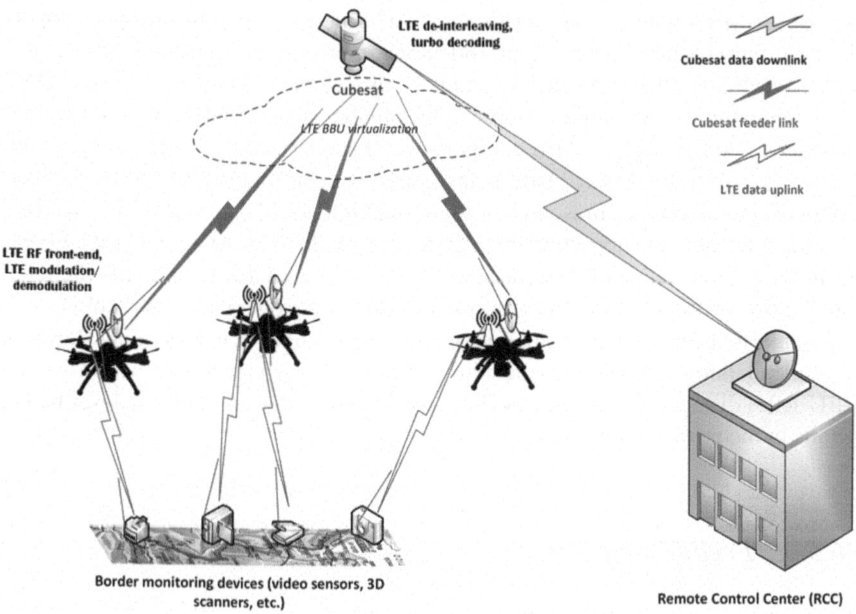

Fig. 26.8 Architecture of the virtualized border monitoring system exploiting the cubesats

cubesat. The BBU performs in real-time the turbo decoding and dispatches the decoded data to the remote control center in downlink at a convenient rate.

The L2 management considers a HARQ mechanism relying on FEC turbo coding only, without any ACK/NACK mechanism. Such a choice has been motivated in order to avoid throughput starvation due to ACK/NACK looping. Such an issue is not critical in terrestrial links, but it may become serious in case of information forwarding through satellite links, whose latency is higher.

We target the implementation of the cubesat transceiver for CRAN splitting, according to the CPRI standard [4]. CPRI is designed for fiber connections, but we can adapt the specifications to the cubesat connection. The channel coding used by CPRI is Reed-Solomon (RS) codes. The RRH-BBU satellite transceiver should be implemented with commercial components.

The mathematical expression linking the sampling rate of the analog front-end AFE to the net bit-rate achievable by the transmitter is the following:

$$R_b = \frac{S_R}{f_{ov}}\gamma \tag{26.1}$$

where *SR* is the AFE sampling rate expressed in Msamples/sec, f_{ov} is the over-sampling factor, expressed in samples/symbol, and γ is the number of information bits carried by each symbol. In [5], the MAX19713 component, commercialized by Maxim Integrated (San Jose, CA) is used for the implementation of a 60 Mb/s cubesat transceiver. The MAX19713 AFE is capable of supporting a sampling rate of 45 MS/s with 12 bits ADC and DAC converters [6]. In order to support the minimum required data rate of 180 Mb/s, imposing an oversampling rate of 3 samples/symbol – that is needed to guarantee the correct working of ADC and DAC [5] – the parameter γ should assume values higher than 12 bit/symbol. Clearly, it is better to consider other AFE components, characterized by higher sample rate. Having a look to the Texas Instruments' shelf, we can find the AFE5816 product, originally designed for ultrasound systems and high-speed data acquisition systems [7], that is capable of supporting up to 80 MS/sec with 12 bit ADC and DAC. Fixing again $fov = 3$, the value of γ decreases to 6.75 bit/symbol that is still quite high, but much more affordable from the viewpoint of the link budget parameterization.

Thanks to the analytical evaluation of BER at the output of RS decoding shown in [6], the link budget should provide a link BER < 10^{-12}, in the worst case of longest BBU-RRH distance. Such values will be confirmed by the simulations, described in details in the next section.

26.5.3 Preliminary Results

Two SIMULINK-based simulators have been employed in order to simulate the cubesat orbits and the cubestat link, respectively. Some relevant results are presented in the following figures, where Fig. 26.9 shows the allowed number of turbo

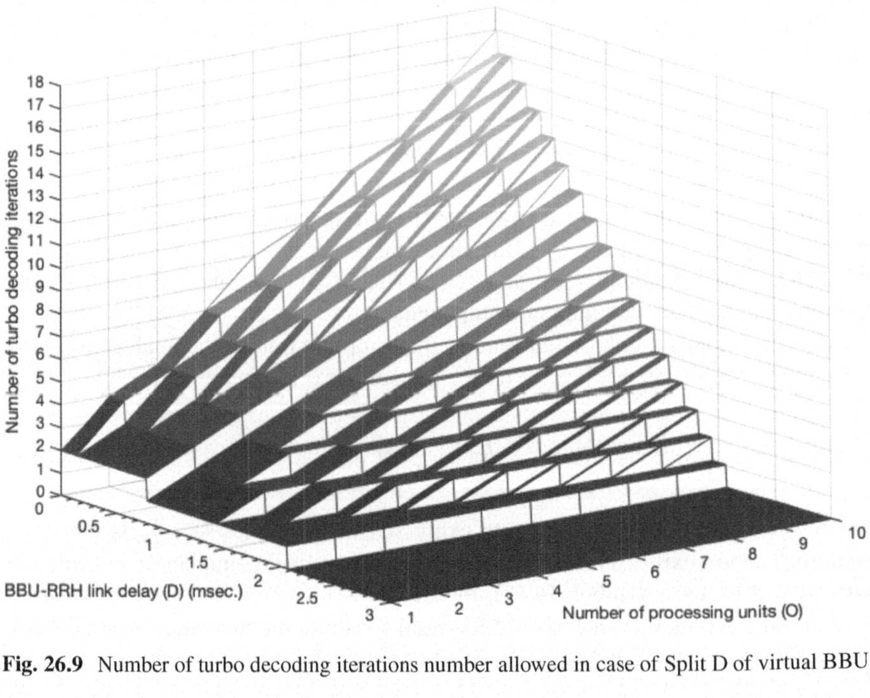

Fig. 26.9 Number of turbo decoding iterations number allowed in case of Split D of virtual BBU

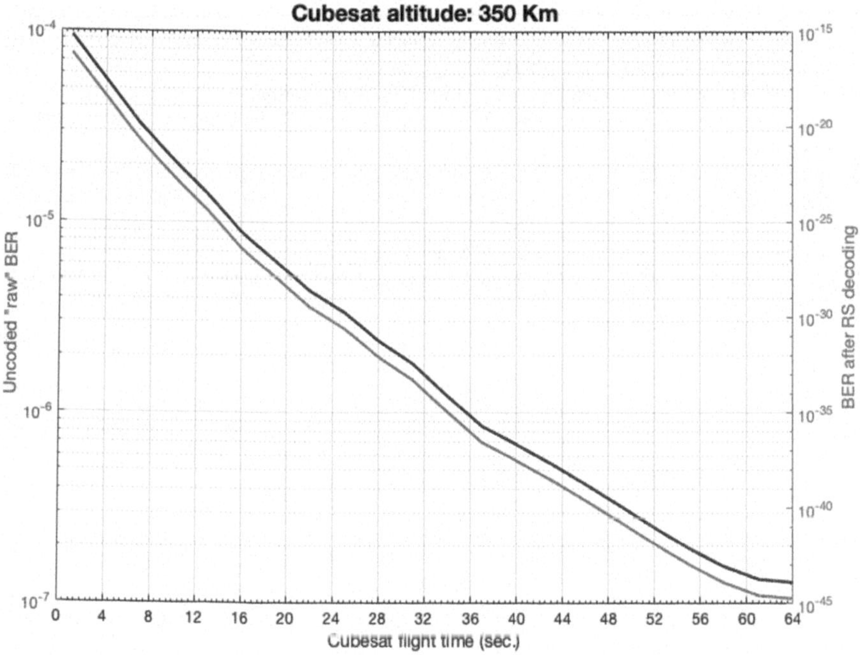

Fig. 26.10 Raw BER performance (blue curve and left axis) and BER after RS decoding (analytical lower bound – orange curve and right axis) vs. cubesat flight time in the worst case of 350 Km altitude and $k_O = 1$

iteration allowed by using the Split D architecture, while Fig. 26.10 represent the corresponding BER.

26.6 Conclusions

This paper presented the ongoing activities to design a flexible border monitoring architecture based on the usage of aerial platforms (UAVs, blimps, cubesats), developed in the framework of the NATO SPS G5428 DAVOSS project. Future activities will be focused on implementing and measuring the actual performance of the architecture on a real scenario.

References

1. Bassoli R, Sacchi C, Granelli F, Ashkenazi I (2019) A virtualized border control system based on UAVs: design and energy efficiency considerations. In: 2019 IEEE Aerospace conference, Big Sky, MT, USA, pp 1–11
2. Dorling K, Heinrichs J, Messier GG, Magierowski S (2017) Vehicle routing problems for drone delivery. IEEE Trans Syst Man Cybern Syst 47:70–85

3. Auer G, Giannini V, Godor I, Skillermark P, Olsson M, Imran MA, Sabella D, Gonzalez MJ, Desset C, Blume O (2011) Cellular energy efficiency evaluation framework. In: IEEE 73rd vehicular technology conference (VTC Spring), pp 1–6
4. Ericsson AB, Huawei Technologies Co. Ltd, NEC Corporation, Alcatel Lucent and Nokia Networks (2015) CPRI Specification V7.0 – Common Public Radio Interface (CPRI); Interface Specification
5. Butters B, Raad R (2014) A 2.4 GHz high data rate radio for picosatellites. In: 2014 8th international conference on telecommunication systems services and applications (TSSA), pp 1–6
6. Maxim Integrated. MAX19713 10-Bit, 45Msps, Full-Duplex Analog Front-End, 2006
7. Texas Instruments. AFE5816 16-Channel Ultrasound AFE With 90- mW/Channel Power, 9 2017. Rev. 1

Chapter 27
Nato SPS Cluster Workshop on Advanced Technologies: Conclusions

Claudio Palestini, Deniz Beten, Ettore Marchesoni, and Marie-Anne Brouillon

Abstract The objective of the Science for Peace and Security (SPS) Programme is to promote dialogue and practical cooperation between NATO members and partner nations based on scientific research, technological innovation and knowledge exchange. As a result, understanding and forecasting evolving trends is paramount for SPS to ensure the promotion of activities that reflect both the ambition of the scientific community as well as the real-world security challenges.

27.1 A Multi-domain Workshop

The NATO SPS Cluster Workshop on Advanced Technologies was an opportunity for researchers and co-directors from NATO and partner nations to further strengthen dialogue, to present their activities, to highlight future trends and to provide feedback on how SPS can contribute to the scientific and technical development in the field of security-related advanced technologies.

It was decided to arrange the Workshop in a single open session and to expose the researchers to several topics, in order to promote as much as possible networking, generation of new ideas, cross-domain innovation, etc. Sharing information, experiences and ideas was the main *motif* of the Workshop; recognizing that creativity and innovation require exposure to different ideas and out-of-the-box thinking, the SPS Programme tried to push its researchers in uncharted but innovative waters. Discussions revolved around the projects' achievements and participants could appreciate scientific and technological developments while engaging with other researchers in constructive dialogues and offering observations and suggestions.

In line with the objectives of NATO and the SPS Programme, the projects developed technologies for the security of communication systems, to enhance their resilience and to prevent network intrusion or detect attacks. Cutting-edge

C. Palestini (✉) · D. Beten · E. Marchesoni · M.-A. Brouillon
NATO Emerging Security Challenges Division (ESCD), Brussels, Belgium
e-mail: palestini.claudio@hq.nato.int

C. Palestini (ed.), *Advanced Technologies for Security Applications*, NATO Science for Peace and Security Series B: Physics and Biophysics,
https://doi.org/10.1007/978-94-024-2021-0_27

production and synthetization techniques were explored to exploit advanced materials' properties in a number of fields, from electromagnetic protection to nuclear screening. Innovative sensors and detectors were also developed for border security, situational awareness, monitoring of critical areas and infrastructures, and the detection of hazardous materials. Finally, unmanned and autonomous systems were designed to support a number of security missions in multiple environments. Based on their research efforts and knowledge, participants shared their thoughts on potential future actions and areas of interest in order to advance the scientific and technical state-of-the-art.

27.2 Recommendations for Future Developments

A number of recommendations were collected and will be used to shape upcoming SPS calls for proposals and activities in the field of security-related advanced technologies. Among these recommendations, the most relevant are related to the following topics:

Artificial Intelligence (AI) and Machine Learning AI is becoming an area of strategic importance and a game-changer in future societal dynamics, from economic development to geopolitics and security. Participants in the Workshop highlighted the implications, opportunities and risks of AI, recalling that many of the current SPS projects already make use of AI in a number of areas (target recognition, signal optimization, data mining, object detection and classification, etc.). Looking into the future, the need for the scientific community to provide advice and solutions to make AI secure, fair, ethical, and trustworthy was recognized. The importance of data availability, quality and integrity as critical elements for success was also recalled. As such, participants recommended to enhance further cooperation in this domain, with an emphasis on data and information sharing among the scientific community.

Unmanned Systems and Autonomy Unmanned systems are already on the verge of the commercial and consumer market. In the Gartner Hype Cycle for Emerging Technologies 2019,[1] technologies like autonomous driving, light cargo delivery drones and decentralized autonomous organizations are in the "Innovation trigger" and "Peak of Inflated Expectations" categories, meaning that there is an intense scientific and media interest at this stage, with their potential mainstream market adoption expected within the next decade. On the other hand, the use of autonomous systems, especially in the field of security, raises some questions and doubts, bringing the need for a profound understanding and general awareness on the

[1] 5 Trends Appear on the Gartner Hype Cycle for Emerging Technologies, 2019 – https://www.gartner.com/smarterwithgartner/5-trends-appear-on-the-gartner-hype-cycle-for-emerging-technologies-2019/

full spectrum of opportunities offered and, at the same time, on the risks and implications of the potential misuse by terrorists and adversaries. To this end, it was recommended to analyze and continue gathering the scientific community's perspectives in these technologies and their security implications, with particular focus on policy, ethical and legal aspects.

Quantum Technology Application of quantum computing is leading to the development of powerful machines capable to solve complex problems which are outside the realm of today's computers' capabilities. In the next couple of decades, this could lead to the cracking of current cryptographic algorithms,[2] making obsolete all current IT security systems, and to the need for new encryption schemes and technologies. Accordingly, several SPS projects have tackled the development of new cryptographic algorithms capable to resist against attacks by quantum computers (also known as post-quantum cryptography) and have contributed to international standardization efforts.[3] Other projects have developed Quantum Key Distribution (QKD) systems, i.e. using quantum technology to establish secure links free from the risk of eavesdropping attacks. The experts recognized the need to combine the two efforts as a future priority and to bring the two communities (the quantum physicists and the cryptologists) together in order to promote novel concepts and breakthrough ideas.

Technology Convergence All projects demonstrated and acknowledged the current trends of integrating and combining multiple scientific disciplines to form new and innovative technologies. For example, sensor fusion and communication are enablers for autonomous systems; similarly, innovative materials are the basis for more capable sensors. This trend will become even more radical in the near future, when new opportunities offered by biotechnologies will open unprecedented possibilities. In this scenario, cooperation between different scientific areas should be encouraged through the promotion of large inter-disciplinary projects (i.e. integration of engineering, biotechnologies, physical sciences, data science, computation, life sciences, social sciences, etc.) as novel security applications will necessarily follow this pattern.

27.3 Keeping SPS Abreast of a Developing Scientific and Security Landscape

This book offers a snapshot on how SPS is currently contributing to the international scientific community's efforts in these domains, and a view on how its role may be shaped in the future.

[2]How a quantum computer could break 2048-bit RSA encryption in 8 h https://www.technologyreview.com/s/613596/how-a-quantum-computer-could-break-2048-bit-rsa-encryption-in-8-hours/

[3]SPS projects contribute, for example, to the US National Institute of Standards and Technology (NIST) Post-Quantum Cryptography Standardization Process

The overall objective of SPS is to promote dialogue and practical cooperation between NATO members and partner nations based on scientific research, technological innovation and knowledge exchange. As a result, understanding and forecasting evolving trends is paramount for SPS to ensure the promotion of activities that reflect both the ambition of the scientific community as well as the real-world security challenges posed by new and disruptive technologies. Events like the SPS Cluster Workshop on Advanced Technologies contribute to keeping the Programme's activities abreast of an ever developing scientific and security landscape.

For the Alliance, technological disruption has been the driving force that started NATO's involvement in the scientific domain. With scientific progress and the evolution of security threats, technological disruption is also bound to be part of NATO's future. Its involvement in the scientific domain and support to activities aimed at understanding and preventing emerging security challenges is therefore essential to the Alliance's adaptation and modernization.

Annexes

Annex I: List of Projects Involved in the Workshop

Communication systems	
G5461	Large Scale Collaborative Detection and Location of Threats in the Electromagnetic Space (SOCRATES)
G5482	Public Safety COmmUNication in ConTExt Related to Terror Attacks (Counter-Terror)
G5269	Flash Crowds Management via Virtualized Network Resources (FALCON)
G5319	Threat Predict: From Global Social and Technical Big Data to Cyber Threat Forecast
G5263	Analysis, Design and Implementation of an End-to-End 400 km QKD Link
G5448	Quantum-safe Authenticated Group Key Establishment
G5485	Secure Quantum Communication Undersea Link

Advanced materials	
G5215	Engineering Silicon Carbide for Enhanced Border and Port Security (E-SiCure)
G5120	Infrared Transparent Ceramic Windows for High-speed Vehicles
G5140	Advanced Nanotechnologies For Multivariate Sensor Fabrication
G5453	Radiation Hard UV Detectors against Terrorist Threats
G5580	Creation of New Generation Titanium Diboride Composite Armour Material
G5030	Titanium Armour with Gradient Structure: Advanced Technology for Fabrication

C. Palestini (ed.), *Advanced Technologies for Security Applications*, NATO Science for Peace and Security Series B: Physics and Biophysics,
https://doi.org/10.1007/978-94-024-2021-0

Sensors and detectors	
G5267	Maritime Security – Multistatic and Multiband Coherent Radar Fleet for Border Security (SOLE)
G5465	Noise Imaging Radar Network for Covert Air and Maritime Border Security (NORMA)
G5248	Compact Eye-Safe Lidar Source for Airborne Laser Scanning (CALIBER)
G4840	Microelectronic 3D Imaging and Neuromorphic Recognition for Autonomous UAVs
G5437	Wide InTegration of sensor Networks to Enable Smart Surveillance (WITNESS)
G5244	Graphene / Polymer based Sensor
G5351	Nanocomposites Based Photonic Crystal Sensors of Biological and Chemical Agents
G5373	Hand-held Gamma Detector based on High-Pressure Xenon Gas

Unmanned and autonomous systems	
G5176	Agile Tyre Mobility for Severe Terrain Environments
G5293	Autonomous Platform for Securing Marine Infrastructures
G5322	High Altitude Balloon-Borne Radar
G5428	Dynamic Architecture based on UAVs Monitoring for Border Security and Safety
G5568	Mobile Adaptive/Reactive Counter Unmanned Aerial System (MARCUS)$_*$

Annex II: Science for Peace and Security (SPS) Grant Mechanisms

1. SPS Grant Mechanisms

The SPS Programme develops and implements practical cooperation and enhances dialogue between NATO nations and Partner countries through capacity-building and security-related science technology and innovation.

All Programme activities contribute towards the Alliance's strategic objectives, have a clear link to security and respond to at least one of the SPS Key Priorities as mentioned below.

The SPS Programme supports collaboration through four established grant mechanisms:

- Research & Development Multi-Year Project (MYP): With a typical duration of 2–3 years, the MYP grant mechanism enables scientists from NATO and its Partner countries to collaborate on applied research and development (R&D) and capacity building projects that result in new civil science advancements with practical application in the security and defence fields. It provides equipment, training and support to young scientists.
- Advanced Research Workshop (ARW): With a duration of 2–5 days, the ARW grant mechanism allows for advanced-level discussions among scientists and experts from different countries in order to identify direction for future actions to address security challenges.
- Advanced Study Institute (ASI): With a duration of at least 7 working days, the ASI grant mechanism offers high-level tutorial courses organized to convey the latest developments in science and innovation to an advanced (PhD level) audience.
- Advanced Training Course (ATC): With a duration of 5–7 working days, the ATC grant mechanism enables specialists in NATO member countries to share their security-related expertise with trainees from NATO's Partner countries.

Each activity is led by project directors from at least one NATO country and one Partner country.

2. SPS Key Priorities

SPS Key Priorities are based on NATO's Strategic Concept as agreed by Allies in Lisbon in November 2010 and the Strategic Objectives of NATO's Partner Relations as agreed in Berlin in April 2011, without any indication of priority ranking.

All SPS activities funded under the SPS Programme must address the SPS Key Priorities listed below and must have a clear link to security and to NATO's strategic objectives.

1. **Facilitate mutually beneficial cooperation on issues of common interest, including international efforts to meet emerging security challenges**

   (a) **Counter-Terrorism**

   (i) Methods for the protection of critical infrastructure, supplies and personnel;
   (ii) Human factors in the defence against terrorism;
   (iii) Detection technologies against the terrorist threat for explosive devices and other illicit activities;
   (iv) Risk management, best practices and technologies in response to terrorism.

   (b) **Energy Security**

   (i) Innovative energy solutions for the military; battlefield energy solutions; renewable energy solutions with military applications;
   (ii) Energy infrastructure security;
   (iii) Maritime aspects of energy security;
   (iv) Technological aspects of energy security.

   (c) **Cyber Defence**

   (i) Critical infrastructure protection, including sharing of best practices, capacity building and policies;
   (ii) Support in developing cyber defence capabilities, including new technologies and support to the construction of information technology infrastructure;
   (iii) Cyber defence situation awareness.

   (d) **Defence against Chemical, Biological, Radiological, and Nuclear (CBRN) Agents**

   (i) Methods and technology regarding the protection against, diagnosing effects, detection, decontamination, destruction, disposal and containment of CBRN agents;
   (ii) Risk management and recovery strategies and technologies;
   (iii) Medical countermeasures against CBRN agents.

   (e) **Environmental Security**

   (i) Security issues arising from key environmental and resource constraints, including health risks, climate change, water scarcity and increasing energy needs, which have the potential to significantly affect NATO's planning and operations;
   (ii) Disaster forecast and prevention of natural catastrophes;
   (iii) Defence-related environmental issues.

2. **Enhance support for NATO-led operations and missions**

   (i) Provision of civilian support through SPS Key Priorities;

(ii) Provision of access to information through internet connectivity as in the SILK-Afghanistan Programme;
(iii) Cultural and social aspects in military operations and missions;
(iv) Enhancing cooperation with other international actors.

3. **Enhance awareness of security developments including through early warning, with a view to preventing crises**

(a) **Security-related Advanced Technology**

(i) Emerging technologies including nanotechnology, optical technology, micro satellites, metallurgy and the development of Unmanned Aerial Vehicle (UAV) platforms.

(b) **Border and Port Security**

(i) Border and port security technology;
(ii) Cross-border communication systems and data fusion;
(iii) Expert advice and assessments of border security needs and best practice.

(c) **Mine and Unexploded Ordnance Detection and Clearance**

(i) Development and provision of advanced technologies, methodologies and best practice;
(ii) Solutions to counter improvised explosive devices (IED).

(d) **Human and Social Aspects of Security related to NATO's strategic objectives**

4. **Any related project clearly linked to a threat to security not otherwise defined in these priorities mayalso be considered for funding under the SPS Programme.**

The manufacturer's authorised representative in the EU is Springer Nature Customer Service Centre GmbH, Europaplatz 3, 69115 Heidelberg, Germany. If you have any concerns regarding our products, please contact ProductSafety@springernature.com

Printed and bound by CPI Group (UK) Ltd, Croydon, CR0 4YY

30/09/2026

02225559-0013